D1144538

WITHDRAWN
STOCK

Proceedings of the Ninth FEBS Meeting

General Editor G. GÁRDOS

Volume 31

PROTEINS OF CONTRACTILE SYSTEMS

Volume 32

MECHANISM OF ACTION AND REGULATION OF ENZYMES

Volume 33

BIOCHEMISTRY OF THE CELL NUCLEUS
MECHANISM AND REGULATION OF GENE EXPRESSION

Volume 34

POST-SYNTHETIC MODIFICATION OF MACROMOLECULES

Volume 35

BIOMEMBRANES Structure and Function

Volume 36

ANTIBODY STRUCTURE AND MOLECULAR IMMUNOLOGY

Volume 37

PROPERTIES OF PURIFIED CHOLINERGIC AND ADRENERGIC RECEPTORS

BIOMEMBRANES:

Structure and function

Proceedings of the Ninth FEBS Meeting

General Editor G. GÁRDOS

Volume 31

PROTEINS OF CONTRACTILE SYSTEMS

Volume 32

MECHANISM OF ACTION AND REGULATION OF ENZYMES

Volume 33

BIOCHEMISTRY OF THE CELL NUCLEUS
MECHANISM AND REGULATION OF GENE EXPRESSION

Volume 34

POST-SYNTHETIC MODIFICATION OF MACROMOLECULES

Volume 35

BIOMEMBRANES Structure and Function

Volume 36

ANTIBODY STRUCTURE AND MOLECULAR IMMUNOLOGY

Volume 37

PROPERTIES OF PURIFIED CHOLINERGIC AND ADRENERGIC RECEPTORS

FEBPBY

FEDERATION OF EUROPEAN BIOCHEMICAL SOCIETIES
NINTH MEETING, BUDAPEST, 1974

BIOMEMBRANES:
Structure and Function

Volume 35

Editors
G. GÁRDOS, *Budapest*
ILMA SZÁSZ, *Budapest*

1975
NORTH-HOLLAND / AMERICAN ELSEVIER

ISBN North-Holland
Series: 0 7204 4300 8
Volume: 0 7204 4335 0
ISBN American Elsevier: 0 444 10935 8

A joint publication of North-Holland Publishing Company-Amsterdam-London
with Akadémiai Kiadó, Budapest

Sole distributors for the USA and Canada:
American Elsevier Publishing Company, Inc.
52 Vanderbilt Avenue
New York, N.Y 10017

Printed in Hungary

LIST OF CONTENTS

Mitochondrial Bioenergetics

INTRODUCTION

This volume comprises the papers presented at the Symposium on "Biomembranes: Structure and Function" of the 9th FEBS Meeting held on August 25–30, 1974 in Budapest, Hungary. The Symposium was organized by Professors A. Kotyk and G. Gárdos. Herewith we should like to express our gratitude to Prof. A. Kotyk for all his kind activities and useful advice in organization.

Parallel with the true understanding of the paramount importance of biomembranes during the last decade, the interest in their structure and function has increased enormously. This symposium was expected to give a cross-section of the progress in this more and more active and diverse field of research. With this in mind the organizers had invited many outstanding experts of this special field to present their recent results and – in addition – to outline their view about the general standing of the question. Attention was focussed on a wide variety of plasma membranes, on mitochondria and lipid bilayers. The results of up-to-date physical and chemical structure research as well as dynamics and molecular mechanisms of membrane transport processes were discussed with special regard to energy-coupling. Nevertheless, it was impossible to cover the whole vast field of membrane research. About 300 short communications and poster presentations joining the Symposium, however, gave a further opportunity to go into the details of the mentioned and related topics.

We hope that in this way the Symposium gave a deeper insight into the problems. Let us further hope that this volume will refresh our memory and remind us of facts and ideas presented and discussed at the Symposium, and it will promote further effective work in the "hot" area of membrane research.

Budapest, October 1974

G. Gárdos
Ilma Szász

Membrane Constituents and Ultrastructure

LIPID-PROTEIN INTERACTION IN MODEL SYSTEMS AND BIOMEMBRANES

L.L.M. van Deenen

Department of Biochemistry, University of Utrecht,
The Netherlands

INTRODUCTION

A brief report on lipids and biomembranes cannot give proper and balanced credit to the many directions of research initiated during the past decade in this vastly expanding field. The chemical characterization of the lipid components has entered the stage of completion. It is possible now to provide a detailed catalogue of the lipid constituents of many membranes. Some biological interfaces display a relatively simple lipid composition, other membranes appear to contain several hundreds of different molecular species. The enzymatic control of the structure and composition of lipids and their dynamic behaviour in membranes represents an extensive chapter in biochemistry in its own. Organic chemists have already synthesized many complex lipids, thus making available well-defined components which cannot be isolated as single molecular species. A great variety of techniques has been utilized to ascertain relatioships between chemical structure and many - but not all - lipid constituents. These studies demonstrated that several properties of biological membranes are dictated to a great extent by the nature of the lipids. Of highest importance are studies on organized lipid systems which resemble in various respects the properties of the lipid core of biomembranes. Many penetrating studies on artificial lipid membranes allowed a simulation of permeability processes including both simple diffusion and carrier-mediated transport. The ionic selectivity produced by macrocyclic compounds in artificial and natural membranes represent a landmark in membrane research. Reconstitution of functionally active membrane systems made up from lipids and enzymes attracts the attention of many biochemists today. Studies on the interaction of protein and lipids, using well-defined components and well-controlled conditions are aimed to understand the protein-lipid association in membranes in molecular detail.

The concept that lipids in biological membranes are arranged to a significant extent in bilayers has survived for nearly fifty years.

The idea that the majority of membrane proteins is located at both surfaces of the lipid bilayer involving principally electrostatic attractions between lipids and proteins has been questioned severely during the past decade. Many alternative models were proposed and claimed to give a better representation of membrane structure. In several of these proposals membrane proteins were considered to be concentrated nearly exclusively in the center of the membrane. As a result of many discussions about these models, the importance of hydrophobic interactions between the hydrocarbon chains of lipids and apolar regions of the proteins was more generally realized. During the past three years we have entered a period of compromise in which the presence of both peripheral proteins and proteins which penetrate or span a lipid bilayer are visualized. This more realistic attitude recognizes that considerable quantitative variations on a general architectural theme may exist between membranes of distinct biological origins, which membranes often have differences in function and in chemical make-up. It is further envisaged that even in adjacent regions within one membrane different molecular arrangements may occur, this in relation to the function of the sites concerned. The dynamic and flexible character of membranes is now accepted by nearly all workers in this field.

The present contribution summarizes mainly some recent studies from the author's laboratory.

TOPOGRAPHY OF PHOSPHOLIPIDS IN THE HUMAN ERYTHROCYTE MEMBRANE.

Phospholipid asymmetry. The lipid composition of erythrocyte membranes has been studied in great detail (van Deenen and de Gier, 1974). Recent studies strongly suggest that a non-uniform distribution of phospholipid classes exists between the exterior and interior region of the human erythrocyte membrane. Labelling of intact erythrocytes with relatively non-permeant agents showed that very few phosphatidylserine and phosphatidylethanolamine molecules reacted (Bretscher, 1972; Gordesky and Marinetti, 1973). On the other hand extensive labelling of these phospholipids occurred when erythrocyte ghosts were exposed to such reagents. Although other explanations were not ruled out completely these results support the idea that phosphatidylethanolamine and phosphatidylserine are mainly located on the inner half of the lipid-bilayer of the human erythrocyte. A preferential location of phosphatidylcholine and sphingomyelin on the outer half of the bilayer was detected by experiments with phospholipases. Limits of space do not permit an extensive discussion of the many studies made in this area (Zwaal *et al.*, 1973) and results are given of only one recent investigation combining phospholipase action and freeze-etch electron microscopy (Verkley *et al.*, 1973). Phospholipase A_2 from *Naja naja* hydrolyses 65-70% of phosphatidylcholine, but no other phosphoglycerides of the intact human red cell membrane. In contrast action of this phospholipase on ghosts produces a complete breakdown of all glycerophospholipids including phosphatidylethanolamine and phosphatidylserine. A sphingomyelinase (Staphylococcus aureus) hydrolyses 80-85% of sphingomyelin of the intact erythrocytes, without cell lysis. Treatment of cells successively with phospholipase A_2 and sphingomyelinase gives a degradation of nearly 50% of the total phospholipids and fragile but intact cells remain. About 75% of the total phosphatidylcholine and 80% of sphingomyelin were found to be hydrolysed; the hydrolysis of phosphatidylethanolamine was limited to 20% and no degradation of phosphatidylserine was detectable. A treatment of erythrocyte ghosts with this combination of enzymes leads to 100% hydrolysis of these major phospholipid classes. The most simple explanation of these results implies that in intact cells the enzymatic degradation is restricted to the phospholipids present in the outer region of the membrane and that the enzymatic action on the ghosts leads in addition to the hydrolysis of phospholipids present in the inner region of the membrane. Freeze-etch electron microscopy supported this view. Sphingomyelinase action on intact cells produced small spheres (75 Å and 200 Å in diameter) which were localized on the outer fracture face; corresponding pits appeared on the inner fraction face. It is concluded that the outer monolayer contains about 50% of the total phospholipids of the human red cell membrane representing about 75% of the total lecithin, 80% of the sphingomyelin; 20% of the phosphatidylethanolamine and nearly none of the phosphatidylserine of this membrane. Apparently, the inner monolayer is composed of phosphatidylserine and phosphatidylethanolamine and a smaller fraction of the choline containing phospholipids.

Lipid packing and phospholipase action on erythrocytes. It is of interest to note that a number of pure phospholipases (e.g. phospholipase A_2 from pig pancreas and Crotalus adamanteus and phospholipase C from Bacillus cereus) fail to attack their phospholipid substrates when present in the membrane of the intact red cell. On the other hand pure phospholipases from other sources (e.g. phospholipase A_2 from Naja naja and bee venom and sphingomyelinase from Staphylococcus aureus) are capable to hydrolyse the phospholipids present in the (outer region) of the membrane. Several explanations for the different behaviour of the various phospholipases can be offered (Zwaal et al., 1973; van Deenen and de Gier, 1974). For instance, if one assumes that the polar headgroups of the phospholipids at the membrane surface are masked completely one could envisage that some phospholipase preparations first cause rearrangements gaining them access to their substrates. Alternatively, the phospholipids at the surface may be at least in part available but their alignment and packing may be such that with some phospholipases an enzyme-substrate complex cannot be formed. It is known from many studies on the action of lipolytic enzymes on monomolecular and bimolecular lipid films that the tightness of the packing of the lipid molecules is an important parameter for enzymatic hydrolysis. Indeed, recent experiments (Demel et al., 1974) suggest that the distinction in action between various phospholipases on the erythrocyte membrane may be related with the packing of the lipids. It was observed that those phospholipases which do not act on the intact erythrocyte fail also to hydrolyse monomolecular films adjusted at pressures above 30 dynes/cm. In contrast phospholipase A_2 from Naja naja and sphingomyelinase (S.aureus) injected underneath lecithin and sphingomyelin monolayers exert their action even above this pressure. Before making final conclusions it is necessary to assay an extensive series of (pure) phospholipases on both erythrocytes and lipid monolayers to ascertain the validity of the proposed relationship. These results may allow to assess, with reasonable preciseness, the virtual surface pressure of the lipids in the outer layer of the erythrocyte membrane. This parameter is of importance for an evaluation of the pionering experiments of Gorter and Grendel (1925) which led them to propose a bimolecular lipid leaflet for the erythrocyte membrane.

Metabolic asymmetry. Although the mature red cell is limited in its phospholipid metabolism when compared with other cells several processes are active in the renewal of phospholipid constituents in this membrane (van Deenen and de Gier, 1974). One mechanism involves the incorporation of fatty acids into lysophospholipids which are either generated in the membrane or supplied by the serum. Incubation of human erythrocytes with radio-active fatty acids results in the formation of labelled lecithin in the membrane. Treatment of the erythrocytes with phospholipase A from Naja naja results in the degradation of about 65% of the lecithin of the cells. As discussed above, this part of the lecithin is considered to be located in the outer monolayer of the erythrocyte membrane. Renooy et al. (1974) found that the great majority of the labelled lecithin is not degraded. Hence, it was concluded that the incorporation of labelled fatty acids into lecithin occurs preferentially into lecithins located at the inner surface of the membrane. These results support the concept of Shohet (1970) that the compartment involved in the incorporation of fatty acids into phosphoglycerides is located "deeper" in the membrane. It is tempting to speculate that the exchange of intact lecithin molecules which occurs between erythrocytes and serum lipoproteins involves lecithin molecules of the outer region of the membrane mainly.

Lipid requirement of (Na^+ + K^+) stimulated ATPase. Membranes perform many of their physiological functions in an asymmetric manner indicating already a firm degree of sideness in architecture. As demonstrated for the erythrocyte this feature involves not only the arrangment of lipids but the distribution of proteins as well (for reviews see Wallach, 1972; Bretscher, 1973; Zwaal et al., 1973). At present it is not known how in the erythrocyte membrane the asymmetric distribution of lipids is linked to that of proteins. In order to obtain some information about the protein-lipid partners at both sides of the membrane experiments are carried out in the author's laboratory to ascertain whether isolated proteins exhibit a preferential binding to specific lipid classes, present in monomolecular layers. As discussed below, this technique gave information about specific interactions between lipids and proteins in the myelin membrane. Furthermore, it is highly likely that proteins which span the erythrocyte membrane have regions of different lipid affinity, which are associated with distinct lipid classes at both sides of the membrane. In this respect it is of interest to devote some attention to the $(Na^+ + K^+)$-stimulated ATP-ase of human erythrocytes. The lipid dependency of this enzyme has been extensively investigated and rather conflicting opinions have been expressed about the nature of the type of phospholipid required for enzymatic activity. Although there is *in vitro* not an absolute specificity for a given phospholipid class. Roelofsen and van Deenen (1973) provided evidence that *in vivo* in the human erythrocyte the $(Na^+ + K^+)$-stimulated ATP-ase activity depends on its association with phosphatidylserine molecules. Enzymatic decarboxylation of all phosphatidylserine of erythrocyte ghosts gave a complete loss of this ATP-ase activity. Restoration of full activity of erythrocyte ghosts treated with phospholipase C and subsequently extracted with dry ether was obtained by addition of phosphatidylserine. An asymmetric alignment of this enzyme has been demonstrated (Whittam, 1967) and the active centre of the $(Na^+ + K^+)$-stimulated ATP-ase appears to be located at the inner surface of the erythrocyte membrane (Marchesi and Palade, 1967). This asymmetry is in line with the localization of phosphatidylserine at the inside of the erythrocyte membrane. This view was supported by the observation that treatment of erythrocytes with phospholipase A_2 (*Naja naja*) and sphingomyelinase (which causes a non-lytic hydrolysis of the phospholipids in the outer membrane surface), did not inactivate the $(Na^+ + K^+)$-stimulated ATP-ase. A small fraction of the total phosphatidylserine of the erythrocyte membrane is associated with this enzyme and essential for a proper function of the active centre of the $(Na^+ + K^+)$-ATP-ase at the inside of the membrane.

ON THE MOLECULAR ARCHITECTURE OF THE MYELIN MEMBRANE.

Specific protein-lipid interaction. The major constituents of the myelin membrane have been characterized. Myelin contains only three major proteins: the A_1 basic protein, Folch-Lees proteolipid and Wolfgram protein. About 80% of this membrane consists of lipids: phospholipids, cerebrosides and cholesterol. Studies with the A_1 basic protein and Folch-Lees protein have attempted to answer the question whether these particular myelin proteins show some degree of specificity in their interactions with lipids. Such interactions were examined using the monolayer technique (Demel *et al.*, 1973; London *et al.*, 1974). A monomolecular film of a defined lipid is formed at the air-water interface and adjusted at a given initial pressure. When a protein is injected into the subsolution penetration into the lipid monolayer will increase the surface pressure; when the surface pressure is kept constant an increase of surface area can be measured. It is also possible to add radio-active protein and to detect a change in surface radio-activity. Measurement of the change in surface pressure showed a pronounced interaction of A_1 basic protein with negatively charged lipids of myelin such as cerebroside sulphate and less for phosphatidylserine, while neutral lipids such as phosphatidylcholine, cholesterol and cerebroside showed markedly less affinity. Measurement of surface radio-activity using ^{131}I-labelled A_1 protein showed that the amount of A_1 basic protein bound to cerebroside sulphate in the monolayer parallels the surface pressure increase observed. Effects of different solutes and an increase of surface area at constant pressure indicated that both ionic and hydrophobic forces are involved in the interaction of A_1 basic protein and lipids (Demel *et al.*, 1973).

The Folch-Lees proteolipid showed a much broader spectrum of affinity than does the A_1 protein and caused an increase in surface pressure for a variety of myelin lipids (London *et al.*, 1974). Most remarkable was an interaction of Folch-Lees protein with cholesterol, which was found to depend on the sterol structure. The differences in lipid affinity between A_1 basic protein and Folch-Lees protein were substantiated by competition experiments. A subsequent injection of A_1 basic protein and Folch-Lees protein underneath a cholesterol monolayer demonstrated that the Folch-Lees protein binds to cholesterol under displacement of A_1 protein from the interface. In similar experiments on monolayers of cerebroside sulphate the A_1 basic protein revealed the highest affinity for this myelin lipid. These experiments lend further support to the concept of specificity in lipid-protein interaction of the myelin membrane *in vivo*.

Asymmetry of the myelin membrane. Detailed X-ray analysis of the complexes formed between the A_1 basic protein and various myelin lipids revealed another clue for understanding of the ultrastructure of this membrane (Mateu *et al.*, 1973). The myelin basic protein gave with the acidic lipid fraction, consisting of various negatively charged phospholipids and cerebroside sulphate, a complex lamellar phase which contains two lipid bilayers in its unit cell. The electron density profile of the complex phase of basic protein-acidic lipid fraction shows two hydrocarbon layers of different thickness. The structure of the complex was deduced to contain two symmetrical lipid bilayers. One of the bilayers appears to be of the liquid type and contains phospholipids mainly; the other bilayer consists of cerebroside sulphate with a considerable fraction of the chains in a more rigid conformation. An asymmetric peak in the polar region suggests that the basic protein molecules interact differently with the segregated phospholipids and cerebroside sulphates. These studies of Mateu *et* al. (1973) demonstrate a specific affinity of the A_1 basic protein for the cerebroside sulphate which is in agreement with the experiments on monomolecular films (Demel et al., 1973). It will be of interest to extend such X-ray studies to the Folch-Lees protein and to ascertain whether this protein preferentially associates with cholesterol as suggested by the investigation on monomolecular films (London et al., 1974). Studies on the structure of the intact myelin membrane indicated that this membrane is asymmetric; both protein and lipid asymmetry have been proposed (compare e.g. Dickinson et al., 1970; Blaurock, 1970; Caspar and Kirschner, 1971, Kirschner and Caspar, 1972). Extrapolation of the results discussed above strongly suggest that the specific protein-lipid association observed in vitro is responsible for both lipid and protein asymmetry in the myelin membrane. The striking differences in the affinity of Folch-Lees protein and the A_1 basic protein for lipids suggest a structure of distinct protein-lipid layers in myelin which has to be elucidated in greater detail by further experimentation.

Regions of A_1 basic proteins which interact with lipids. The general problem whether membrane proteins have particular regions that have a higher affinity for lipids than others was studied with A_1 myelin basic protein after its specific interaction with monolayers of cerebroside sulphate. The known amino acid sequence of this protein (Eylar et al., 1971; Carnegie, 1971) made it attractive to investigate the protection of various regions of the protein from the hydrolytic action of trypsin after formation of ^{131}I-A_1 basic protein-lipid complexes. Trypsin digestion of the A_1 basic protein after penetration of the lipid monolayer gave a reduction of about 50% of surface pressure and surface radio-activity (London et al., 1973). This suggests that protein hydrolysis might be incomplete as a result of protection by the lipids of sites available to enzymatic hydrolysis in the basic protein alone. A comparison of the peptide maps obtained with and without protein-lipid interaction showed that linkages in the N-terminal part of the protein (position 20-113) were preserved in the protein-lipid complex. A hypothetical model for the A_1 basic protein-lipid complex representing the specific lipid binding sites of the protein and involving ionic and hydrophobic interactions was proposed by London et al. (1973).

RELATIONS OF LIPID-LIPID, LIPID-CATION AND LIPID-PROTEIN INTERACTION WITH MEMBRANE PROPERTIES.

Many physical properties of biomembranes are dictated by the physical characteristics of the lipids which depend on the chemical structure of these lipid components. Excellent correlations have been found between permeability properties of liposomes and natural membranes (de Gier et al., 1972; van Deenen et al., 1972). The consequences of induction of chemical variations in lipid structure of both artificial and natural membranes could be predicted in many cases on the basis of the behaviour of single compounds in a most simple model system viz. a monomolecular film. Such correlations are mostly qualitative because in natural membranes various forms of lipid-lipid, lipid-cation and lipid-protein interaction create a more complicated situation.

Fatty acid structure and passive transport. The effect of systematic alterations in fatty acid composition and permeation of non-electrolytes (glycerol and erythritol) through the cell membrane of *Acholeplasma laidlawii* demonstrated that chain length, branching and degree of unsaturation of the hydrocarbon chains modifies passive permeability in a similar manner as in liposomes (McElhaney *et al.*, 1973 de Kruyff *et al.*, 1973a). It was deduced that the diffusion of these non-electrolytes through the *A.laidlawii* cell membrane at temperatures in the region of transition of membrane lipids from the gel to liquid-crystalline state occurs preferentially in those parts of the membrane where the fatty acids are still in the liquid state and have the highest thermal mobility.

Comparable results on fatty acid composition and non-electrolyte diffusion were obtained with an unsaturated fatty acid requiring auxotroph of *Escherichia coli* (Haest *et al.*, 1972a).

Lipid phase transition and membrane leakage. Thermotrophic transitions in biomembranes have been studied in great detail (Steim, 1972). The temperature induced transition of lipids from the liquid crystalline to the gel phase effects the properties of the membrane considerable. In *A.laidlawii* cells and *E.coli* cells with all their lipids in the gel phase a high degree of fragility towards mechanical forces was observed. In the case of *A.laidlawii* cells no damage of the membrane occurs as such by passing the transition temperature (de Kruyff *et al.*, 1973a). However, in membranes of an unsaturated fatty acid-requiring mutant of *E.coli* a rapid cation release was observed when the cells were rapidly cooled below the transition temperature of the membrane lipids. This cold shock induced permeability turned out to be somewhat selective in as much as the spontaneous leak of K^+ was not accompanied by a release of cytoplasmic enzymes (Haest *et al.*, 1972a). It is noteworthy that the change in permeability induced by lipid phase transition in *E.coli* membranes can also be demonstrated in liposomes made of synthetic phospholipids. These artificial bilayers revealed a comparable release of K^+ when exposed to temperatures below that of liquid crystalline-gel phase transition as detected by differential scanning calorimetry (Haest *et al.*, 1972a). Recently, an increase in ion permeability of liposomes in the phase transition was reported also by Papahadjopoulos *et al.*, (1973). For the explanation of the observed alteration of the barrier properties of the lipid-membranes it is of interest to quote results of the X-ray studies (Luzzati *et al.*, 1972) and freeze-etch electron microscopy on organized lipid system (Ververgaert et al., 1973). Upon lowering of the temperature various phospholipids pass from the liquid-crystalline state into "undulated" bilayers in which the molecules are rigid with stiffened chains and tilted with respect to the normal of the bilayer plane. The transition to this more close packing of the hydrocarbon chains may permit ion to penetrate the hydrophobic barrier. Regions of band patterns, which may reflect similarly structure lipid areas, were also observed below the transition temperature of A.laidlawii and E.coli membranes (Verkleij et al., 1972).

Fatty acid composition and valinomycin mediated transport. Studies on valinomycin induced transport of Rb^+ and K^+ showed that the structure of the fatty acid constituents of phospholipids in liposomal membranes (in addition to the phospholipid polar headgroup) determines the rate of carrier mediated transport (de Gier et al., 1972). Studies on A.laidlawii cells with different fatty acid composition demonstrated that the conclusions previously taken for the model systems are also valid for a natural membrane (van der Neut-Kok et al., 1974). Further studies on liposomes indicate that the higher actions of valinomycin in the more unsaturated systems is not due to a higher affinity of the ionophore for the membrane but is the consequence of a greater mobility of the cation-anion-carrier complex (Blok et al., 1974). The chemical nature of the immediate lipid environment of a membrane carrier is believed to be important for the effectiveness of the transport system. A regulatory function of lipids on the turnover rate of carrier mediated transport is possible.

Cholesterol-phospholipid interaction. Continuous attention over the years has been given to the effect of cholesterol on biological membranes. The so-termed condensing effect of cholesterol on most phospholipid species in the liquid-crystalline state causes a reduction of permeability (passive and carrier-mediated) of artificial and natural membranes (compare e.g. van Deenen et al., 1972). On the other hand cholesterol may effect the packing of phospholipid species in the gel-state, and prevent for some phospholipid species (e.g. dipalmitoylglycerophosphorylcholine) the formation of crystalline-gel like regions. An extensive monolayer and differential scanning calorimetry study on a variety of sterols and phospholipids gave further information about the effect of the polar headgroup (de Kruyff et al. 1973b). In agreement with various studies from this laboratory on model systems it was found in A.laidlawii that only sterols having a 3β-OH group, a planar ring and a hydrophobic side chain at C_{17} were capable to induce permeability alterations in this natural membrane (de Kruyff, 1973a). The effect of cholesterol on the phase transition of codispersions of pairs of synthetic lecithins with different fatty acid chain length and degree of unsaturation was studied by D.S.C. (de Kruyff et al., 1973b, 1974). In these mixtures cholesterol interacts preferentially with the lecithin species having the lowest transition temperature. At temperatures at which phase separation occurs cholesterol will certainly not be distributed randomly. The preferential interaction of cholesterol with the phospholipid species in the liquid-crystalline state could also be visualized by freeze-etch electron microscopy (Verkleij *et al.*, 1974b). A non-random distribution of cholesterol may occur in several biological membranes.

Effects of polar headgroups of phospholipids. A most pronounced difference in structure and properties is displayed by two related bacterial phospholipids, the negatively charged phosphatidylglycerol and the positively charged lysylphosphatidylglycerol. Because in *S.aureus* the ratio of these two lipids is highly dependent on the environmental pH it is possible to study the effect of phospholipids on membrane properties such as non-electrolyte diffusion and valinomycin mediated cation transport (Haest *et al.*, 1972b). The permeability of intact cells appeared to increase with increasing lysylphosphatidylglycerol to phosphatidylglycerol ratio. The non-electrolyte permeability of liposomes prepared with lysylphosphatidylglycerol was higher than those prepared with phosphatidylglycerol as was expected on the basis of the behaviour of monomolecular layers, demonstrating a larger area per molecule for lysylphosphatidylglycerol than for glycerol. However, the relations found in this case between natural membranes and model systems may have been explained in a somewhat oversimplified manner. In a recent study Tocanne *et al.* (1974a) found that a phosphatidylglycerol-lysylphosphatidylglycerol mixture at a ratio 1:1 forms a strong condensed state and both at the air-water interface and in bulk the phospholipid-phospholipid interaction gives rise to a new system.

Measurements of the action of valinomycin showed that there is a striking reduction in the action of this carrier in the cell membrane when the ratio of positively to negatively charged phospholipid increased (Haest *et al.*, 1972b). A similar behaviour was found when liposomes of different phospholipid composition were studied.

Effects of phospholipid-cation interaction on lipid phase transition. Much attention has been given to cation binding to phospholipids and the relevance of the interaction of monovalent and divalent cation with lipid monolayers and bilayers has been generally recognized. This area of research may have received a new impetus by the recent demonstration that the transition of phospholipids from the gel to the liquid-crystalline state is drastically influenced by pH and the presence of divalent cation (Träuble and Eibl, 1974; Verkleij *et al.*, 1974a; Tocanne *et al.*, 1974b). The first mentioned authors demonstrated by means of a fluorescent probe technique the effect of pH and several cations on the transition temperature of phosphatidic acid and other negatively charged phospholipids. Provocative biological implications were proposed. The investigations in the author's laboratory were carried out on phosphatidylglycerol, by monolayers, D.S.C. and freeze-etch electron microscopy. Even at low Ca^{++} concentration

(Ca^{++}; phosphatidylglycerol 1/100) the transition temperature is markedly raised. At equivalent ratios a jump of about 70° in the transition temperature occurs and a structure of tightly packed bilayers wrapped in cylinder could be observed. Recent studies showed that Mg^{++} also causes an upward shift in the transition temperature, an increase of the heat content, a decrease of the molecular area of phosphatidylglycerol while cilindrical studies could be observed as well (Ververgaert et al., 1974). The interaction of polar headgroups of negatively charged phospholipids with divalent cations is transmitted to the apolar region of the bilayers causing a limitation of the mobility of the hydrocarbon chains. Similar effects can be noted with positively charged phospholipids and anions, and during the interaction between positively and negatively charged phospholipids.

Effects of lipid-protein interaction on lipid phase transition. On ordering effect of protein on the lipid core can be envisaged when ionic forces are predominantly involved in lipid-protein association. Butler et al. (1973) made a spin probe and X-ray study on the effects of proteins on bilayers of brain lipids. They observed that only proteins which have a net positive charge effected the molecular orientation of the lipids. With the aid of D.S.C. it was demonstrated that the transition temperature of negatively charged phospholipid is considerably increased by A_1 basic protein from myelin (Verkleij et al., 1974a). The interaction of membrane proteins with other components in principle may also effect the physical state of lipids. Further studies are also required to study how the formation of tricomplexes of protein-lipid-cation effect the organization of lipids.

Effect of lipid phase transition on lipid-protein interaction. The visualization of the transition from the liquid-crystalline to the gel phase of lipids in liposomes (Verkleij et al., 1972; Ververgaert et al., 1973) stimulated the study to detect possible phase transitions in natural membranes by freeze-etch electronmicroscopy. It was found by Verkleij et al. (1972) that particles which can be seen on freeze-fracture faces of many membranes (compare Branton et al., 1972) aggregate upon cooling of A.laidlawii cells below the temperature of phase transition of the membrane lipids. It was argued that the condensation of the hydrocarbon chain of the membrane lipids results in a squeezing out of protein parts which were embedded in the apolar part of lipid core. A similar reorganization was also observed in the fracture faces of Tetrahymena after lowering the temperature (Speth and Wunderlich, 1973). That the extent of particle aggregation is correlated with the degree of lipid phase transition was demonstrated through the use of cell membranes in which the fatty acid composition had been varied viz A.laidlawii (James and Branton, 1973) and a mutant of E.coli (Haest et al., 1974). In the latter study particle aggregation induced by lipid phase transition was, however, not seen in the fracture faces of some bacterial species (e.g. S.aureus), although differential scanning calorimetry and breaks in the Arrhenius plots of membrane bound enzymes indicated that such transitions occurred. The non-appearance of particle aggregation could be explained by the presence of branched chain fatty acid in the membrane lipids of these bacteria. When these fatty acids were incorporated into A.laidlawii the membrane failed to give the reorganization induced phase transition. It is considered likely that a rather loose packing of the lipids containing branched fatty acid constituents in the gel phase gives space for protein parts to remain localized in the hydrocarbon region (Haest et al., 1974).

REFERENCES

Blaurock, A.E. (1970). J. Mol. Biol., 56, 35
Blok, M.C., de Gier, J. and van Deenen, L.L.M. (1974). Biochim. Biophys. Acta, in the press
Branton, D., Elgsaeter, A. and James, R. (1972). in: Proc. 8th FEBS Meeting, Amsterdam. North Holland American Elsevier, Amsterdam, New York vol. 28, 165
Bretscher, M.S. (1972). J. Mol. Biol. 71, 523
Bretscher, M.S. (1973) Science 181, 622
Butler, K.W., Hanson, A.W., Smith, J.C.P. and Schneider, H. (1973). Can. J. Biochem. 51, 980
Carnegie, P.R. (1971). Biochem. J. 123, 57
Caspar, D.L.D. and Kirschner, D.A. (1971) Nature, New Biol. 231, 46
van Deenen, L.L.M. and de Gier, J. (1974). in: "The Red Blood Cell" ed. D.M. Surgenor, Academic Press, New York and London, vol. I, p. 147
van Deenen, L.L.M., de Gier, J. and Demel, R.A. (1972). in: "Current Trends in the Biochemistry of Lipids". J. Ganguly and R.M.S. Smellie, eds. Academic Press, London, p. 377
Demel, R.A., London, Y., Geurts van Kessel, W.S.M., Vossenberg, F.G.A. and van Deenen, L.L.M. (1973). Biochim. Biophys. Acta 311, 507
Demel, R.A., Geurts van Kessel, W.S.M., Zwaal, R.F.A., Roelofsen, B. and van Deenen, L.L.M. (1974). paper in preparation
Dickinson, J.P., Jones, K.M., Aparicio, S.R. and Lumsden, C.E. (1970). Nature 227, 1133
de Gier, J., Haest, C.W.M., van der Neut-Kok, E.C.M., Mandersloot, J.G. and van Deenen, L.L.M. (1972). in: Proc. 8th FEBS Meeting Amsterdam, North-Holland-American Elsevier, Amsterdam, New York, vol. 28, p. 263
Gordesky, S.E. and Marinetti, G.V. (1973). Biochem. Biophys. Res. Commun. 50, 1027
Gorter, E. and Grendel, F. (1925). J. Exp. Med. 41, 439
Haest, C.W.M., de Gier, J., van Es, G.A., Verkleij, A.J. and van Deenen, L.L.M. (1972a). Biochim. Biophys. Acta 288, 43
Haest, C.W.M., de Gier, J., Op den Kamp, J.A.F., Bartels, P. and van

Deenen, L.L.M. (1972b). Biochim. Biophys. Acta 255, 720
Haest. C.W.M., Verkleij, A.J., de Gier, J., Scheek, R., Ververgaert, P.H.J., van Deenen, L.L.M. (1974). Biochim. Biophys. Acta 356, 17
James, R. and Branton, D. (1973). Biochim. Biophys. Acta 323, 378
Kirschner, D.A. and Caspar, D.L.D. (1972). Ann. N.Y. Acad. Sci. 195, 309
de Kruyff, B., de Greef, W.J., van Eyk, R.V.W., Demel, R.A. and van Deenen, L.L.M. (1973a). Biochim. Biophys. Acta 298, 479
de Kruyff, B., Demel, R.A., Slotboom, A.J., van Deenen, L.L.M. and Rosenthal, A.F. (1973b). Biochim. Biophys. Acta 307, 1
de Kruyff, B., van Dijck, P.W.M., Demel, R.A., Schuyff, A., Brants, F. and van Deenen, L.L.M. (1974). Biochim. Biophys. Acta 356, 1
London, Y.; Demel, R.A., Geurts van Kessel, W.S.M., Vossenberg, F.G.A. and van Deenen, L.L.M. (1973). Biochim. Biophys. Acta 311, 520
London, Y., Demel, R.A., Geurts van Kessel, W.S.M., Zahler, P. and van Deenen, L.L.M. (1974). Biochim. Biophys. Acta 332, 69
Luzzati, V., Tardieu, A., Gulik-Krzywicki, T., Mateu, L., Rouch, J.L., Shechter, E., Cabre, M. and Caron, F. (1972). Proc. 8th FEBS Meeting, Amsterdam, North-Holland-American Elsevier, Amsterdam, New York, vol. 28, p. 173
Mateu, L., Luzzati, V., London, Y., Gould, R.M., Vosseberg, F.G.A. and Olive, J. (1973). J. Mol. Biol. 75, 697
Marchesi, V.T. and Palade, G.E. (1967). J. Cell. Biol. 35, 385
McElhaney, R.N., de Gier, J. and van der Neut-Kok, E.C.M. (1973). Biochim. Biophys. Acta 298, 500
van der Neut-Kok, E.C.M., de Gier, J., Middelbeek, E.J. and van Deenen, L.L.M. (1974). Biochim. Biophys. Acta 332, 97
Papahadjopoulos, D., Jacobson, K., Niz, S. and Isac, T. (1973). Biochim. Biophys. Acta 311, 330
Renooy, W., van Golde, L.M.G., Zwaal, R.F.A., Roelofsen, B. and van Deenen, L.L.M. (1974). Biochim. Biophys. Acta, in the press
Roelofsen, B. and van Deenen, L.L.M. (1973). Eur. J. Biochem. 40, 245
Steim, J.M. (1972). Proc. 8th FEBS Meeting, Amsterdam, North-Holland-American Elsevier, Amsterdam, New York, vol. 28, p. 185
Shohet, S.B. (1970). J. Clin. Invest. 49, 1668
Speth, V. and Wunderlich, F. (1973). Biochim. Biophys. Acta 291, 621
Träuble, H., Eibl. H. (1974). Proc. Nat. Ac. Sci. U.S. 71. 214
Tocanne, J.F., Ververgaert, P.H.J. Th., Verkleij, A.J., and van Deenen, L.L.M. (1974a). Chem. Phys. Lipids 12, 220
Tocanne, J.F., Ververgaert, P.H.J.Th., Verkleij, A.J. and van Deenen, L.L.M. (1974b). Chem. Phys. Lipids 12, 201
Verkleij, A.J., Ververgaert, P.H.J.Th., van Deenen, L.L.M. and Elbers, P.F. (1972). Biochim. Biophys. Acta 288, 326
Verkleij, A.J., Zwaal, R.F.A., Roelofsen, B., Comfurius, P., Kastelijn, D. and van Deenen, L.L.M. (1973). Biochim. Biophys. Acta 323, 178
Verkleij, A.J., de Kruyff, B., Ververgaert, P.H.J.Th., Tocanne, J.F.

and van Deenen, L.L.M. (1974a). Biochim. Biophys. Acta 339, 432
Verkleij, A.J., Ververgaert, P.H.J.Th., de Kruyff, B. and van Deenen, L.L.M. (1974b). Biochim. Biophys. Acta in the press
Ververgaert, P.H.J.Th., Verkleij, A.J., Elbers, P.F. and van Deenen, L.L.M. (1973). Biochim. Biophys. Acta 311, 320
Ververgaert, P.H.J.Th., de Kruyff, B., Verkleij, A.J., Tocanne, J.F. and van Deenen, L.L.M. (1974) paper submitted
Wallach, D.F.H. (1972). Biochim. Biophys. Acta 265, 61
Whittam, R. (1967). in: "The Molecular Mechanisms of Active Transport" eds. G.C. Quarton, T. Melneclink and F.O. Schmitt, (Rockefeller Press, New York) p. 313
Zwaal, R.F.A., Roelofsen, B. and Colley, C.M. (1973). Biochim. Biophys. Acta 300, 159

PROTEIN-LIPID INTERACTIONS IN ERYTHROCYTE MEMBRANES STRUCTURAL AND DYNAMIC ASPECTS

P. ZAHLER, R.F.A. ZWAAL, R. KRAMER

Theodor Kocher-Institute, University of Berne
Central Laboratory, Transfusion Service
Swiss Red Cross

INTRODUCTION

The elucidation of membrane structure is certainly still far from being resolved, although important progress has been achieved within the last 10 years. Thus there now is no doubt that membrane proteins are localized in many alternative ways with respect to the lipid bilayer (Fig. 1).

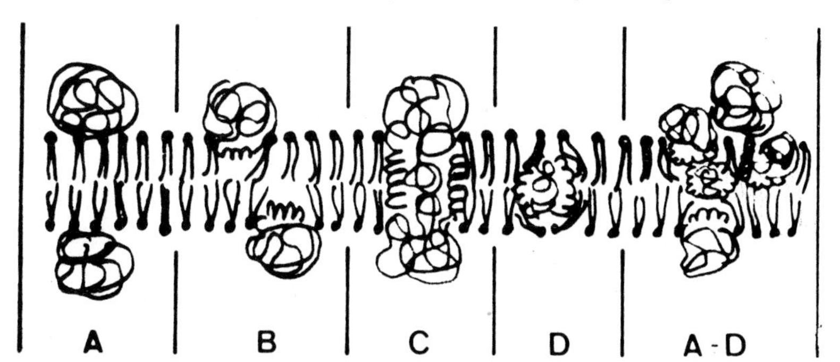

Fig. 1 Possible localization of membrane proteins with respect to the lipid bilayer

A: Membrane associated proteins interacting with the negatively charged surface of the bilayer by predominantly electrostatic bounds.

B: Penetrating amphipatic proteins interacting with the lipids both hydrophobically and electrostatically; localization inside or outside.

C: Proteins extending completely through the bilayer exposing two extreme parts of the protein molecule which have a hydrophilic periphery intercepted by a hydrophobic part.

D: Proteins fully included within the bilayer and consequently exposing a totally hydrophobic surface of their molecule.

A-D: Possible arrangement of functional units where all possibilities may be realized within a well defined quarternary structure.

Some proteins are attached to the polar surface as it was proposed by the "unit membrane-theory" (A), some penetrate the bilayer from either side (B), some pass through the hydrophobic barrier (C) and some are fully included (D). Protein-protein interactions may then lead to the kinds of complexes as indicated in A-D.

With the exception of attached proteins[1] it is quite clear that protein-lipid interactions also occur in purely hydrophobic areas of the membrane, implying that membrane proteins must contain hydrophobic regions on their surfaces allowing apropriate interaction with the hydrocarbon portions of the lipid molecules. Although it is known from oligomeric proteins that hydrophobic protein-protein interactions may play an important role also in non-membrane proteins it seems that the special character of "membrane-proteins" [2] may be defined by the necessity of considerable regions of the protein periphery showing a predominantly hydrophobic arrangement of amino acid side chains. In principle there exist 3 ways in which proteins may fulfill such a necessity for hydrophobic surface areas.

a) Membrane proteins may contain a higher proportion of amino acids with hydrophobic side chains as compared to the water-soluble proteins. In fact several authors have studied amino acid compositions of mixtures and also of single membrane proteins and have found relatively large portions of amino acids with hydrophobic side chains (Rosenberg and Guidotti, 1964; Vanderkooi and Capaldi, 1972; Capaldi and Vanderkooi, 1972).

b) The membrane proteins, although showing an amino acid composition similar to non-membrane proteins, may contain segments, some being predominantly hydrophobic, others being mainly hydrophilic, thus leading to amphipatic molecules. The major glycoprotein from red cells may be such an example (Marchesi et al., 1972; Segrest et al., 1972; Segrest et al., 1973). Other investigators have described the splitting of hydrophilic portions of membrane proteins by proteolytic enzymes such as trypsin or papain leaving behind a hydrophobic segment anchored in the membrane core (Jackson and Seaman, 1972; Winzler et al., 1967; Winzler, 1969). These findings may be interpreted as favouring this alternative concept of amphipatic protein molecules.

c) A third possibility may be realized by a special folding of a polypeptide with a "normal" amino acid composition to give a tertiary structure in which predominantly hydrophobic side chains are exposed to the exterior while in the interior of the protein molecule hydrophilic regions would predominate. α-helical segments are known in many cases to have hydrophilic and hydrophobic half sides and some data indicate that such hydrophobic half sides exposed on the surface of the protein molecule would represent the binding sites towards lipid molecules of the membrane bilayer (Wallach and Zahler, 1966; Zahler and Weibel, 1970; Jackson et al., 1973).

---

1 The following alternative terms are currently found in the literature: Membrane associated proteins, extrinsic proteins, loosely-bound proteins, water-soluble membrane proteins, shockable proteins (bacterial membranes)

2 Alternative terms: Intrinsic proteins, strongly-bound proteins, core proteins, hydrophobic membrane proteins.

Very little is known about the way membrane proteins are inserted into the membrane after being synthesized inside the cell. It is difficult to conceive a mechanism allowing the transport of such hydrophobic molecules from the rough endoplasmatic reticulum to the plasma membrane through the cytoplasm without the proteins immediately aggregating or without binding lipid instantly and assembling to form membranes right away. One should therefore postulate the synthesis of "pre-membrane proteins", which are soluble and monomeric in the cytoplasm and which travel to the plasma membrane where they could interact, penetrate and be modified chemically (e.g. glycosylated); only then would they be refolded to form membrane proteins, exposing hydrophobic areas and interacting with the membrane lipids. An alternative possibility would imply that membrane vesicles, formed next to the loci of biosynthesis of the proteins, are transported to the surface membrane, to which they could then fuse. Such a mechanism however would conflict with the lipid asymmetry of membrane fractions (Zwaal et al., 1973; Gordesky and Marinetti, 1973). The continuity of internal and plasma membrane as postulated on the basis of electron microscopy also leads to difficulties in understanding lipid asymmetry.

The control of the composition of membrane lipids and their asymmetric disposition is strongly connected to the problem of protein-lipid interaction. We have shown that the extremely low content in phosphatidyl-choline of sheep red cell membranes may partly be the result of a preferential binding of sphingomyelin by the apoproteins of this membrane type (Kramer et al., 1972). These experiments were based on the recombination technique using 2-chloroethanol for solubilizing the membranes and isolating the lipidfree membrane proteins. The fact that such a preference could be shown after treatment with this helicogenic solvent strongly supports the hypothesis that the α-helical segments might be the sites of interaction with the lipid molecule. The experiments also proved that membrane proteins penetrate into the bilayer because distinction of the two choline-containing phospholipids can only be made in sections II and III of the molecule, which both are known to reside in the interior of a bilayer (Fig. 2).

During the recombination experiments related to the preferential binding of sphingomyelin by the apoproteins of sheep red cell membranes we consistently observed the appearance of lyso-phosphatidyl-choline after the binding dialysis, suggesting the presence of a phospholipase A. This finding was unexpected in two ways:

First, our recombination technique, based on the use of 2-chloroethanol as solubilizing agent, led to an inactivation of all enzymatic membrane proteins, as far as they had been tested (Zahler and Weibel, 1970). If the appearance of lyso-phosphatidyl-choline was the result of a catalytic breakdown, this would mean that the phospholipase A in question escaped denaturation by 2-chloroethanol or, alternatively, that this membrane protein was reactivated during the transfer to an aqueous medium in the presence of lipid.

Second, most of the current work on phospholipase A in cell membranes indicates that such an enzyme seems to be completely absent from red cell membranes generally (Robertson and Lands, 1964; Oliveira and Vaughan, 1964; Munder et al., 1965; Mulder et al., 1965; Mulder and van Deenen, 1965). However Paysant et al., 1967 have described various experiments which indicated that red cell membranes from rats have phospholipase A activity

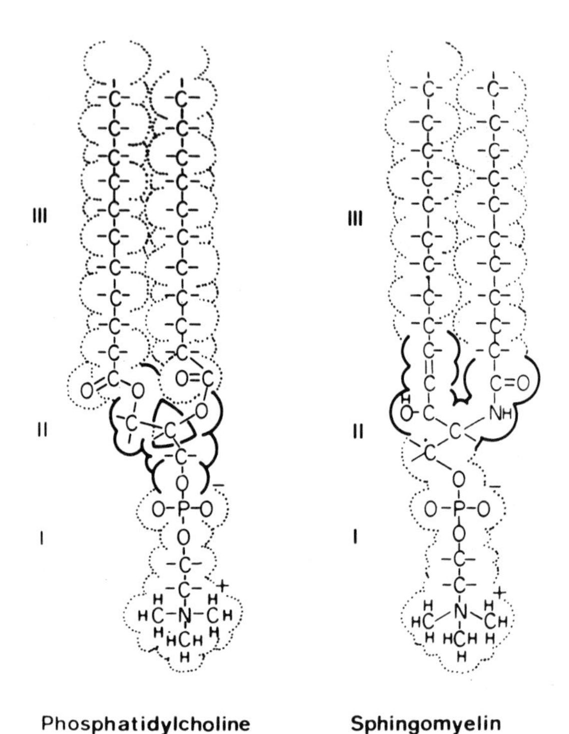

Fig. 2 Comparison of the structure of phosphatidyl-choline and sphingomyelin.
I Charged region in both cases with the identical phosphorylcholine group.
II Polar region of two ester bonds in phosphatidyl-choline and the amide bond and hydroxyl group in sphingomyelin.
III Hydrophobic region of the hydrocarbon chains differing in phosphatidyl-choline and sphingomyelin.

towards phosphatidyl-glycerol from bacteria. Later they could show (Paysant et al., 1970) that human erythrocytes contain a similar enzyme which also splits phosphatidyl-ethanolamine and that there was a considerable increase in activity after treatment of red cell haemolysate with trypsin. This proteolytic treatment also resulted in a breakdown of phosphatidyl-choline.

The possibility of the existence of a phospholipase A in the membrane of sheep red cells and its possible regulatory function in relation to the extraordinary lipid composition of this membrane type induced a detailed study on the identity and properties of this enzyme with the following results.

RESULTS AND DISCUSSION

An enzymatic system in sheep red cell membranes which catalyses the hydrolysis of exogenous phosphatidyl-choline into free fatty acids and 1-acyl-phosphatidyl-choline could be demonstrated.
(Phospholipase A_2, phosphatide acyl hydrolase, E.C. 3.1.1.4) (Kramer et al., 1974).

With incubation times longer than one hour lyso-phosphatidyl-choline is further degraded into free fatty acid and water-soluble products (Fig. 3 and Fig. 5). Most likely, two enzymes are involved, a phospholipase A_2 and a lyso-phospholipase; this is in agreement with previous findings of Mulder et al., 1965 and Munder et al., 1965 showing the existence of a lyso-phospholipase in erythrocyte haemolysate of human, rabbit and ox. Some general characteristics of the phospholipase A in sheep erythrocyte membranes do agree with those of well known phospholipases A from other sources (Crot. ad. venom, E.coli, post heparin plasma, pig pancreas, rat spleen, rat brain and rat liver). The enzyme has an alkaline pH-optimum at pH 8 and it is likely that it is specific for the fatty acid at the 2-position (Fig. 4). Enzyme reaction is only obtained in the presence of Ca ions and is greatly stimulated by detergents. Moreover, similarities are observed in properties concerning the protein nature of the phospholipase A. Thus, the enzyme is affected neither by handling nor by freezing and thawing. It also shows a remarkable stability against heat and the denaturing action of surface active compounds. In spite of these similarities, however, we do not observe that the phospholipase A of sheep red cell membranes is completely identical to one of the well described phospholipases from other sources.

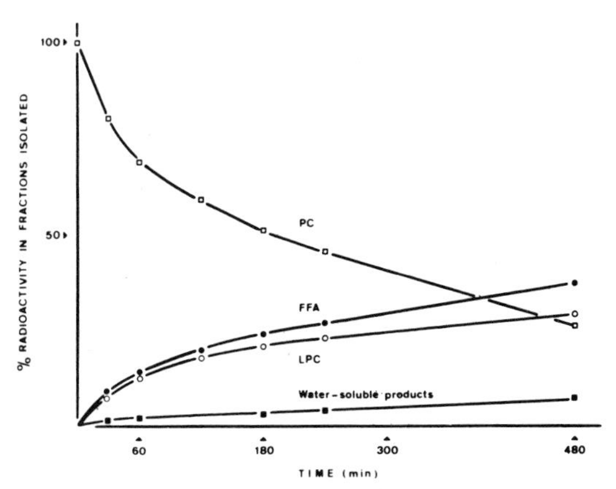

Fig. 3 Phospholipase A activity of sheep red cell membranes. Time dependence of the degradation of phosphatidyl-choline (□——□) and formation of the lyso-phosphatidyl-choline (○——○), free fatty acids (●——●) and water-soluble products (■——■).
Assay system: 0.05 M glycyl-glycin pH 8.0, 8 mM $CaCl_2$, 3 mg/ml Triton-X-100, 1mg phosphatidyl-choline and $5x10^{-3}$ nmoles ^{14}C-phosphatidyl-choline in a total volume of 0.5 ml.

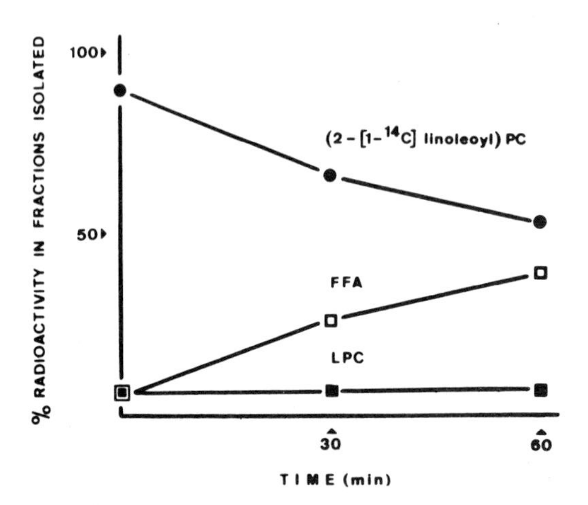

Fig. 4 Identification of the phospholipase A_2 specifity of the enzyme from sheep red cell membrane. Phosphatidyl-choline labeled with ^{14}C-linoleic acid in the 2-position (2-[1-^{14}C] linoleoyl) phosphatidyl-choline) was used as substrate. No label can be found in the lyso-compound. FFA = free fatty acid, LPC = lyso-phosphatidyl-choline.

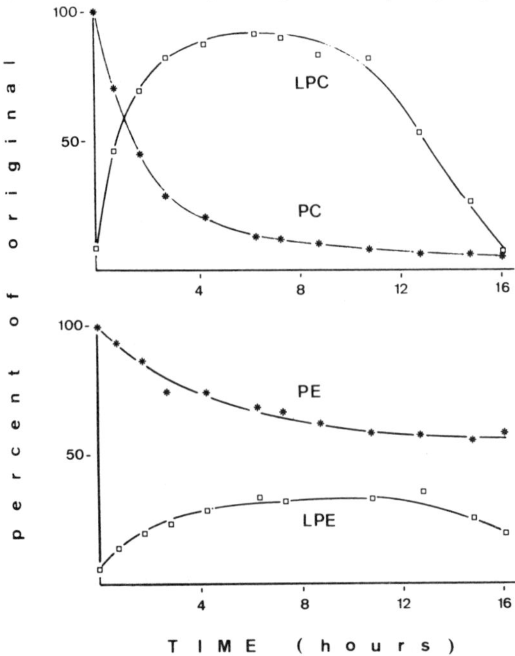

Fig. 5 Degradation of glycerophospholipids by sheep red cell membrane phospholipase A. Abbreviations: PC, phosphatidyl-choline; LPC, lyso-phosphatidyl-choline; PE, phosphatidyl-ethanolamine; LPE, lyso-phosphatidyl-ethanolamine.

As many other phospholipases A the enzyme is influenced by the addition of detergents. Accordingly, surface active compounds, such as deoxycholate and Triton X-100, cause a marked stimulation of the enzyme at low concentration (up to 3 mg/ml) and become slightly inhibitory when larger quantities are added to the reaction mixture. In contrast, dodecylsulfate moderately stimulates in a narrow concentration range (up to 0.5 mg/ml) and strongly inhibits the enzyme at higher concentrations. Our studies demonstrate the complex interaction of detergents with the membrane-bound phospholipase A, the overall effect being dependent on the type of detergent and its concentration. As previously discussed detergents may influence the rate of hydrolysis by altering the conformation or the packing of the substrate (Magee et al., 1962 and Shankland, 1970; de Haas et al., 1971). The marked variations in activation of phospholipase A in erythrocyte membrane apoproteins (or recombinate) as compared to stroma recall the "allotopic properties" of other membrane-associated enzymes (Coleman, 1973). Thus, when the characteristic microenvironment of membrane-bound enzymes is altered, differences in specific properties are often apparent (Scandella and Kornberg, 1971; Racker, 1967).

Complete inhibition by EDTA and subsequent activation of the enzyme by Ca ions are taken to indicate that metal ions are an absolute prerequisite for enzymatic activity. The stimulation of the phospholipase A by Ca ions, and to a smaller degree by Mg ions, indicates that $Ca^{\bullet\bullet}$ is not specifically required for the enzymatic reaction and suggests that $Mg^{\bullet\bullet}$ may partly substitute for it.

Organic solvents produce a depression in enzymatic activity, which is more pronounced with solvents of increased solubilizing power for membranes, such as 2-chloroethanol and butanol. However, further studies with 2-chloroethanol indicated that the inhibitory action of this solvent is not due to an unspecific denaturation of the enzyme protein. 2-chloroethanol is known to be effective for the total solubilization of erythrocyte membranes by separating lipids from proteins (Zahler and Wallach, 1967), but has always caused loss of enzymatic activity (Zahler et al., 1967). According to our results, the phospholipase behaves differently from other membrane-bound enzymes, in that after removal of 2-chloroethanol by dialysis it partially regains activity. As stated, 2-chloroethanol promotes the formation of α -helix conformation (Wallach and Zahler, 1966) and apparently converts the enzyme to a form which renatures rapidly when transferred to water.

An increasing number of membrane-bound enzymes have been shown to require lipid in order to carry out their functions (Coleman, 1973). Although phospholipase A activity is observed in isolated membrane proteins, the possibility that lipids are necessary for correct functioning of the enzyme cannot be excluded. Phosphatidyl-choline, present as a substrate in the incubation mixture, may cause a lipid reactivation of the enzyme. The role of lipids in the action of phospholipase A is now being investigated.

Dodecylsulfate is known to destroy rapidly the enzymatic activities of biological membranes by unfolding the polypeptide chains of the proteins. However, the phospholipase A is not irreversibely inactivated by its denaturating effect; although the enzyme is completely inhibited in the presence of dodecylsulfate at levels above 2 mg/ml, after removal of the denaturating agent, activity is restored. It has been established that after treatment of red cell ghosts with urea at a concentration of 8 M subsequent elimination of the denaturing agent, partial restoration of enzyme activity occurs. On the other hand, the enzyme is destroyed by dithiothreitol, a reagent used for cleavage of disulfide bridges. Obviously, the enzyme is sensitive to disulfide reduction.

When erythrocyte ghosts from various mammalian species were assayed for phospholipase activity at pH 8 in the presence of Triton X-100 and $CaCl_2$, breakdown of phosphatidyl-choline was found exclusively in sheep, ox and goat red cell ghosts (Table 1)

Table 1

Phospholipase A activities in various mammalian erythrocyte ghosts

Values indicate percentage degradation of endogenous phospholipids per 6 hour incubation at 37°C. Ruminant red cell ghosts were mixed, prior to incubation, with equal amounts of human red cell ghosts or an equivalent amount of sonicated egg phosphatidyl-choline. Degradation of phosphatidyl-serine is probably absent in all ghosts, but this could not always be established unambigously.

Species	Phosphatidyl-choline	Phosphatidyl-ethanolamine
Sheep	30 - 95	30 - 45
Ox	60 - 80	5 - 15
Goat	70 - 75	5 - 20
Rabbit	-	10 - 20
Human	-	-
Pig	-	5 - 20
Dog	-	-
Rat	-	35 - 50

It is emphasized that this lecithinase activity of ruminant red cell ghosts was detected on exogenous phosphatidyl-choline as substrate, since these ghosts contain very little phosphatidyl-choline (Van Deenen and de Gier, 1964; Nelson, 1967). No significant differences in degradation were observed when phosphatidyl-choline was supplied as egg phosphatidyl-choline, as total lipid extract of human erythrocyte ghosts, or as intact human erythrocyte ghosts mixed with an equal amount of the ruminant red cell ghosts. In all cases incubations in the presence of EDTA did not produce any significant phospholipid breakdown as compared to non-incubated ghosts. With the exception of human and dog, all the mammalian red cell ghosts tested produced breakdown of phosphatidyl-ethanolamine, although to a lower extent than the phosphatidyl-choline breakdown with ruminant erythrocyte ghosts. Degradation of phosphatidyl-choline or phosphatidyl-ethanolamine was always paralleled by the formation of the corresponding lyso-derivates. In some cases phosphatidyl-serine was reduced relative to sphingomyelin, but no lyso-phosphatidyl-serine was produced. Moreover, this occurrence appeared to be completely arbitrary and uncontrollable, and is probably due to a lower extractability of phosphatidyl-serine from ghosts incubated in the presence of Ca^{++}, in spite of adding EDTA prior to lipid extraction. It has been suggested by Paysant et al., 1970 that lysates of human erythrocytes exhibit a weak phospholipase A activity which is enhanced by trypsin treatment. Under our experimental conditions, however, pretreatment of human ghosts with trypsin, as described by Paysant et al., 1970 did not produce any detectable phospholipid degradation.

As stated above, degradation of phosphatidyl-ethanolamine was accompanied by the production of their lyso-derivates. This is shown for sheep erythrocyte membranes in Fig. 5.

Prolonged incubation periods produced a decrease in the lyso-compounds, probably due to lyso-phospholipase activity which is known to be present in other red cell membranes (Mulder et al., 1965; Munder et al., 1965; Ferber et al., 1968). At present it is unclear why this phospholipase activity seems to be enhanced in a later stage of the incubation. Although lyso-phosphatidyl-choline was nearly completely degraded, a residual amount of lyso-phosphatidyl-ethanolamine was always observed. The plasmalogen content of phosphatidyl-ethanolamine appeared to be approximately 25 % by weight. Since this content was not changed in the phosphatidyl-ethanolamine fraction after degradation, it is likely, that the residual lyso-phosphatidyl-ethanolamine is formed from the original alk-1-enyl compound and hence is not susceptible to lyso-phospholipase activity.

In further experiments, attention was focused on the degradation of phosphatidyl-choline by phospholipase A of ruminant erythrocyte membranes. In order to establish whether this activity is caused by a membrane-bound enzyme, both ox and sheep erythrocyte ghosts were extracted with 1 mM EDTA. In both cases, 15 - 20 % of the membrane proteins were extracted but no significant lecithinase activity could be detected in this fraction. On the other hand, the sediment after EDTA extraction showed a marked lecithinase activity, although approximately 20 % inactivation (as compared to the original ghosts) was observed.

Treatment of ox or sheep erythrocyte ghosts with pronase almost completely prevented degradation of phosphatidyl-choline. Similar results were obtained when intact cells were treated with pronase, suggesting that the majority of the lecithinase activity is located at the outside of the cell. (Fig. 6)

In order to detect whether this enzyme plays a role in phosphatidyl-choline turnover and renewal in intact ruminant erythrocytes, both pronase-treated and control cells (derived from sheep) were incubated with equal volumes of heat-treated serum to which (^{14}C)-phosphatidyl-choline was added. As shown in Fig. 7 A pronase treated cells incorporated approximately three times as much radioactive phosphatidyl-choline as control cells. In the case of control cells the incorporation was probably due to exchange between cold phosphatidyl-choline in cells and radioactive phosphatidyl-choline in serum, since the phosphatidyl-choline content of the membrane was not altered as compared to freshly isolated ghosts (0.7-8.9 % of the total phospholipids). On the other hand, the phosphatidyl-choline of pronase treated cells incubated with serum was definitely increased (2.5-3.0 %)

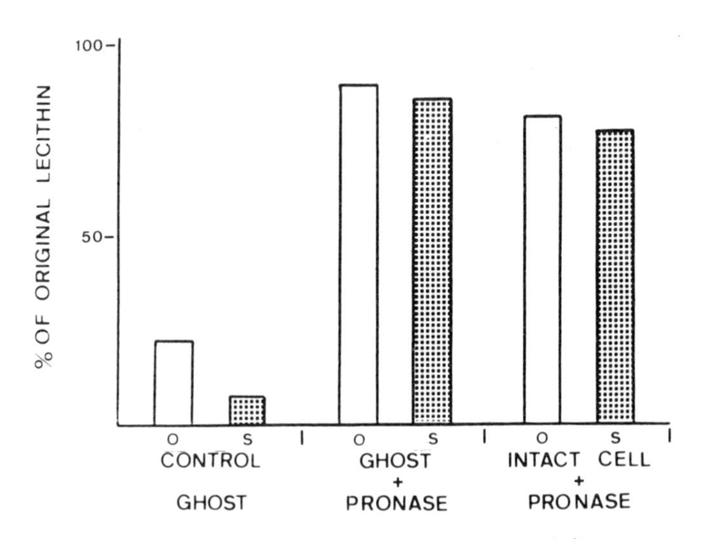

Fig. 6 Localization of the phospholipase A from ox (o) and sheep (s) red cell membranes. Degradation of phosphatidyl-choline by ghosts is shown after treatment of the ghosts or the intact cells prior to the ghost preparation with pronase. The inactivation of the enzyme in intact cells indicates its localization at the outside of the membrane.

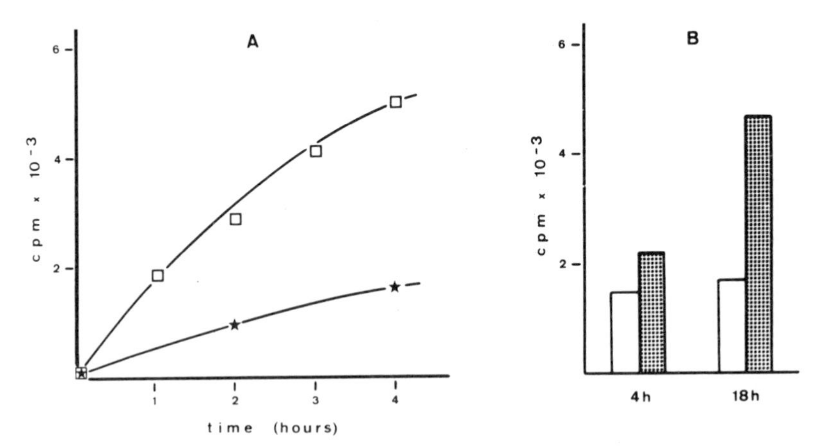

Fig. 7 A, B Incorporation of radioactive phosphatidyl-choline from serum into sheep red cells. Ordinate indicates counts/min in phosphatidyl-choline per mg of total membrane phospholipids.

A Cells treated with pronase prior to incubation procedure (□—□), control cells (✱—✱).

B Cells incubated with serum in the presence of EDTA (shaded bars) compared to control cells (open bars).

5. Pronase treatment of intact ox and sheep erythrocytes largely inactivated lecithinase activity, suggesting that the enzyme is present at the outside of the membrane. Incubation of serum with sheep red cells in which lecithinase was inactivated by pronase or inhibited by EDTA resulted in a slow net incorporation of lecithin from serum into cells.

6. It is concluded that lecithinase activities may be generally present at the outside of ruminant erythrocytes. It is speculated that this enzyme might play a role (together with other proteins) in maintaining the low lecithin content of the membrane.

REFERENCES

Capaldi, R.A. and Vanderkooi, G. (1972). Proc. Nat. Acad. Sci. (USA) 69, 930.
Coleman, R. (1973). Biochem. Biophys. Acta 300, 1.
Van Deenen, L.L.M. and de Gier, J. (1964). In: "The Red Cell" (Bishop, C. and Surgenor, D.M., eds) Academic Press, New York, Chapter VII, 243.
Ferber, E., Munder, P.G., Kohlschütter, A. and Fischer, H. (1968) European J. Biochem. 5, 395.
Gordesky, St.E. and Marinetti, C.V. (1973). Biochem. and Biophys. Res. Com. 50, 1027.
De Haas, G.H., Bonsen, P.P.M., Pieterson, W.A. and van Deenen, L.L.M. (1971). Biochem. Biophys. Acta 239, 252.
Jackson, L.T. and Seaman, G.V.F. (1972). Biochemistry 11, 44.
Jackson, R.L., Morrisett, J.D. and Gotto, A.M.Jr. (1973). In:"Protides of Biological Fluids"(ed. by Peeters, H., Brugge)p. 227.
Kramer, R., Schlatter, Ch. and Zahler, P. (1972). Biochem. Biophys. Acta 282, 146.
Kramer, R., Jungi, B. and Zahler, P. (1974). In press BBA Biochem. Biophys. Acta.
Magee, W.L., Gallai-Hatchard, J., Sanders, H. and Thompson, R.H.S. (1962). Biochem. J. 83, 17.
Marchesi, V.T., Tillack, T.W., Jackson, R.L., Segrest, J.P. and Scott, R.E. (1972). Proc. Nat. Acad. Sci. (USA) 69, 1445.
Mulder, E., van den Berg, J.W.O. and van Deenen, L.L.M. (1965). Biochem. Biophys. Acta 106, 118.
Mulder, E. and van Deenen, L.L.M. (1965). Biochem. Biophys. Acta 106, 348.
Munder, P.G., Ferber, E. and Fischer, H. (1965). Z. Naturforschg. 20 b, 1048.
Nelson, G.J. (1967). Biochem. Biophys. Acta 144, 221.
Oliveira, M.M. and Vaughan, M. (1964). J. of Lipid Res. 5, 156.
Paysant, M., Delbauffe, D., Wald, R. and Polonovski, J. (1967). Bull. Soc. Chim. Biol. 49, 169.
Paysant, M., Bitran, M., Wald, R. and Polonovski, J. (1970). Bull. Soc. Chim. Biol. 52, 1257.
Racker, E. (1967). Fed. Proc. 26, 1335.
Robertson, A.F. and Lands, W.E.M. (1964). J. Lipid Res. 5, 88.
Rosenberg, S.A. and Guidotti, G. (1969). In:"Red Cell Membrane".(Jamieson, G.A. and Greenwalt, T.J., eds.) Lippincott, J.B. Co. 93, 109.
Scandella, C.J. and Kornberg, A. (1971). Biochemistry 10, 4447.
Segrest, J.P., Jackson, R.L. and Marchesi, V.T. (1972). Biochem. and Biophys. Res. Comm. 49, 964.
Segrest, J.P., Kahane, I., Jackson, R.L. and Marchesi, V.T. (1973). Arch. of Biochem. and Biophys. 155, 167.
Shankland, W. (1970). Chem. Phys. Lipids 4, 109.
Vanderkooi, G. and Capaldi, R.A. (1972). Ann. N.Y. Acad. Sci. 195, 135.
Wallach, D.F.H. and Zahler, P.H. (1966). Proc. Nat. Acad. Sci. (USA) 56, 1552.
Winzler, R.J., Harris, E.D., Pekas, D.J., Johnson, C.A. and Weber, P. (1967). Biochemistry 6, 2195.

Winzler, R.J. (1969). In:"Red Cell Membrane" (Jamieson, G.A. and Greenwalt, T.J., eds) Lippincott, T.J. 157, 171.
Zahler, P.H. and Wallach, D.F.H. (1967). Biochim. Biophys. Acta 135, 371.
Zahler, P.H., Wallach, D.H.F. and Lüscher, E.F. (1967). Protides of the Biol. Fluids 15, 69.
Zahler, P.H. and Weibel, E.R. (1970). Biochim. Biophys. Acta 219, 320.
Zwaal, R.F.A., Roelofsen, B. and Colley, C.M. (1973). Biochim. Biophys. Acta 300, 159.

USE OF SHIFT and BROADENING REAGENTS IN THE NMR INVESTIGATION OF MEMBRANES

L.D.BERGELSON and V.F.BYSTROV

Shemyakin Institute of Bioorganic Chemistry
Academy of Science of the USSR
Moscow, USSR

INTRODUCTION

Magnetic resonance methods are of great use in elucidating the organization of lipids in membranes as well as the relative motions of atoms and groups within the lipid molecules.

Unfortunately when conventional proton nuclear magnetic resonance (nmr) is used the signals from the lipids on the inside and the outside of the bilayer are usually overlapping despite the differences in their chemical environment. Resolution can be improved by increasing field strength with stronger magnets or by using pulse Fourier transformed spectrometry. Both methods require sophisticated and highly expensive equipment.

A much cheaper way is to use paramagnetic ions capable of forming weakly bound complexes with the substances under investigation. The unpaired electrons of the paramagnetic ions produce a strong magnetic field in their nearest neighbourhoud. When this microscopic field is superimposed on the external magnetic field of the nmr spectrometer it can markedly improve its resolving power.

In 1970 we introduced the use of paramagnetic ions, which form complexes with the polar head groups of phospholipids, in the nmr investigation of sonicated phospholipid vesicles (liposomes)(Bergelson, 1970). Our work (Bystrov et al., 1971, 1972; Barsukov et al., 1972, 1973, 1974; Shapiro et al., 1974) as well as that of others who subsequently tried this method (Finer et al., 1971; Levine et al., 1973; Fernandez and Cerbon, 1973; Michaelson, 1973; Andrews et al., 1973; Hauser and Barratt, 1973) has demonstrated that the paramagnetic probes considerably increase the potentialities of nmr for the study of membrane systems.

RESULTS AND DISCUSSION

According to their effect on the nmr signals of phospholipid vesicles the paramagnetic substances may be divided into two groups: the shift reagents and the broadening reagents.

To the first group belong the lanthanides, for example the salts of Eu^{3+} and Pr^{3+}. These ions, by coordination to the phosphate moiety are capable of changing the chemical shifts of the atoms forming the polar head group of the phospholipid molecules. Since the cations normally can not penetrate the bilayer, only the outward facing nuclei experience an isotropic shift due to the paramagnetic ions while the inner nuclei remain unshifted. Thus when Eu^{3+} is added to a dispersion of sonicated phosphatidylcholine (PC) the hydrocarbon chain signals in the ^{1}H nmr spectrum remain unaltered, while the signal of the choline N-methyl groups is splitted into two components: one unshifted (originating from the internal lecithin molecules of the bilayer) and the other shiftened upfield - from the outward facing molecules, which are in contact with the paramagnetic ions (Fig. 1c). Pr^{3+} shifts the outer signal in the opposite direction, i.e.downfield.

The broadening agents, e.g. salts of Mn, Fe, Co and Cu have a different action. On addition of $MnSO_4$ the choline N-methyl signal remains at its initial position, but becomes less intense because the signal from the outward facing lecithins which are in contact with Mn^{2+} becomes broadened beyond detection (Fig. 1b). The intensity of the remaining "inner" signal is always about 1/3 of the intensity of the initial signal, which correlates well with the relative intensity of the two signals splitted by Eu^{3+} or Pr^{3+}.

The paramagnetic ions may be used not only in the proton magnetic resonance of phospholipid vesicles, but also in ^{13}C-

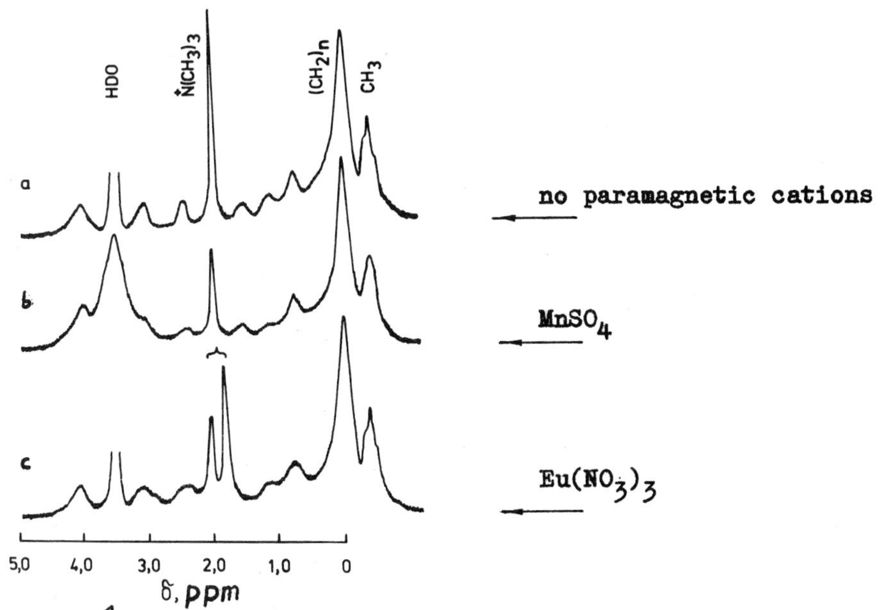

Fig. 1. ^{1}H-nmr spectra of sonicated egg lecithin liposomes (20% in D_2O).

and phosphorus resonances. Fig. 2 shows a part of the ^{13}C nmr spectrum of PC - liposomes in the absence and in the presence of external added Pr^{3+}. Here not only the N-methyl signals but also those of other C-atoms of the polar parts of the lipid molecules are splitted each into an "outside" (shifted) and

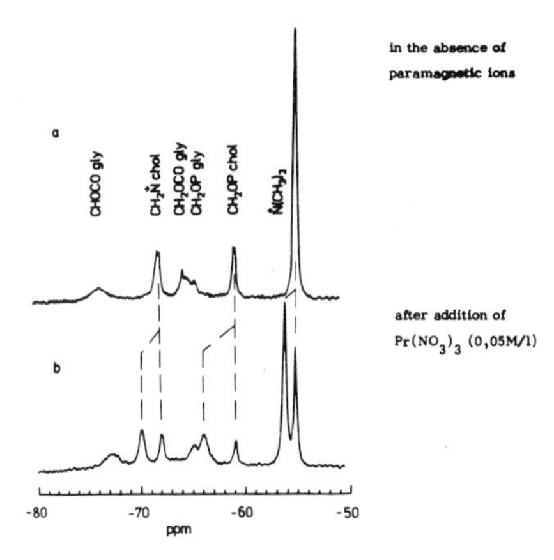

Fig. 2. ^{13}C-nmr spectra of egg lecithin liposomes (20% in D_2O).

"inside" (unshifted) component.

Addition of Pr^{3+} to PC-liposomes shifts also the phosphorus signal from the outward facing phosphate groups (Fig. 3B). The magnitude of the induced shift depends on the number of binding sites which is determined by the paramagnetic ion to phospholipid ratio and by the overall electrical charge of the liposomes. As can be seen from Fig. 3C the induced shift increases with increasing anion concentration in the external solution. This is because the absorption of the nitrate ions on the liposome surface increases its negative charge.

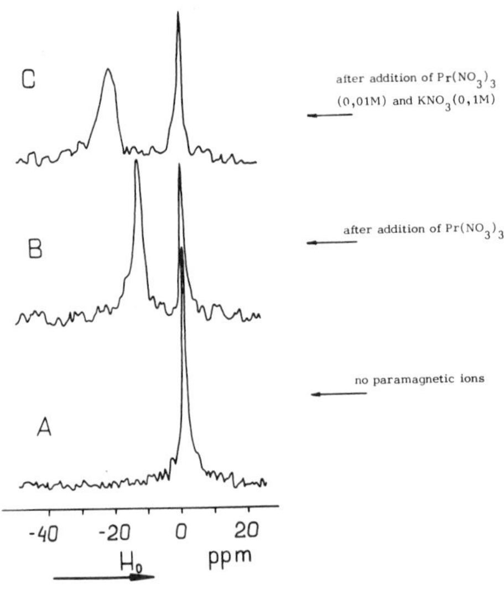

Fig. 3. ^{31}P-nmr spectra of egg lecithin liposomes (20% in D_2O)

In this way the lipids at the outer surface of a membrane easily can be distinguished from those located at the inner surface. In biological membranes the various phospholipid classes are appearently distributed asymmetrically between the two surfaces of the membrane. The above approach appears to be a promising method to discover such asymmetry.

The use of shift reagents enables one also to exploit nmr for the <u>in situ</u> analysis of the lipid composition of the outer surface of vesicular membranes and for the detection of lateral heterogeneity in membrane systems. The dependence of the induced shift on the number of binding sites makes it highly sensitive to the phospholipid composition of the bilayer.

Table 1.

Dependence of Pr^{3+} induced shifts of the $\overset{+}{N}(CH_3)$signal on the lipid composition of liposomes

Sample	$\Delta\delta^1H$	$\Delta\delta^{13}C$
Phosphatidylcholine	0.42	0.28
Phosphatidylcholine-phosphatidylserine (3:1)	-	0.48
Phosphatidylcholine-phosphatidylinositol (3:1)	0.84	0.42
Phosphatidylcholine-cholesterol (1:1)	0.28	-
Phosphatidylcholine-sphingomyelin (1:1)	0.55	

0.03 M/l PC; 0.01 M/l $Pr(NO_3)_3$ were added to the external solution after sonication

$\Delta\delta$= difference of chemical shifts in the absence and in the presence of Pr^{3+}

As can be seen from the table the acidic phospholipids phosphatidylserine and phosphatidylinositol (PI) markedly increase the induced shift of the NMe_3 signal. Evidently this is due to the increased negative charge of the mixed liposomes i.e. to the increase in the number of binding sites. Sphingomyelin also increases the induced shift, which may be attributed to its free hydroxyl group serving as an additional binding site for the paramagnetic ions. The introduction of cholesterol decreases the induced shift because of formation of PC-cholesterol associates, leading to a decrease of the number of available phosphate groups. By this technique it was shown that in mixed cholesterol - egg lecithin liposomes the cholesterol molecules are distributed asymmetrically in favor of the inner layer (Ching-Hsien Huang et al., 1974).

Fig. 4 gives an other example of the possibilities of this method. It shows the dependence of the induced shift in the spectra of mixed sphingomyelin-PC-liposomes on their component ratio. In this manner the composition of the outer surface of mixed liposomes may be determined in situ simply by measurement of the chemical shifts.

Thereupon the information provided by the paramagnetic probes is by no means exhausted. From the ratio of the "inner" and "outer" signals one can calculate the number of outside and inside facing molecules and the average size of the vesicules. The persistance of the two signals confirm that the membranes under investigation form indeed sealed cation impermeable vesicles.

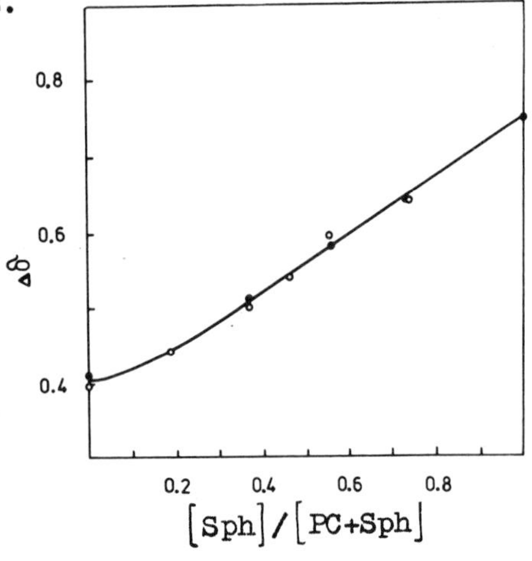

Fig. 4. Dependence of the induced shift of the "outer" $\overset{+}{N}(CH_3)$ signals of mixed sphingomyelin-phosphatidylcholine liposomes on their molecular composition.
• at 29°; ∘ at 63°
0.01 M/l $Pr(NO_3)_3$ was added after sonication.

This is a very reliable criterion of vesicularity. We are using it permanently in the investigation of sonicated dispersious of lipids other than PC. In this way it was proved that aqueous dispersions obtained on sonication of PI, phosphatidylglycerol or phosphatidylserine consist indeed of small vesicles the interior of which is tightly sealed from the outside.

Fig. 5 gives an example of the use of shift reagents in the control of permeability. When liposomes composed of PC and lyso-PC are incubated with Pr^{3+} the "inside" signal shifts, and finally merges with the outer one, showing that such liposomes are leaky. The leakage can be followed in time by measurement of the induced shift at various intervals. This method

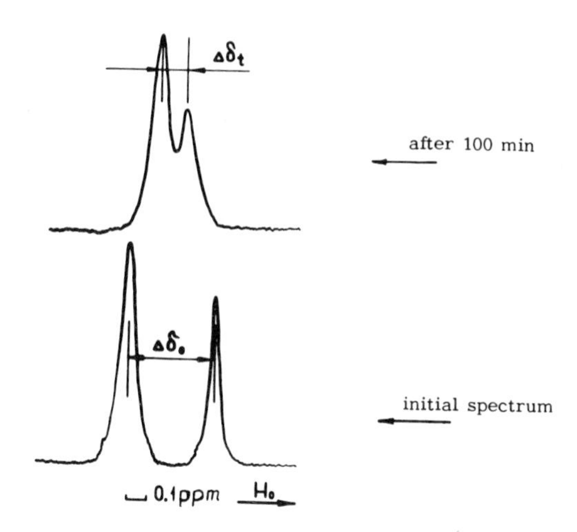

Fig. 5. Time dependent change of the "inner" $\overset{+}{N}(CH_3)_3$ signal in the ^{1}H-nmr spectrum of (4:1) PC-lyso-PC liposomes on incubation with $Pr(NO_3)_3$(0.01 M/l) at 33°.

has succesfully been used for the control of permeability induced by an ionophoric antibiotic (Fernandez and Cerbon, 1973).

Other uses of the paramagnetic probes include the study of the diffusion of water through the bilayer, the investigation of the inside-outside transitions (or so called "flip-flop") of lipids in the liposomal membrane and the determination of the conformation of the polar head groups of the outward facing phospholipids molecules. Examination of the water signal in the nmr spectra of PC-micelles in C_6D_6 containing D_2O in the presence of paramagnetic ions revealed that water is diffusing freely through the micelle surface, the lifetime of water molecules in the internal aqueous phase being shorter than 10^{-2} s (Bystrov et al., 1971). Measurements carried out at 80° C confirmed that even at elevated temperatures phospholipid flip-flop is a very rare event (Bergelson et al., 1970). It was also shown that lyso-PC enhances the rate of flip-flop and that the latter is a cooperative process requiring participation of several phopsholipid molecules (Barsukov et al., 1973). Based on this finding we suggested that lysolecithin induced flip-flop proceeds through formation of small phospholipid aggregates slowly rotating within the liposomal membrane.

Next I should like to describe the use of the paramagnetic hydrophylic probes for the investigation of intermembrane phospholipid exchange. Such an exchange has been established by in vitro experiments for subcellular fractions of the liver and some other tissues (Dawson, 1973). It seems to be an important step in the biosynthesis and repair of cell membranes. Two types of vehicles for the intermembrane transfer of phospholipids may be envisaged.

One of them are the lysophosphatides, which solubilize phospholipids by forming small micelles having a high critical micelle concentration. The second type of vehicle are a special group of proteins, the so called phospholipid exchanging proteins. It is likely that in the cell a variety of lipid exchanging proteins exists, each of which has the ability to stimulate the exchange of a single phospholipid or a limited number of phospholipids. Two of such proteins, which specifically induce the intermembrane exchange of PC and PI have recently been isolated by van Deenens group in Utrecht (Wirtz et al., 1972; Helmkamp et al., 1974).

The procedure mostly used for studying the phospholipid exchange activity of such proteins is to incubate labelled and non labelled subcellular particles of different density, to separate them and to estimate the specific activity of the individual lipid fractions. This is a very cumbersome task especially if one has to investigate the kinetics of the exchange process. The paramagnetic hydrophylic probes enabled us to elaborate a completely different approach for the assay of phospholipid transfer activity.

According to our method two populatious of liposomes made up from different phospholipids are incubated, together with the protein under investigation. The alterations in the phospholipid composition of the liposomes during the exchange process are followed by observing the changes in the nmr spectra after adding paramagnetic ions. This method has a number of advantages. It is a one step procedure, it requires no radioactive label, no separation of membrane fractions or phospholipids and it should be useful in kinetic measurements since it is simple, accurate and fast. We employed it for assaying the PI transfer activity of a water soluble protein fraction from beef liver.

The assay method was based on the fact that even minute amounts of PI are considerably increasing the paramagnetic ion induced shift in the nmr spectra of PC-liposomes (see table 1 and Fig. 6). In this way less than 2% of PI can be detected on the outer surface of PC-liposomes.

As can be seen from Fig. 6C the ^{1}H-nmr **spectrum** of mixtures consisting of cosonicated PC-PI-liposomes and of pure PC-liposomes reveals 3 $\overset{+}{N}Me_3$ signals: one frome the inside facing molecules, another from the outside molecules of the pure PC-liposomes which are in contact with Pr_3^+ and a third one from the outward facing choline groups of the mixed PC-PI-liposomes which are in contact both with Pr^{3+} and PI.

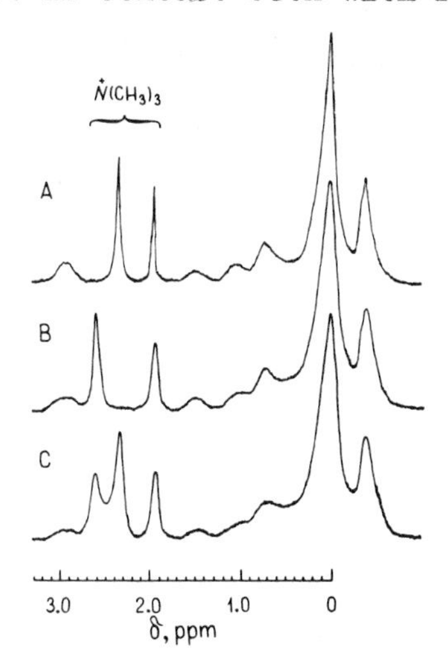

Fig. 6. ^{1}H-nmr spectra of different liposome populations (external addition of 0.01 M/l $Pr(NO_3)_3$ after sonication). A- pure PC-liposomes; B - cosonicated 7:3 PC-PI-liposomes. C - 1:1 mixture of samples A and B.

The position of these three signals does not change on coincubation of the two liposome populations for many hours. However when small amounts of lyso-PC were present the two outside signals drew together and finally merged completely. This indicated that now all the outward facing molecules were in the same magnetic environment, i.e. that PI and PC had been redistributed between the two types of vesicles. The persistance of the "inside" signal testified that during the short incubation period the liposomes retained their vesicular structure. These results demonstrated clearly that lyso-PC is indeed capable of stimulating the intermembrane exchange of lipids.

The induced shifts and the ratio of the "inside" and "outside" $\overset{+}{N}Me_3$ signals in the ^{1}H-nmr spectrum of the liposomes produced by lyso-PC induced phospholipid exchange almost coincided with those in the spectrum of co-sonicated PC-PI (3:1) liposomes. It follows that under these conditions the transferred phospholipid molecules are introduced both into the outer and the inner surface of the bilayer.

In contrast to the non-selective exchange induced by lyso-PC the exchange stimulated by the phospholipid exchanging proteins seems to be highly specific. For an investigation of this process mixtures of PC and PI vesicles were incubated with a post-microsomal supernatant fraction from rat liver (see table 2).

Table 2.
Stimulation of PI incorporation into PC liposomes by rat liver lipid exchanging protein

Incubation time (hr)	$\Delta\delta$ ppm	
	without protein	with protein
0	0.292	0.298
2	0.303	0.418
3	0.314	0.420
4	0.314	0.437

PC (0.03 M) and PI (0.01 M) liposomes were incubated at 50° with protein (5 mg/ml). $Eu(NO_3)_3$ (10^{-3} M/l) added before measurement.

As can be seen from the table the NMe_3 signal of the outward facing molecules gradually shifted upfield, demonstrating thet PI was transferred by the protein from the PI-liposomes to the PC-vesicles.

In contrast to the lyso-PC induced exchange the protein stimulated PI transfer leads to formation of an asymmetric bilayer. A comparison of the ^{31}P-nmr spectra of PC-PI-liposomes obtained by incubation of separately sonicated PC and PI liposomes with the protein and of cosonicated PC-PI-liposomes showed the induced shift to be much larger in the former sample (Fig. 7). Evidently the concentration of outward facing PI molecules is essentially higher in the liposomes obtained by protein stimulated transfer. A simple calculation showed that about 90% of the transferred PI is located at the outer surface of the bilayer.

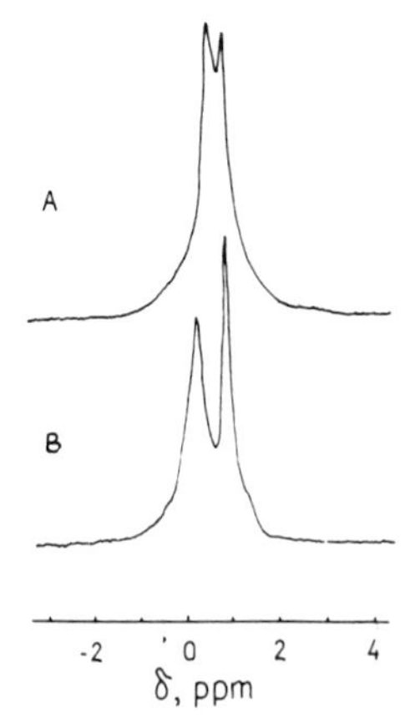

Fig. 7. ^{31}P-nmr spectra of 98:2 PC-PI-liposomes: A - obtained from PC and PI by cosonication; B - obtained by incubation of separately sonicated PC and PI liposomes with lipid exchange protein (5 mg/ml). $Pr(NO_3)_3$ (10^{-3} M/l) was added to samples A and B before recording the spectra.

The reverse process - i.e. the transfer of PC from the PC liposomes into the PI vesicles - should lead to the appearance of an even more shifted signal - from the newly introduced PC - molecules, which are in contact with very large amounts of PI. Such signals were never observed, which indicates that the protein induced exchange proceeds as a one way traffic, at least in the presence of excess PI. It was indeed found that the PC - exchanging protein from beef liver is inhibited by acidic phospholipids.

As to the mechanism of the action of the protein it is important to note that the protein does not disturb the vesicularity of the liposomes since the "inside" signal persists and is not changed in the presence of the protein. The protein does also not penetrate through the bilayer since the signals from the CH_2 groups of the hydrophobic chains are not changed in the presence of the protein. Consequently the protein is not acting as a liposome - destroying detergent.

This was further confirmed by the fact that the phosphorus resonance of pure PC - vesicles splits in the presence of the protein into two components (Fig. 8). On external addition of Pr^{3+} only the highfield component is affected. This permits to attribute the highfield component to the outward facing and the lowfield component to the "inside" molecules. It thus seems that the protein is adsorbed on the outer surface of the liposomes, but is not destroying their integrity.

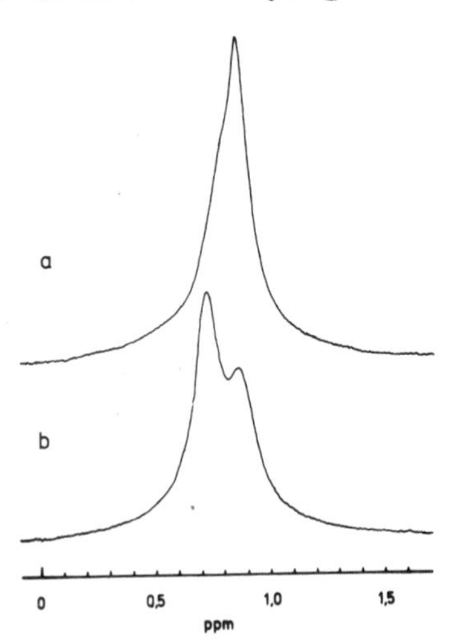

Fig. 8. ^{31}P-nmr spectrum of PC-liposomes (9% in D_2O). a - without protein; b - after addition of lipid exchange protein (5 mg/ml).

On this basis the mechanism of the action of the

PI - exchanging protein may be imagined as consisting of the following steps: adsorption of the protein on the outer surface of a PI liposome; insertion of the protein into the PI bilayer; detachement of the lipoprotein complex from the PI-liposomes; adsorption of the lipoprotein on the outer surface of a PC-liposome and incorporation of the PI molecule into its outer monolayer.

The fact that the protein induced exchange affects only the outer surface of the bilayer enabled us to use the exchange procedure for the production of tailor-made liposomes with an asymmetric distribution of the phospholipids across the membrane for instance of mixed composition liposome species containing the cholinephosphatides only in the outer monolayer and the aminophospholipids predominantly in the inner monolayer. This exactly is the situation in the bilayer regions of the erythrocyte membrane. The advent of such asymmetric liposomes may be considered as a step forward on the long way from the liposome model to natural membrane systems.

REFERENCES

Andrews, S.B., Faller, J.W., Gilliam, J.M. and Barnett, R.J. (1973). Proc. Natl. Acad. Sci. US 70, 1814.

Barsukov, L.I., Shapiro, Yu.E., Viktorov, A.V., Bystrov, V.F. and Bergelson, L.D. (1973). Dokl. Akad. Nauk SSSR 208, 717; 211, 847.

Bergelson, L.D., Barsukov, L.I., Dubrovina, N.I. and Bystrov, V.F. (1970). Dokl. Akad. Nauk SSSR 194, 708.

Bystrov, V.F., Dubrovina, N.I., Barsukov, L.I. and Bergelson, L.D. (1971). Chem. Phys. Lipids 6, 343.

Bystrov, V.F., Shapiro, Yu.E., Viktorov, A.V., Barsukov, L.I. and Bergelson, L.D. (1972). FEBS Letters 25, 337.

Ching-Hsien Huang, Sipe, J.P. and Martin, B. (1974). Proc. Natl. Acad. Sci. US 71, 359.

Dawson, R.M.C. (1973). Sub-Cell. Biochem. 2, 69.

Fernandez, M.S. and Cerbon, J. (1973). Biochim. Biophys. Acta 298, 8.

Finer, E.G., Flook, A.G. and Hauser, H. (1971). FEBS Letters 18, 331.

Helmkamp, Jr.,G.M., Harvey, M.S., Wirtz, K.W.A., van Deenen, L.L.M. (1974). 9th FEBS Meeting, Budapest, Abstr. of Communications, p. 366.

Levine, Y.K., Lee, A.G., Birdsall, N.J.M., Metcalfe, J.C. and Robinson, J.P. (1973). Biochim. Biophys. Acta 291, 592.

Michaelson, D.M., Horwitz, A.F. and Klein, M.P. (1973). Biochemistry (USA) 12, 2637.

Shapiro, Yu.E., Viktorov, A.V., Volkova, V.I., Barsukov, L.I., Bystrov, V.F. and Bergelson, L.D. (1974). Chem. Phys. Lipids, in press.

Wirtz, K.W.A., Kamp, H.H. and van Deenen, L.L.M. (1972). Biochim. Biophys. Acta 274, 606.

FUNCTIONAL EFFECTS OF CROSS-LINKING OXIDATION-REDUCTION ENZYMES OF MEMBRANES AND HUMAN ERYTHROCYTES

LESTER PACKER

Department of Physiology-Anatomy, University of California at Berkeley, and Energy and Environments Division, Lawrence Berkeley Laboratory, Berkeley, California, 94720, USA

INTRODUCTION

Under physiological conditions, biological membranes manifest lateral and perpendicular mobility of membrane components. In an attempt to evaluate the relative importance of such mobility parameters for membrane functions, we have been developing a strategy involving the use of bifunctional cross-linking reagents, particularly imidoesters to immobilize oxidation-reduction components of membranes in organelles and also in red blood cells. These studies are relevant to two problems which will be considered in this article. 1) The mechanism of energy coupling in inner mitochondrial and chloroplast membranes; and 2) The mechanism of how a conformational change first initiated in deoxy-hemoglobin S causes the sickling of human erythrocytes.

RESULTS

Our approach involves the use of bifunctional imidoesters of various chain lengths. These amidating reagents have been widely used for cross-linking of proteins and act rapidly at slightly alkaline pH[1]. Unlike bifunctional aldehydes, bifunctional imidoesters do not appreciably change the net charge of proteins, as they bring an amino group with them for each ε or N-terminal amino group with which they react as shown in fig. 1. The reagents used for our studies have been primarily the 6 carbon chain dimethyl adipimidate (DMA) and the 8 carbon chain dimethyl suberimidate (DMS).

Immobilization of Mitochondrial Membranes

Rat liver mitochondria or inner membranes made from mitochondria, i.e. submitochondrial preparations (SMP) have been studied (in collaboration with Dr. H. Tinberg, C. Lee, and J. Maguire) in regard to retention of functions following treatment with imidoesters. Figures 2 and 3 show that treatment of mitochondria with 50 mM DMS results in extensive cross-linking as judged by formation and disappearance of bands on SDS-acrylamide gels and in the complete loss of osmotic sensitivity (fig. 3) of preparations following such treatment. Thus it can be anticipated that significant lateral and perpendicular movements of protein components would not be expected in these preparations. Nevertheless, they retain to a considerable degree, both in whole mitochondria and SMP preparations, the capacity to generate electron transport[2]. Retention of electron transport functions of the

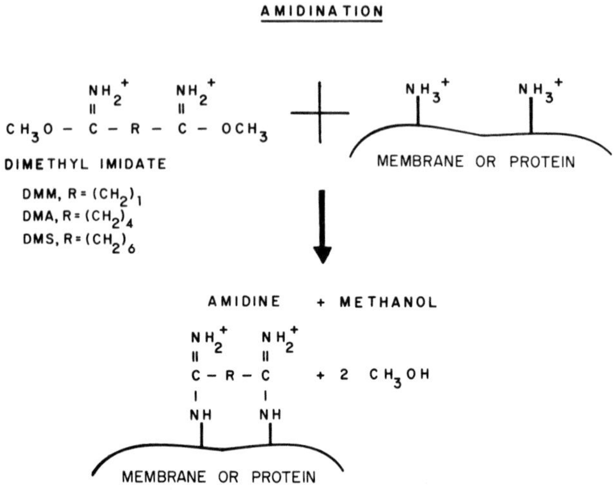

AMIDINATION

$$CH_3O-\overset{\overset{NH_2^+}{\|}}{C}-R-\overset{\overset{NH_2^+}{\|}}{C}-OCH_3 \quad + \quad \underbrace{NH_3^+ \qquad NH_3^+}_{\text{MEMBRANE OR PROTEIN}}$$

DIMETHYL IMIDATE

DMM, $R=(CH_2)_1$
DMA, $R=(CH_2)_4$
DMS, $R=(CH_2)_6$

↓

AMIDINE + METHANOL

$$\underbrace{\overset{\overset{NH_2^+}{\|}}{C}(-NH)-R-\overset{\overset{NH_2^+}{\|}}{C}(-NH)}_{\text{MEMBRANE OR PROTEIN}} \quad + \; 2\; CH_3OH$$

Fig. 1. Scheme for amidation reactions.

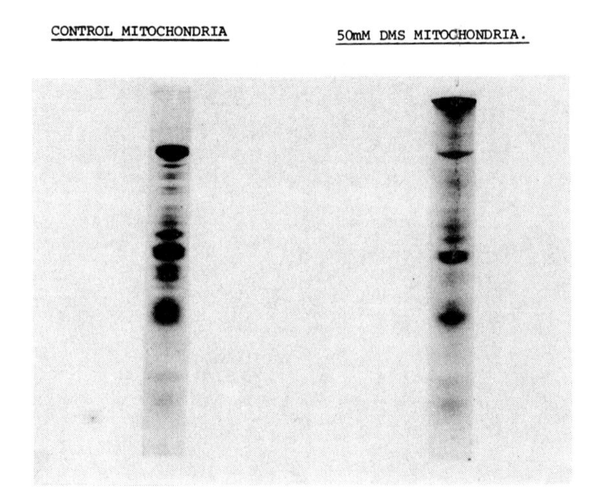

Fig. 2. SDS-Polyacrylamide gel electrophoresis of whole and dimethyl suberimidate treated rat liver mitochondria.

respiratory chain is shown in figure 4. Note that certain activities are stimulated (like NADH-DCIP reductase) and others are progressively inhibited (like NADH oxidase), and yet others reach a constant level of inhibition (like) Ascorbate-TMPD oxidase). In other experiments we have found that although the initial activity may be diminished in these preparations by DMS treatment, certain activities such as Ascorbate-TMPD oxidase are stabilized by immobilization treatment they retain activity upon storage of the preparation for long periods of time. Furthermore, electron transport can be coupled to energy-linked H^+ transport as demonstrated by studies employing both direct measurements of pH shifts following deliverance of an oxygen pulse to an anaerobic suspension of either mitochondria or as monitored by following quenching of 9-amino acridine fluorescence.

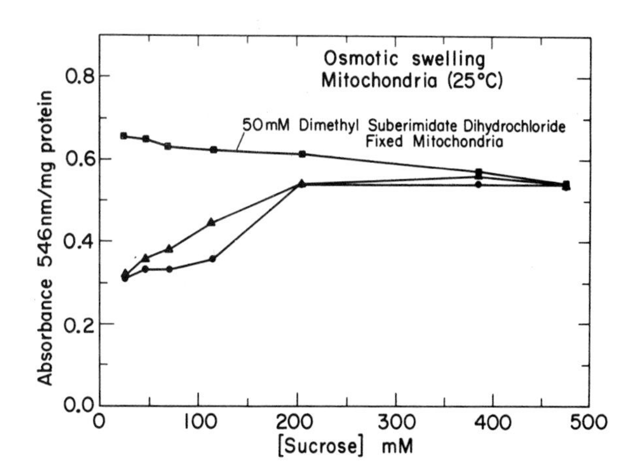

Fig. 3. Loss of osmotic swelling of rat liver mitochondria following treatment with dimethyl suberimidate.

Immobilization of Chloroplast Membranes

Experiments have been carried out in collaboration with J. Torres-Pereira, P. Chang and S. Hansen[3,4] with chloroplast membranes where the light induced electron transport dependent formation of H^+ gradients is being evaluated using the technique of the quenching of acridine dyes. Studies in Kraayenhof's laboratory[5] established that the light induced fluorescence quenching and subsequent dark decay of acridine compounds in chloroplasts could be attributed to the quenching of the acridine dye which occurs as a result of binding of the dye to the membrane, hence it serves as a monitor of the membrane pH gradient. Subsequent studies by Schuldiner, Rottenberg and Avron[6] established that quenching largely reflected the formation of a pH gradient across the thylakoid membrane. The loss of this gradient results from the equilibration of H^+ (as for K^+) as the pH gradient is dissipated in darkness. A most interesting observation of Kraayenhof[5] was that at low temperature the rate of decay of the ion gradient was inhibited, i.e. there was a "freezing" of the energized ion gradient at lower temperatures.

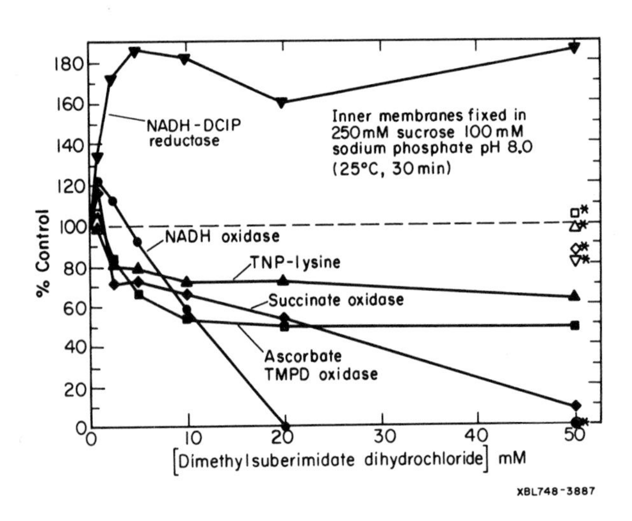

Fig. 4. Effect on several mitochondrial electron transport systems with increasing DMS concentration. TNP-lysine represents the amount of unreacted primary amines after treatment with DMS. NADH-2,6 dichlorophenol-indophenol (DCIP) reductase is the only enzyme stimulated significantly by DMS. Asterisks represent enzyme activities after treatment with 50 mM hydrolyzed DMS.

An example of this type of result is shown in some of our experiments in fig. 5 which demonstrates the time course of atebrin fluorescence quenching and its reversibility at 20° and 1°C. At low temperature there is no appreciable affect on the ability of the chloroplast electron transport system to drive the light dependent energization process. The "frozen" ion gradient at 1°C in the darkness is sensitive to dissipation by the uncoupling agent S-13.

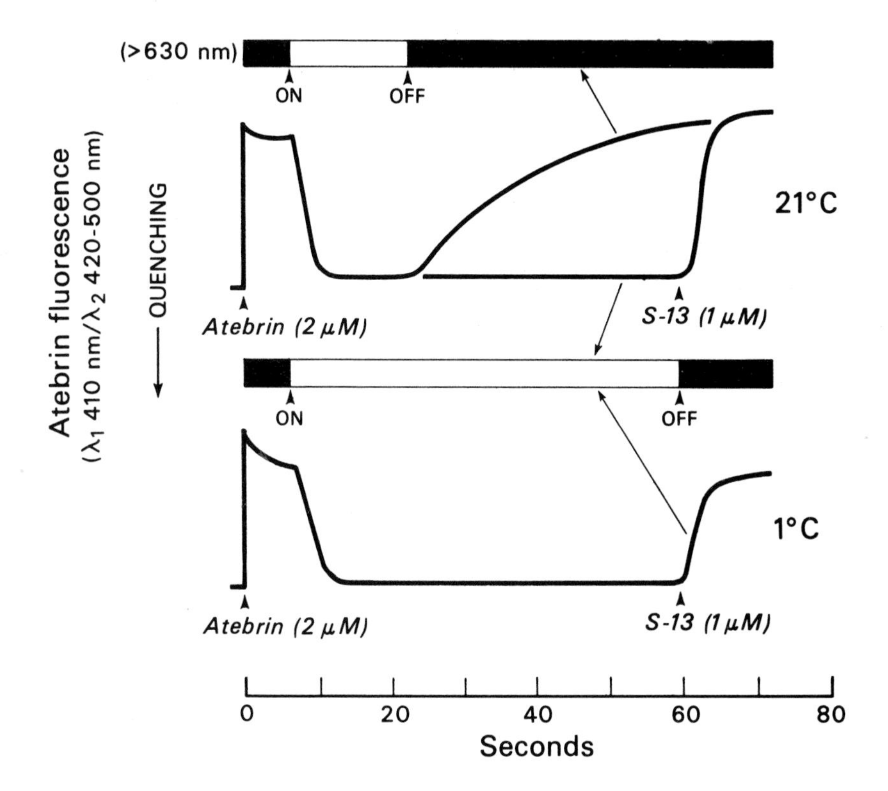

Fig. 5. Light induced atebrin fluorescence quenching in chloroplasts at 1° and 21°C.

We have now found that the addition of bifunctional imidoesters in the μM range to chloroplast membranes do not affect the time course of the quenching of 9-amino acridine or atebrin fluorescence upon illumination of chloroplasts, nor do they affect the extent of the fluorescence quenching. The effect of DMA on the kinetics and extent of light induced fluorescence quenching and its subsequent dark decay are given in fig. 6. There is no effect on the rate or extent to which the fluorescence is quenched, but a progressively increasing inhibition of the decay of quenching in the dark occurs with increasing concentrations of DMA in the μM range. 75 μM DMA treatment completely inhibits the dark decay. Other experiments show that the decay is fully reversible by uncoupling reagents as S-13.

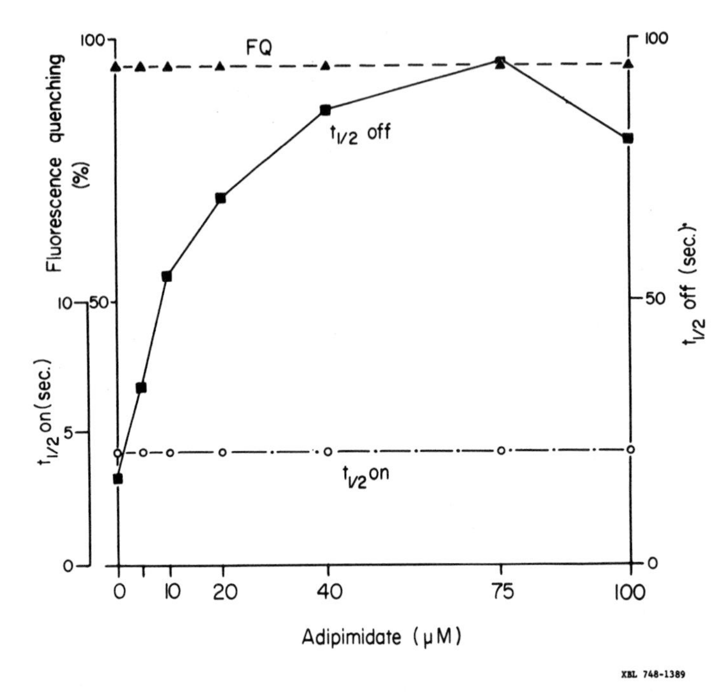

Fig. 6. Influence of dimethyl adipimidate treatment of chloroplasts on light induced atebrin fluorescence quenching.

The longer chain bifunctional imidoester DMS is more effective on an equal concentration basis than DMA. With differing chain lengths these compounds may act as "molecular rulers" cross-linking at critical distances necessary for "freezing" of the energized state. Their effectiveness as "molecular rulers" was further evaluated by manipulating the distances between reactive groups in the membrane by varying the osmolarity of the medium, thus causing thylakoid membranes to contract or expand. Other such experiments showed that there is an optimum range of osmolarities of NaCl which give the most effective range for "freezing" of the energized state.

Yet another action of bifunctional imidoesters in chloroplasts is revealed by studies employing these reagents in the mM range of concentration. In this range of concentration osmotic sensitivity of chloroplast membranes is lost, but chloroplasts still retain Hill reaction activity. Also, they exert a favorable effect on extending the functioning lifetime of light induced 9-amino acridine fluorescence quenching activity in chloroplasts stored continuously at 15°C for many days after isolation from the leaf.

Immobilization of Human Red Blood Cells

Earlier studies on the cross-linking of erythrocytes carried out with bifunctional reagents such as glutaraldehyde revealed that such treatment results in the complete collapse of energy dependent K^+ and H^+ gradients. We first decided to further investigate this problem using imidoesters for treatment of normal human red blood cells. In both normal[7] and sickle erythrocytes[8] we have found the potassium gradients are unaltered by concentrations of DMA which inhibit osmotic lysis. Furthermore, DMA treatment of normal human red cells increases the resistance of the cell to lysis by the nonionic detergent Triton X-100. Normal K^+ gradients were retained after DMA treatment and can be released by adding the ionophore gramicidin D (fig. 7). Under these conditions osmotic fragility of the cells is also markedly inhibited by DMA (fig. 8).

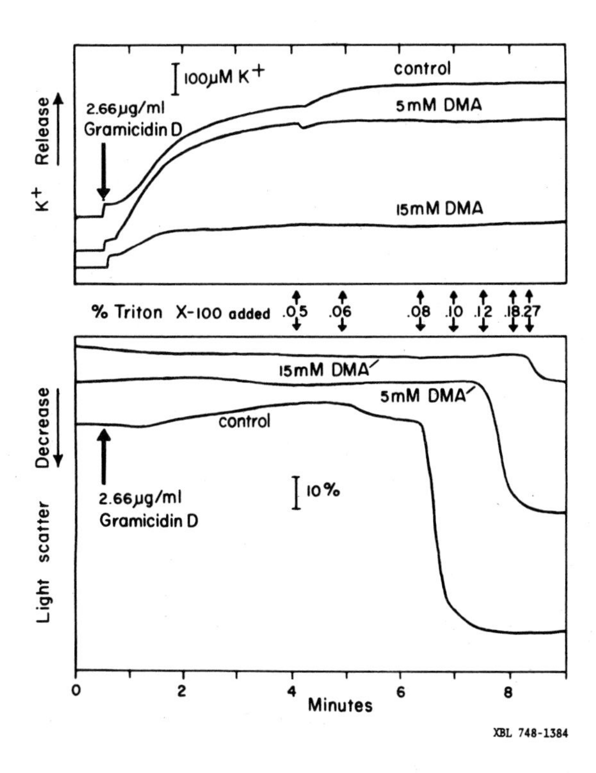

Fig. 7. Light-scattering changes and K^+ release from normal and imidoester-treated erythrocytes.

From these experiments we concluded that cross-linking of the membrane probably occured, and was in part responsible for the resistance to osmotic and Triton lysis. Furthermore, it seemed reasonable to conclude that the retention of the monovalent cation gradients after treatment with bifunctional imidoesters, as contrasted with treatment with bifunctional aldehydes, was probably the result of not appreciably changing the charge on the membrane following treatment.

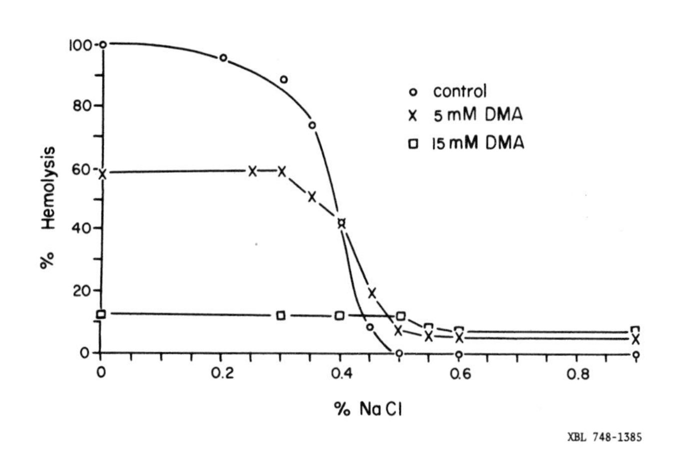

Fig. 8. Osmotic fragility of normal and DMA-treated erythrocytes.

We then decided to determine whether the use of bifunctional imidoesters might provide a strategy for combating the sickling of human erythrocytes that is initiated by the conformational change in deoxy-hemoglobin S that eventually results in a change in cell shape to the sickle form. Sickle erythrocytes were known to lose potassium more rapidly than normal cells[9]. Treatment of sickle erythrocytes with DMA revealed that inhibition of sickling occured under conditions where the K^+ loss was prevented. Moreover, the extent of treatment with low concentrations of DMA was not sufficient to markedly alter cell deformability as judged by studies of whole blood viscosity after cell suspensions were treated with DMA (fig. 9).

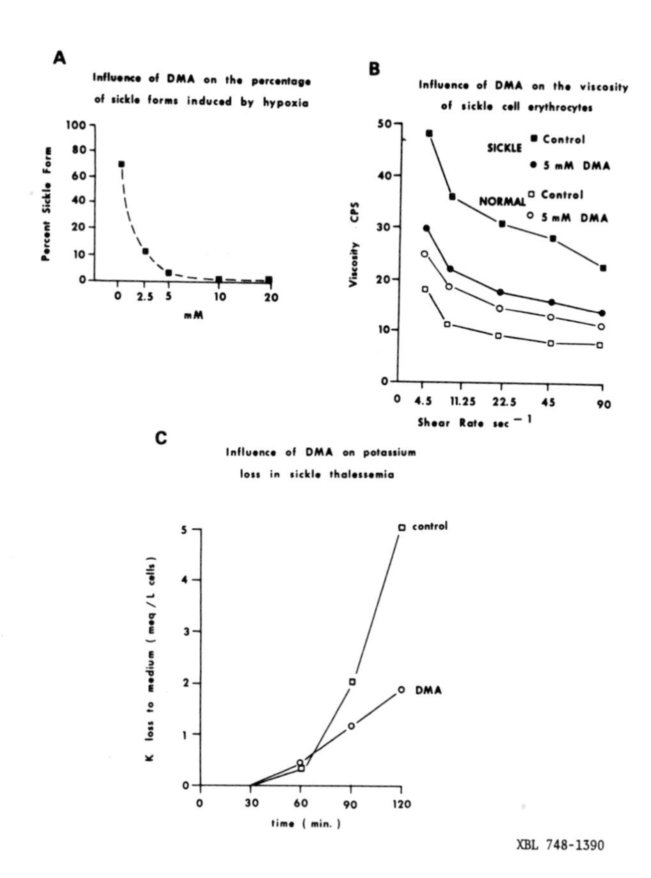

Fig. 9. Influence of DMA treatment on normal and sickle erythrocytes on sickling, potassium loss, and red cell viscosity.

Under conditions which effectively prevent the sickling other experiments[8] have revealed that the capacity of the cells to utilize glucose and produce lactic acid, i.e. glycolysis, and ATP synthesis are not inhibited. Thus suggesting the potential usefulness of these types of reagents for providing a strategy for the combating of sickling in extra-corporeal experiments. However, these studies left many questions unanswered. Earlier studies had already shown that monofunctional alkylating reagents like cyanate can also be used to prevent the sickling of human erythrocytes, but the therapeutic application of this reagent manifests many harmful side effects. We have now found that like cyanate, DMA reacts in erythrocytes to shift the hemoglobin oxygen dissociation curve, and the p50 values are decreased by DMA in sickle erythrocytes in a manner quite similar to that of cyanate. Our studies reveal that the shift in the oxygen dissociaton curve brought about by DMA are independent of hemoglobin-2,3,-diphosphoglycerate levels[8]. And thus it would appear that the effects of bifunctional imidoesters on erythrocytes are of potential importance to the understanding of the molecular "repair" of this disease, and perhaps the treatment of sickle cell anemia. In view of this, we have undertaken to evaluate (in collaboration with E. Bymun) the relative effectiveness of monofunctional and bifunctional imidoesters on the prevention of sickling. Here a crucial question is to what extent is the inhibition of sickling due to chemical modification as contrasted to the cross-linking effects of the reagent.

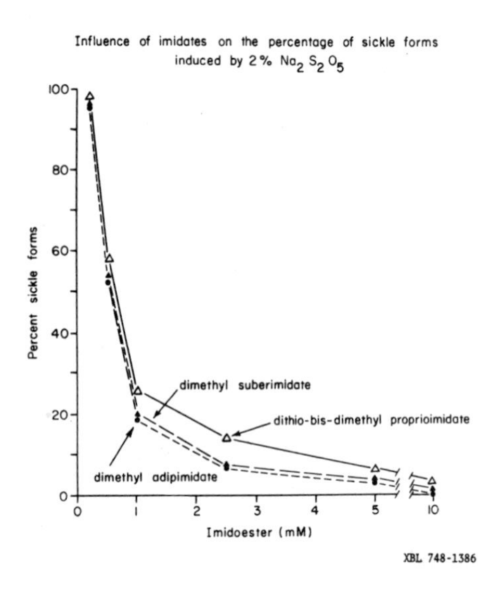

Fig. 10. Influence of imidates on the percentage of sickle forms induced by 92% nitrogen.

The influence of the various imidates on the percentage of sickle forms is shown in fig. 10. The results show that under 92% nitrogen when deoxyhemoglobin S is present in sufficient concentrations to induce sickling, DMA and dimethyl suberimidate (DMS), an 8 carbon chain imidoester, and dithiobis-proprioimidate, a cleavable imidoester, were approximately equally effective; however, monofunctional reagents like 4-methyl-mercaptobutyrimidate and the shorter 4 carbon chain bifunctional imidate, dimethyl malonimidate, are unable to inhibit sickling under these conditions. In other experiments we found that in completely deoxygenated hemoglobin S the influence of DMA, DMS and dithio-bis-dimethyl proprioimidate on sickling are virtually the same as in cells incubated under 92% nitrogen. Hence bifunctional imidoesters inhibit sickling even when hemoglobin S is completely deoxygenated (in contrast to cyanate).

Therefore, the long chain bifunctional imidoesters reported here offer promise for investigating the molecular basis of the defect in hemoglobin S, both at the level of hemoglobin and at the membrane level in sickle erythrocytes. It is premature to speculate which of the many sites (fig. 11) at which bifunctional imidoesters can react in erythrocytes to prevent sickling, reversible sickling, or indeed the development of irreversible sickled forms. The diagram indicates the sites where the relative effectiveness of crosslinking reactions by imidoesters could be investigated, which could be helpful in elucidating the molecular basis for inhibtion of the sickling phenomena.

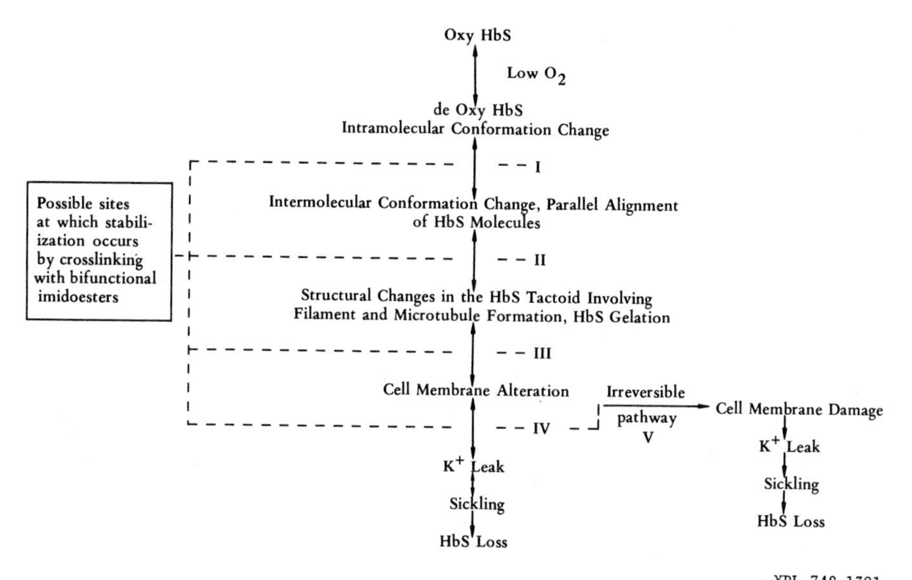

Fig. 11. Possible sites of DMA involved in the inhibition of sickling of human erythrocytes.

CONCLUDING REMARKS

The relative degree with which treatment with bifunctional imidoesters causes inhibition of functions by chemical modification, as contrasted with crosslinking effects, is still unresolved. However, it is clear that if retentions of activity occurs after extensive immobilization, certainly these functions are independent of mobility parameters in the membrane or in the case of the red cell of structural associations such as found in the network of molecules that comprise the hemoglobin tactoid. Chemical immobilization of energy transducing membranes and of whole red blood cells offer some new approaches for investigating some interesting biological problems.

1) Oxidation-reduction functions of mitochondrial and chloroplast membranes and of hemoglobin are relatively unaffected by treatment with bi-

functional imidoesters and indeed certain activities, while initially partially inhibited following immobilization, are stabilized during in vitro storage as extending the functioning lifetime of electron transport enzymes in mitochondrial inner membranes and chloroplast thylakoid membranes. This could have relevance for construction of a "biological fuel cell" or for biological energy conversion by applying principles of "enzyme engineering" to such membrane systems.

Apart from the possible technological applications of stabilized and immobilized electron transport containing membranes, these studies are also of fundamental importance as they address themselves to mechanisms of energy transduction. Further studies on the effects of chemical modification and cross-linking by bifunctional imidoesters on energy linked functions such as transhydrogenase, ATP synthatase-ATPase, the various anion and cation translocation systems, etc., will help to answer the fundamental questions of how the components in the membrane interact so as to partition energy between different modalities and how asymmetric membrane organization plays a role in these processes.

2) In red blood cells extra-corporeal use of the reagents offers a means to prevent the sickling. Although the sites of anti-sickling action is unknown, these reagents provide a probe to investigate the various levels in the structural hierarchy, involving intra and interchain effects, aggregation states of hemoglobin, the role of the membrane, and interaction of hemoglobin with the membrane, in the sickling process. Remarkably key physiological properties of erythrocytes seem unaffected. Hence this "enzyme engineering" approach could provide a strategy for extending the functioning shelf-life in vitro of blood cells.

REFERENCES

Neihaus, W.G. and Wold, F. (1970), Biochem. Biophys. Acta 196, 170-175.

Tinberg, H.M., Melnick, R.L., Maguire, J., and Packer, L. (1974), BBA Library, Vol. 13, Dynamics of Energy-Transducing Membranes, ed. by Ernster, L., Estabrook, R.W., and Slater, E.C. Elsevier Scientific Publishing Co. New York, p. 539.

Takaoki, T., Torres-Pereira, J., and Packer, L. (1974), Biochim. et Biophys. Acta, 352, 260-267.

Packer, L. Torres-Pereira, J., Chang, P., and Hansen, S. (1974), Proceedings of the Third International Congress on Photosynthesis in A.S.P. Biological and Medical Press B.V.

Kraayenhof, R. (1971), Thesis, Unversiteit van Amsterdam, Mondeel-Offsetdrukkerij, Amsterdam.

Schuldiner, S., Rottenberg, H., and Avron, M. (1972), Eur. J. Biochem. 25, 64-70.

Krinsky, N.I., Bymun, E.N., and Packer, L. (1974), Arch. Biochem. Biophys. 160, 350-352.

Lubin, B.H., Pena, V. Mentzer, W.C., Bymun, E., and Packer, L. (1974), Proceedings of the National Academy of Sciences, (in press).

Tosteson, D.C., Carlsen, E., and Dunham, E.T. (1956), J. Gen. Physiol. 39, 31-53.

RED CELL MEMBRANE ALTERATIONS WITH PATHOLOGICAL IMPLICATION

SUSAN R. HOLLÁN, MARIA HASITZ and JUDITH H. BREUER

National Institute of Haematology and Blood Transfusion, Budapest, Hungary and Institute of Experimental Medical Research of the Hungarian Academy of Sciences, Budapest, Hungary

Pathological changes in membrane structure and function can only be understood on the basis of normal membrane characteristics. On the other hand a better insight to membrane changes under pathological conditions may give an insight to the importance of definite structural and functional properties of membranes which bear importance to the cell's normal function and survival. It is not within the scope of this paper to give an overall review of the abundant data of changes of erythrocyte membrane in disease. Arbitrary chosen literary data and some of the recent experimental evidence will be discussed to demonstrate that at the time of our recent knowledge it is not justified to speak about membrane diseases. We are still far from understanding the changes in membrane structure and function at a molecular pathological level. Most of the alterations described so far are rather pathological reaction patterns characterizing very diverse disease conditions.

As all other plasma membranes the erythrocyte membrane acts as a biointerface between extra- and intracellular space, and transports matter and information across this limiting boundary. It serves as stereospecific support of enzyme function and plays a central role in energy conversion, and integration of energy yielding enzyme systems with energy requiring reactions of active transport, shape maintaining function and maintenance of viscosity and deformability. Apart from these functions /common to all biomembranes/ the most important function of the red cell and its limiting membrane is to provide the proper environment for the haemoglobin molecule. The red cell interior is characterized by

1. extremely high protein concentration where protein-protein interactions are maximal, by

2. special cation, anion and nucleotide composition and

3. a very unique condition of oxydation-reduction potential.

As all biomembranes the erythrocyte membrane is composed of protein, lipid, carbohydrate, ions and water. The oligosaccharides are attached to both lipids and proteins. It is now generally accepted that biomembranes have a mosaic structure /Lenard and Singer, 1966; Wallach and Zahler, 1966; Branton et al., 1972/. The special organization of the two main components: the lipids and proteins is still not completely resolved but the only membrane model consistant with all experimental data and with thermodynamic restrictions is the fluid mosaic model /Singer and Nicolson, 1972/. According to this dynamic model the proteins are free to diffuse in a discontinuous lipid matrix. The physical state of the membrane is essentially fluid but with high local concentration of components in a relatively ordered state and locally constrained. Conformational changes of proteins are probably modified in membranes, because the association of proteins with lipids.

Molecular structure and function of membranes change upon binding of suitable ligands. And because macromolecular interactions are very extensive in membranes, cooperative effects are potentially multiplied and may propagate over large areas of the membrane. Changeux and associates /1967/ pointed first to the similarity in the behaviour of membranes with regulatory enzymes. In other words, binding of a structure determining ligand to one of the subunits of equivalent association perturbs other subunits, having other functions resulting in functional pleiotropism. Thus the binding of only a few molecules of a regulatory membrane ligand may perturb the whole membrane lattice. Under physiological conditions external and internal mechanisms maintain ligand binding at a 'poising' level, thus minor biological stimuli can result in major changes in the state of the membrane reflected in triggering and gating and in the "all or none" response in excitable membranes.

Table 1.

List of the most extensively studied membrane ligands

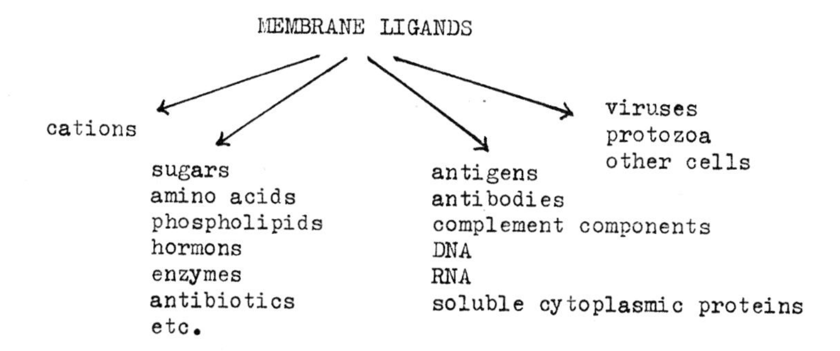

If we run through the list of the most extensively studied membrane ligands /Table 1/ it is selfexplicatory that these wide range of ligands binding to the membrane may perturb the whole structure of the membrane and may produce diverse and multiple pathological manifestations, the reaction pattern of which may be very similar.

The complex nature of the effect of ligand binding is reflected in the well established fact, that a number of agents, especially some organic solvents and general anaesthetics, which at high concentrations haemolyze the red cells, at low levels stabilize their membrane against osmotic and mechanical stress. For each membrane active agent there is a narrow "critical" concentration range, at which the switch from stabilization to disruption occurs.

As mentioned before membranes are asymmetric mosaic lattices, their tangential and transverse polarities are by now well established facts. The asymmetry was first proposed by Bretscher /Bretscher, 1971a, 1971b/ who - by the use of labelling erythrocytes and ghosts with a relatively non-permeant reagent - revealed that two proteins span the membrane and one of these contains most of the membrane glycoproteins. The transverse polarity of the major peptides in human erythrocyte membranes was also derived from other labelling procedures and from the differences in their accessibility to various proteolytic enzymes applied to the external or internal surface of the "inside out" and normally oriented membrane vesicles /Steck et al., 1970, 1971/. This vesicles originated always from purified, haemoglobin-free erythrocyte ghosts. The fact, that in contrast to the haemoglobin-free membranes intact erythrocytes are very resistant to external proteolysis, points to essential differences in their structure.

The distribution of the oligosaccharides differs on the two surfaces of the erythrocyte membrane. Specific sugar residues reacting with plant agglutinins /lectins/, blood groups specific and tumor-specific antigens, virus and other receptor structures are confined to the external surface of the membrane. Antigens exposed to the outer surface of the membrane can be localized by fluorescent or isotopic labelled antibodies. The "capping" of these antigens revealed their free mobility in the plane of the membrane.

Based on the selective degradative action of lipases Van Deenan and associates /Zwaal et al., 1973; Verkleij et al., 1973/ could determine the asymmetric distribution of phospholipids in erythrocyte membranes. The outer layer consists mainly of phosphatidylcholine /PC/ and sphingomyelin /S/, while the inner layer contains only small amounts of these two choline containing phospholipids and the major fraction of phosphatidylethanolamine /PE/ and the total phosphatidylserin /PS/ of the membrane. Their interesting finding that phospholipase C only acts on ghosts, points again to the fact that the membrane structure of haemolyzed red cells differ from that of intact cells.

1-fluoro-2,4-dinitrobenzene /FDNB/ has been used /Gordesky et al., 1973/ to study the availability of amino-containing phospholipids in erythrocyte membrane. It is suggested that 46 % of PE and 12 % of PS of the membrane are free in a lipid bilayer, 27 % and 9 % of these respective lipids are loosely bound to proteins and lost during the preparation of ghosts. 27 % of the PE and 79 % of the PS are tightly bound to core proteins.

It is a generally accepted idea, that out of the three main classes of components - proteins, lipids, oligosaccharides - the proteins play the crucial role in determining the structure of the membrane. And therefore the most important shortcoming of membrane structure studies resides in the fact that in contrast to the well resolved and characterized lipids and oligosaccharides, the structure and function of membrane proteins is still controversary in spite of the tremendeous efforts that have been made to isolate and characterize them. The problem derives from the inability to study membrane proteins by conventional methods. The only available approach has been to use some type of detergent or denaturating solvent to "solubilize" them. In most cases, however, solubilization has been synonymous with fragmentation of the membrane and not dispersion of the protein components. Both "structural proteins" and "miniproteins" turned out to be artefacts. The previously described structural protein preparations reflect properties of a mixture of species. The introduction of the use of the detergent sodiumdodecylsulphate /SDS/ rendered new possibilities in the study of membrane proteins. Major peptides of the membrane can be separated according to their molecular weight by electrophoretic molecular sieving in polyacrylamide gel containing SDS. The patterns are well reproducible and have been extensively studied. About 16-20 bands are consistantly present, out of which 6 to 7 are major components. The size of the polypeptide chains range from about 250.000 to 15.000 daltons. The first description of size distribution was given by Berg /1969/ and confirmed by many investigators. The most successful resolution and characterization of erythrocyte membrane proteins have been accomplished by Rosenberg and Guidotti /1969/ and by Steck et al., 1970, 1971.

There is a large body of information concerning different procedures of solubilization of erythrocyte membrane proteins. We have compared - with my coworkers, Hasitz and Szelényi /Hasitz et al., 1972/ 9 different methods of membrane solubilization /Table 2/.

Without going into details just three points of our conclusion should be mentioned.

First, in good agreement with literary data only one homogeneous fraction could be separated /Fig.1/, the spectrin by the method of Marchesi and Steers, 1968; Tillack et al., 1970.

Table 2

The methods used to the preparation of membrane proteins

Number of samples examined	Solubilizing procedure	Reference
4	Dialysis against 1 mM EDTA and 50 mM β-mercaptoethanol	Tillack et al. /1970/
7	1 M NaI	
50	0.8 M NaCl	Rosenberg, Guidotti /1969/
50	3 % sodium dodecyl sulphate	Berg /1969/
10	5 % saponin	
20	Phenol : urea : acetic acid : distilled water 2w : 1.2w : 1v : 1v	Limber et al. /1970/
100	"Splitting" solution: 8 M urea, 0.16 M β-mercaptoethanol	Azen et al. /1965/
4	Extraction with n-butanol	Zwaal, van Deenen /1968/
7	Pyridine and dialysis against distilled water	Blumenfeld /1968/

Second, in good agreement with Kaplan and Criddle /1971/, we hold the opinion that two types of solvent systems are useful in the study of erythrocyte membrane proteins, those containing urea and those which contain SDS.

The use of SDS-containing solvents has the well-known and aforementioned advantages, but has the disadvantage that the analyses of the resulting polypeptide chains give a minimum estimate of the number of proteins in the membrane studied, because each size class is most probably heterogeneous with respect to the amino acid sequence of the polypeptide chains and because many polypeptides can be present in small, non-detectable amounts /Guidotti, 1972/.

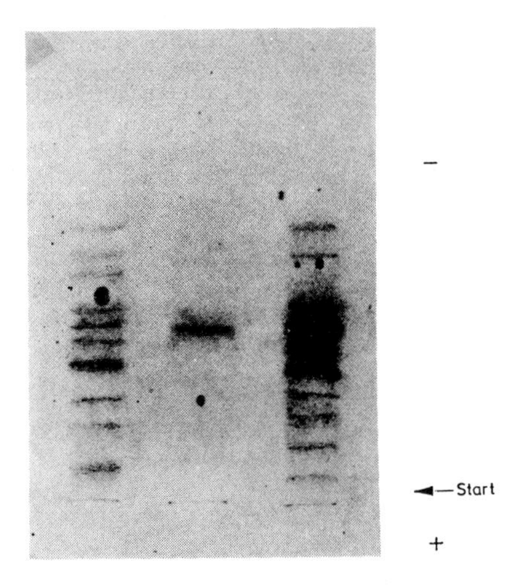

Fig. 1. Results of urea-β-mercaptoethanol /βME/ starch gel electrophoresis /according to Azen et al., 1965/. Middle pattern: spectrin, solubilized from the red cell membrane suspension by dialysis against 1 mM EDTA and 50 mM ME. Left pattern: the residue after dialysis was solubilized by the splitting solution. Right pattern: membrane proteins solubilized by the splitting solution

Urea containing solvent systems fail to solubilize completely the membrane proteins. High molecular weight aggregates may remain at the origin. Although interpretations based on the use of urea gels must be made with extreme care we got well reproducible results by the use of Azen's urea-mercaptoethanol starch gel method /Azen et al., 1965/ and found it advantageous to use it in parallel with the SDS acrylamide gel method.

Third, all methods of membrane solubilization are drastic, and yield denatured proteins, which do not reflect their in situ state in the membrane. This is especially true in pathological situations. We are convinced that the best approach to define a membrane alteration is to provide as many criteria as possible to identify a change with pathological implication.

Another pitfall of the characterization of membrane proteins resides in the unresolved problem of the differentiation of extrinsic and intrinsic membrane components. An essential criterion of a "true" membrane preparation should be the retention of the integral components and the removal of "weekly absorbed" substances. However, recovery of the essential membrane components, i.e. that of proteins and lipids by weight or assay suffers for the time being in a quantitative sense.

Vanderkooi /1972/ defines intrinsic membrane proteins as those which make up part of the membrane continuum and are intimately associated with the lipid. Extrinsic membrane proteins, on the other hand, are those which are associated with the membrane, but do not make up a part of the continuum. But the discrimination for a given protein as intrinsic or extrinsic component is by no means firm. Probably most of the protein species referred to as "membrane proteins" are of the extrinsic or peripheral type, especially in the case of the erythrocyte ghost membrane which contains a large amount of membrane associated protein.

Opinions differ - for example - significantly in the point of view whether haemoglobin is an integral or peripheral part of the red cell membrane. Recent data favour the opinion that haemoglobin--free ghosts do not represent the state of intact red cell membrane. Intact erythrocytes - in contrast to isolated membranes - are very resistant to external action of proteases and phospholipase C. The activity of some membrane bound enzymes, for instance the availability at Na^{+}-K^{+}-sensitive ATPase is influenced by the haemoglobin content of the erythrocyte membrane.

Under well controlled conditions of pH and osmolarity red cell membranes essentially free of haemoglobin can be obtained. But as pointed out by Hanahan /1973/, the slightest changes in the osmolarity of the suspending media of the haemolyzing mixture at pH 7.6 alter very definitely the amount of haemoglobin retained in the isolated membrane. The haemoglobin retention is also highly pH dependent. It is also significantly influenced by the divalent cation content of the membrane.

We have plotted the correlation between membrane bound haemoglobin and the Ca^{2+} content of the membrane /Fig. 2/. Addition of Ca^{2+} or Mg^{2+} to the haemolyzing buffer - especially at pH 7.6 leads to a membrane with increased haemoglobin content. Inclusion of EDTA to the haemolyzing solution results in an important removal of the retained haemoglobin.

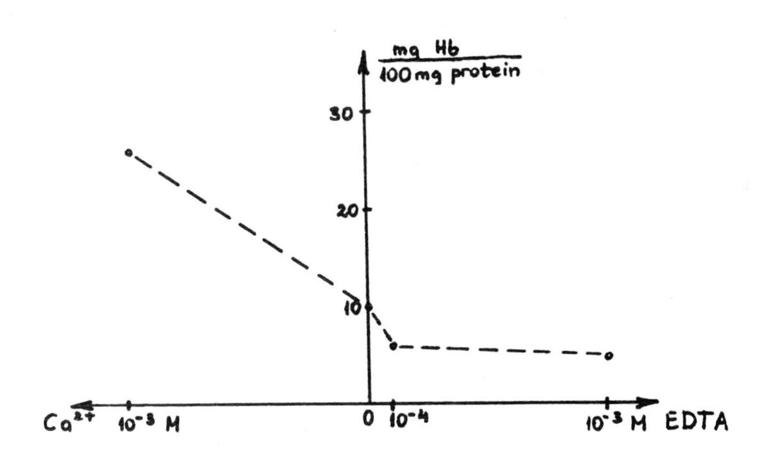

Fig. 2. Correlation between membrane-bound haemoglobin and the Ca^{2+} content of the red cell membrane

<u>In vivo</u> accumulation of red cell Ca^{2+} content also results in increased haemoglobin retention. Membrane Ca^{2+} present in ghosts prepared from normal or ATP-depleted cells represents most of the cell Ca^{2+} content. The Ca^{2+} permeability of the membrane is very low. The intracellular Ca^{2+} pool is kept near zero by the Ca^{2+} pump. Depletion of the Ca^{2+} chelating ATP results in intracellular Ca^{2+} accumulation /Eaton et al., 1973/.

With my associates Szász and Gárdos /1974/, we have studied the $^{45}Ca^{2+}$ penetration into the red cells. The $^{45}Ca^{2+}$ influx into the erythrocytes of patients with different red cell abnormalities differed significantly from the normal. Cells treated <u>in vitro</u> with acetylphenyl-hydrazine /Acphe/ showed differences as well /Fig. 3/.

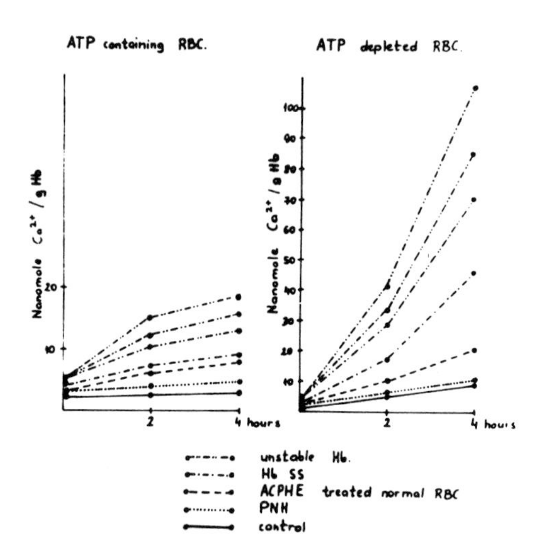

Fig. 3. $^{45}Ca^{2+}$ penetration into the red cells from patients with different types of haemolytic anaemia and into normal human red cells treated Acphe <u>in vitro</u>

The most conspicuous increase of Ca^{2+} influx was measured in red cells from patients with unstable haemoglobin causing Heinz body anaemia. In good agreement with the data of Eaton we have found a definite increase in the Ca^{2+} penetration into sickle cells. A moderate but still significant increase was observed in the Ca^{2+} content of Acphe treated erythrocytes. No definite changes could be detected in red cells from patients with paroxysmal nocturnal haemoglobinuria /PNH/. The observed changes in Ca^{2+} influx were always more pronounced in ATP depleted cells. The increased Ca^{2+} content of sphaerocytes had been described earlier.

As a consequence one can conclude, that ATP depletion, irrespective of whatever its origin is, results in increased Ca^{2+} influx with increased haemoglobin retention, changes in cation permeability, decreased cell elasticity, deformability and viability /Weed, 1970/.

But the structural and functional alterations of red cell membranes in pathological reactions depend not only upon the metabolic state but also upon the age and differentiation of the cell.

We have demonstrated definite differences between the lipid composition of foetal and adult erythrocytes. There are very definite differences between foetal and adult red cells in their phospholipid distribution. Foetal erythrocytes contain less phosphatidyl-ethanolamine and phosphatidyl-choline and more sphingomyelin than their adult counterpart /Table 3/.

Table 3

The lipid composition of cord blood cells

Components	Adult RBC /n= 30/	Foetal cell	
	$/10^{-12}$mg/cell/		
Cholesterol	125,4±3,37+	169,6±22,0	/n=13/
TEFA	199,8±2,72	203,0±24,4	/n=13/
Lipid P	11,8±0,235	11,07±1,43	/n=13/
Phospholipid fractions	/µgP/100 µgP/		
PE	28,68±0,15	27,62±0,28□	/n=11/
PS	15,04±0,16	15,85±0,43	/n=11/
PC	31,04±0,20	27,59±0,39■	/n=11/
S	25,35±0,16	28,74±0,37■	/n=11/

+ : $\bar{x} \pm$ SE
□ : significant deviation from normal adult /$p < 0,05$/
■ : significant deviation from normal adult /$p < 0,01$/

Difference was found between foetal and adult red cells in the fatty acid profile of their total phospholipid. The concentration of linoleic acid /$C_{18:2}$/ was significantly lower, the oleic acid /$C_{18:1}$/ content slightly higher, and the palmitic acid /$C_{16:0}$/ level lower in foetal red cells. Eicosotrienic acid /$C_{20:3}$/

was detected in nearly 50 per cent of the cord blood samples, while in none of the adult cells /Table 4/.

Table 4

The lipid composition of cord blood cells

/Hollán et al., 1972/

Components	Adult /n= 3o/	Foetal /n=21/
Fatty acid profile of the total phospholipid	g/100 g fatty acid	
16 : 0	25,19±0,36[+]	30,78±0,85[■]
16 : 1	1,70±0,17	2,40±0,52
18 : 0	20,72±0,20	18,42±0,44[■]
18 : 1	20,74±0,28	22,92±0,67[■]
18 : 2	12,13±0,26	4,81±0,26[■]
20 : 4	19,70±0,44	19,80±1,31

+ : $\bar{x} \pm$ SE
■ : significant deviation from normal adult /$p < 0,01$/

There is a difference between the membrane protein patterns of foetal and adult cells when the membrane proteins are solubilized by urea-mercaptoethanol according to the method of Azen et al., 1965 /Fig. 4/.

Definite differences could be revealed between the non--electrolyte permeability /Fig. 5/ of foetal and adult erythrocytes /Hollán et al., 1967/. Ethanol-fixed foetal and adult red cells differ significantly in the degree of haemoglobin-elution induced by a treatment with 4 molar urea /Fig. 6/.

We have pointed to the fact that a number of red cell membrane changes found in different types of haemolytic anaemias can be derived from the foetal type cell population produced by the rapidly expanding compensatory erythropoiesis /Hollán and Breuer, 1968; Hollán et al., 1970/.

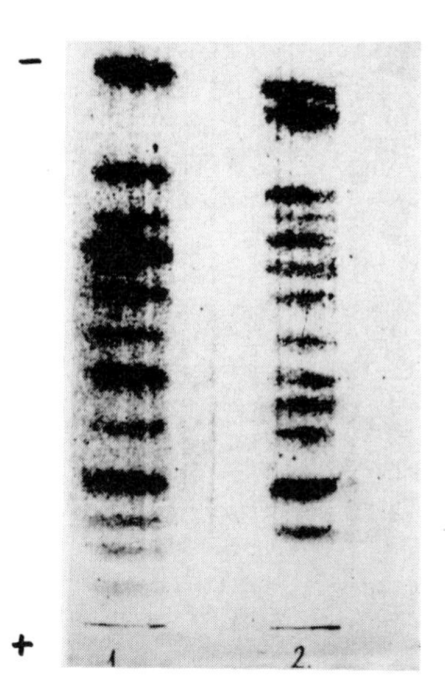

Fig. 4. Results of urea-β ME starch gel electrophoresis /according to Azen et al., 1965/. Left pattern: membrane protein from adult RBC. Right pattern: membrane protein from foetal RBC

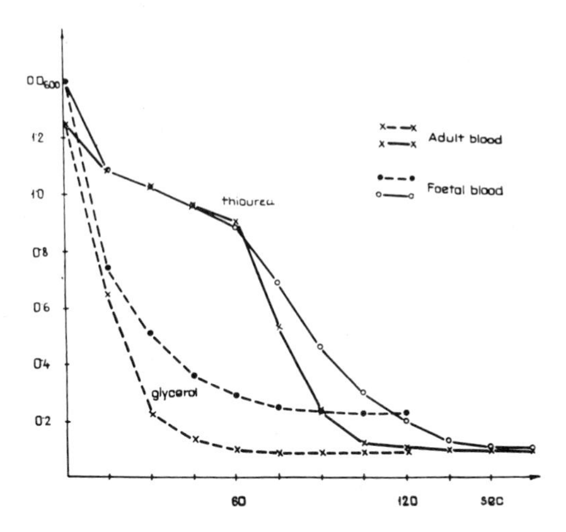

Fig. 5. Rate of haemolysis of adult and foetal erythrocytes in isotonic glycerol and thiourea

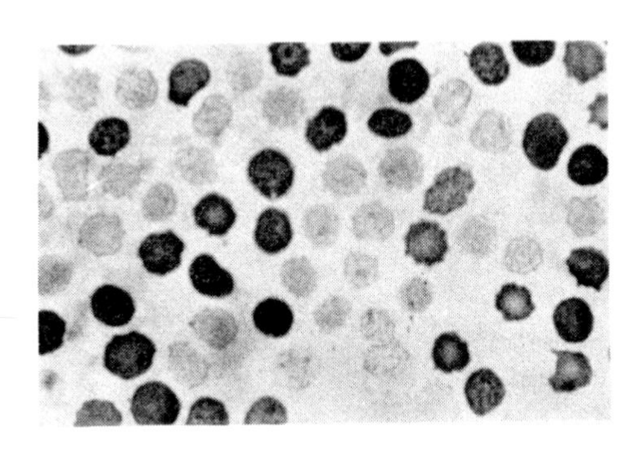

Fig. 6. 4 molar urea treatment /Szelényi and Hollán, 1967/ of ethanol fixed red cells. An artifical mixture of equal volumes of adult and foetal blood

We have found /Szelényi et al., 1972/ the most conspicuous changes in red cells with inclusion /Heinz body/. These membrane alterations differed significantly from those characterizing the foetal cell membrane.

The formation of inclusion bodies in erythrocytes and the concomitant decreased viability, may be caused by various factors such as unstable haemoglobin disease, red cell enzyme deficiency or dose-dependent large amounts of oxydative agents, like Acphe and drugs.

Inclusion body containing erythrocytes from 7 patients with unstable haemoglobin disease and from Acphe treated rabbits were studied. Non-electrolyte permeability was found to be decreased in both type of Heinz-body containing cells /Fig. 7/.

Well reproducible changes could be revealed in the electrophoretic pattern of the inclusion body containing human and rabbit erythrocyte membrane protein, when solubilized by the method of Azen et al. /1965/. The most conspicuous changes were similar in both type of red cells /Fig. 8 and 9/.

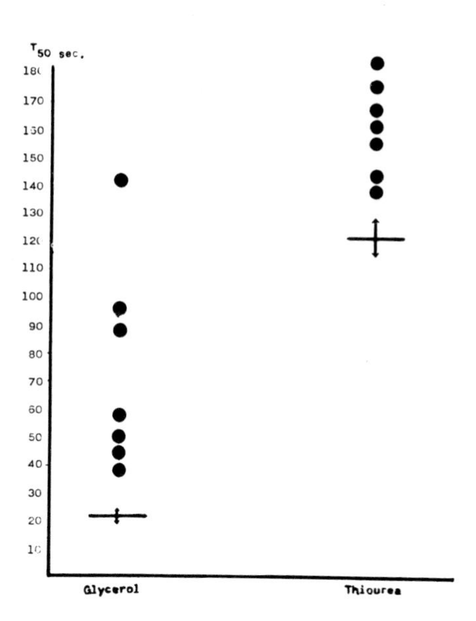

Fig. 7. Rate of penetration of isotonic glycerol and thiourea into the inclusion body containing red cells, as reflected by the increase of the half time of haemolysis. The cells originated from splenectomized patients with unstable haemoglobin disease

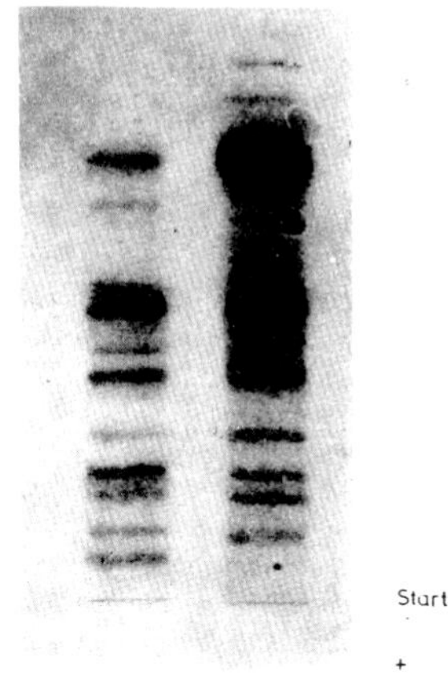

Fig. 8. Results of urea-β ME starch gel electrophoresis /according to Azen et al., 1965/. Membrane proteins from erythrocytes /N/ and from a splenectomized patient /ZB/ with unstable haemoglobin

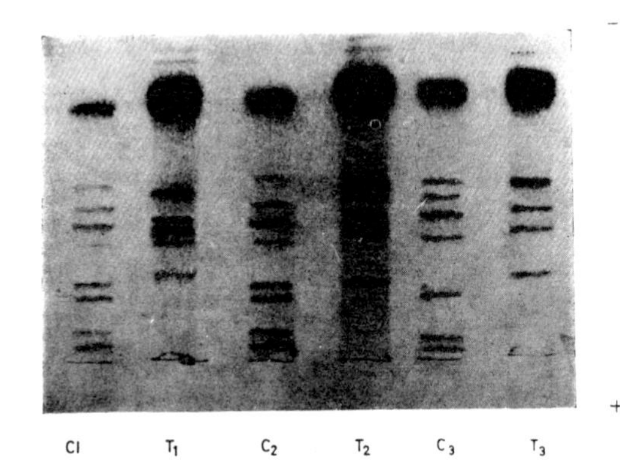

Fig. 9. Results of urea-β ME starch gel electrophoresis /according to Azen et al., 1965/ of membrane proteins of erythrocytes from three rabbits before /C_1 C_2 C_3/ and after /T_1 T_2 T_3/ Acphe treatment in vivo

1./ More globin is bound to the membrane of inclusion body containing cells than to the normal red cell membrane

2./ Two new fast moving protein bands are consistantly present

3./ The fraction or fractions with the slowest electrophoretic mobility are absent.

The SDS-acrylamide gel membrane protein patterns of inclusion body containing red cells did not differ from their normal counterparts.

Striking changes were found in the lipid composition of inclusion body containing erythrocytes. The cholesterol, total esterified fatty acid /TEFA/ and phospholipid content of the red cells of both, Acphe treated rabbits /Table 5/ and of patients with unstable haemoglobin disease /Table 6/ were significantly increased. There was also a definite change in the relative amount of phospholipid classes characterized by an increased phosphatydil--choline/sphingomyelin ratio /Table 6 and 7/.

Table 5

Effect of <u>in vivo</u> Acphe treatment on the red cell main lipid classes of rabbits

No.	Cholesterol		TEFA		Lipid P	
	Before	After	Before	After	Before	After
	10^{-12} mg/ml		10^{-12} mg/ml		10^{-12} mg/ml	
90	76	40	149	1066	14.9	133.0
91	113	185	370	530	21.0	55.2
96	150	340	300	940	29.0	62.0
08	125	110	240	165	13.7	26.0
09	124	130	230	230	12.6	22.1
13	135	153	430	280	42.0	11.3
14	165	172	200	366	15.1	49.5
20	112	305	189	660	11.4	43.2
28	109	310	175	710	10.7	32.0
30	90	178	194	292	8.2	15.5
17	134	230	211	1010	14.6	36.1
Mean	*121.2*	*229.0*	*244.4*	*568.1*	*17.6*	*44.1*
S.D.±	25.4	101.3	132.7	354.3	3.1	10.6
t	3.15		2.94		2.29	
P	<0.01		<0.01		<0.05	

Table 6

Lipid pattern of erythrocytes of splenectomized patients with unstable haemoglobin

Patient	Cholesterol 10^{-12}	TEFA mg/100 ml	Lipid P	PE	PS	L	S
				per cent			
D. M.	210	280	43.0	26.4	15.0	37.2	21.4
H. Gy.	130	230	33.0	31.5	14.8	30.6	23.0
D. F.	143	170	25.0	29.1	16.1	31.8	22.9
P. M.	176	384	38.0	27.4	16.8	32.6	23.1
Sz. J.	390	817	49.0	25.5	15.8	38.1	18.7
Z. B.	230	305	13.0	27.3	13.2	36.9	22.5
R. B.	216	250	11.2	23.8	15.5	39.8	20.8
Mean	*213.6*	*305.1*	*30.5*	*27.3*	*15.3*	*35.3*	*21.7*
S.D.±	86.4	114.8	12.6	2.5	2.8	3.6	2.0
Normal mean	128.9	176.6	11.8	28.6	14.9	31.1	25.6
S.D.±	15.3	24.5	0.9	1.0	1.1	1.4	1.0
t	3.9	4.4	5.9	1.9	0.6	4.7	11.7
P	<0.01	<0.01	<0.01	>0.05	>0.05	<0.01	<0.01

Table 7

Effect of in vivo Acphe treatment of rabbits on the red cell phospholipid fractions /in per cent/

No.	Phosphatidyl-ethanolamine		Phosphatidyl-serine		Phosphatidyl-choline		Sphingomyelin	
	Before	After	Before	After	Before	After	Before	After
90	32.4	33.5	13.7	16.1	34.1	31.9	20.1	18.4
91	33.8	32.1	16.7	15.6	30.0	35.7	19.4	16.6
96	32.4	32.4	17.6	16.4	30.7	34.3	20.1	16.9
08	31.2	31.9	15.4	15.3	32.8	31.9	20.4	20.9
09	32.6	31.6	14.1	15.9	32.8	33.7	20.5	18.7
13	34.3	31.0	14.9	17.1	31.0	34.7	19.7	17.1
14	33.5	32.0	15.5	15.8	30.5	33.7	20.5	18.4
20	31.3	29.5	15.4	16.0	32.7	37.6	20.5	16.9
28	33.4	29.2	14.4	15.7	31.5	35.2	20.5	17.9
30	33.1	31.9	14.3	14.7	33.5	34.1	19.1	19.3
17	31.5	29.1	14.6	15.5	33.0	36.1	28.8	19.1
Mean	*32.7*	*31.3*	*15.1*	*15.8*	*32.0*	*34.6*	*20.1*	*18.2*
S.D.±	0.33	0.45	0.33	0.19	0.43	0.59	0.17	0.41
t	2.69		1.65		3.33		4.19	
P	< 0.01		> 0.05		< 0.01		< 0.01	

The fatty acid composition of the red cell phospholipids of the patients with unstable haemoglobin differed definitely from the normal. There was a significant decrease in linoleic acid /$C_{18:2}$/ and a decrease in the stearic acid/oleic acid ratio /Table 8/. The fatty acid profile of the phospholipids of the rabbit cells did not change upon Acphe treatment.

Membrane changes induced by the attachment of Heinz bodies to the membrane are very diverse and multiple. We have found that it is even reflected in an increased binding of ^{125}I-labelled IgG to the red cell surface /Table 9/.

The mode of attachment of Heinz bodies to the red cell membrane is still unsolved. The binding of the denatured haemoglobin via mixed disulphid bonds was proposed by Jacob et al. /1968/. In contrast to their findings we could confirm the results of Winterbourne and Carrel /1973/ according to which sulfhydryl reducing or blocking agents have no effect on the attachment of the insoluble haemoglobin to the membrane. The reduction of electrostatic interactions in

Table 8

Fatty acid distribution of total phospholipids of splenectomized patients with unstable haemoglobin

Patient	$C_{16:0}$	$C_{16:1}$	$C_{18:0}$	$C_{18:1}$	$C_{18:2}$	$C_{20:4}$
D. M.	26.6	0.6	19.1	22.9	11.2	19.7
H. Gy.	25.2	0.7	20.0	22.6	13.9	17.6
D. F.	24.5	0.6	20.6	22.3	13.4	18.6
P. M.	27.0	0.8	22.3	21.8	7.4	20.7
Sz. J.	26.8	1.3	19.3	20.2	8.9	23.5
Z. B.	26.2	2.2	17.5	22.7	11.7	19.7
R. B.	23.2	0.6	19.9	20.4	9.7	23.2
Mean	*25.64*	*0.97*	*19.81*	*21.88*	*10.88*	*20.43*
S.D.±	1.40	0.72	1.45	1.10	2.38	1.90
Normal mean	25.2	1.7	20.7	20.7	12.1	19.7
S.D.±	2.4	1.1	1.3	1.9	1.8	2.9
t	0.6	3.1	9.0	1.2	2.7	1.1
P	> 0.05	<0.01	< 0.01	>0.05	<0.05	> 0.05

Table 9

The binding of ^{125}I-labelled IgG to inclusion body containing erythrocytes

Origin of red cells	10^{-10} mg IgG/cell	10^{5} mol IgG/cell
Z.B.+	1,17	4,38
P.M.+	1,43	5,32
Normal range	0,65-0,94	0,95-3,8

\+ splenectomized patients with unstable haemoglobin disease

2M NaCl was also found ineffective. These results point rather to a hydrophobic bonding between inclusion bodies and the red cell membrane.

The firm attachment of aggregates of denatured haemoglobin to the internal surface of the membrane produces marked alterations of both, internal and external surface of the membrane. This is also well reflected by the freeze-cleavage studies of Lessin et al. /1972/ according to which rearrangements of the membrane associated particles /MAP-s/ could be revealed in Heinz body containing erythrocytes. Both, aggregation and rarification of the MAP-s characterized the overlying surface of Heinz bodies.

It is by now well established that when exposing the inner hydrophobic matrix of the erythrocyte membrane by freeze-fracture cleaving it presents smooth areas and particles which have an average diameter of 8.5 nm. They apparently consist of protein and are associated with the antigens on the outer surface pointing to their oligosaccharide content.

In freshly prepared ghosts the distribution of the particles appears to be random. Pinto da Silva /1972/ has shown that pH changes of the suspending medium of the cell ghosts induce reversible translational movement of the MAPs. A series of publication of Branton and coworkers /1972/, Greenwalt and Steane /1973/ and from other laboratories revealed that redistribution of the particles follows a wide variety of treatment with very diverse agents, including gentle proteolysis, lipolysis.

The wide variety of conditions eliciting particle aggregation suggests that the redistribution of particles may be explained by charge interactions. The control of particle aggregation may depend upon the manner in which the particles are associated with other proteins probably on the inner surface of the membrane.

The most meaningful conclusion of the studies of particle aggregation is that protein-protein interactions can modify and regulate the whole membrane organization in response to changes in its external and internal aqueous environment. This has a paramount importance in different pathological conditions and different cell--cell interactions and in the action of different membrane active ligands. This is because electrokinetic and immunologic properties of the cell surface, receptor functions and adhaesivity, recognition, motility and communication of cells are probably highly dependent upon these membrane particles.

SUMMARY

Changes in membrane structure and function are not yet understood on a molecular pathological level. Most of the alterations described so far are rather pathological reaction patterns characterizing very diverse disease conditions.

Molecular structure and function of membranes change upon binding of suitable ligands. And because of the very extensive macromolecular interactions, cooperative effects are multiplied and may perturb the whole membrane lattice.

The best approach to define a membrane alteration is to provide as many criteria as possible to identify a change with pathological implication.

The use of urea containing solvent systems in parallel with the SDS acrylamide gel method was found advantageous in characterizing developmental and pathological membrane protein alterations.

A number of red cell membrane changes found in different types of haemolytic anaemias can be derived from the foetal type cell population produced by the rapidly expanding compensatory erythropoiesis.

The differentiation between extrinsic and intrinsic membrane proteins is extremely difficult. Recent data favour the opinion that haemoglobin forms an integral part of the red cell membrane.

A wide range of red cell defects especially those inducing membrane changes: like inclusion body formation or sickling and all those which cause a depletion of the Ca-chelating ATP elicit an increased Ca^{2+} influx which in turn leads to a more firm attachment of haemoglobin to the membrane and to very significant changes in the membrane structure and biophysical properties, such as increased cation permeability and osmotic fragility, decreased elasticity, deformability and viability.

Changes in the organization of a membrane may be elicited by either extrinsic changes in the cell environment or changes originating from the cell interior. The attachment of aggregated denatured haemoglobin to the internal surface of the membrane elicits significant changes in the lipid composition, in the membrane protein pattern and the non-electrolyte permeability of the cell. The binding of the Heinz bodies to the membrane induces aggregation of the membrane particles and increases the binding of IgG to the external surface of the cell.

ACKNOWLEDGEMENTS

Thanks are due to Dr. J.G. Szelényi, Dr. I. Szász and Dr. G. Gárdos, for some of these experiments.

REFERENCES

Azen, E.A., Orr, S. and Smithies, O. /1965/. J. Lab. clin. Med. 65, 440.

Berg, H.C. /1969/. Biochim. biophys. Acta /Amst./ 183, 65.

Blumenfeld, O.O. /1968/. Biochim. biophys. Acta /Amst./ 30, 200.

Branton, D., Elgsaeter, A. and James, R. /1972/. In: Mitochondria, biomembranes, eds.: S.G. van den Bergh, P.Borst, L.L.M. van Deenen, J.C. Riemersma, E.C. Slater and J.M. Tager /North--Holland Pub. Co., Amsterdam/ p. 165.

Bretscher, M.S. /1971a/. J. Molec. Biol. 58, 775.

Bretscher, M.S. /1971b/. J. Molec. Biol. 59, 351.

Changeux, J.P., Tung, Y. and Kittel, C. /1967/. Proc. nat. Acad. Sci. /Wash./ 57, 335.

Eaton, J.W., Shelton, T.D., Swofford, H.S., Kolpin, C.E. and Jacob, H.S. /1973/. Nature /London/ 246, 105.

Gordesky, S.E., Marinetti, G.V. and Segal, G.B. /1973/. J.Membrane Biol. 14, 229.

Greenwalt, T.J. and Steane, E.A. /1973/. Brit. J. Haemat. 25, 227.

Guidotti, G. /1972/. Arch. intern. Med. 129, 194.

Hanahan, D.J. /1973/. Biochim. biophys. Acta /Amst./ 300, 319.

Hasitz, M., Szelényi, J.G., Hollán, S.R. and Baumann, M. /1972/. Haematologia 6, 249.

Hollán, S.R. and Breuer, J.H. /1968/. Acta med. Acad. Sci. hung. 25, 421.

Hollán, S.R., Breuer, J.H. and Szelényi, J.G. /1972/. Haematologia 6, 217.

Hollán, S.R., Charlesworth, D., Szelényi, J.G., Miltényi, M. and Lehmann, H. /1970/. Haematologia 4, 141.

Hollán, S.R., Szelényi, J.G., Breuer, J.H., Medgyesi, G.A. and Sőtér, V.N. /1967/. Haematologia 4, 409.

Jacob, H.S., Brain, N.C. and Dacie, V. /1968/. J. clin. Invest. 47, 2664.

Kaplan, D.M. and Criddle, S.R. /1971/. Physiol. Rev. 51, 249.

Lenard, J. and Singer, S. /1966/. Proc. nat. Acad. Sci. /Wash./ 56, 1828.

Lessin, L.S., Jensen, W.N. and Klug, P. /1972/. Arch. intern. Med. 129, 306.

Limber, G.K., Davies, R.F. and Bakerman, S. /1970/. Blood 36, 111.

Marchesi, V.T. and Steers, E. Jr. /1968/. Science 159, 203.

Pinto da Silva, P. /1972/. J. Cell Biol. 53, 777.

Rosenberg, S.A. and Guidotti, G. /1969/. J. biol. Chem. 244, 5118

Singer, S.J. and Nicolson, G.L. /1972/. Science 175, 720.

Steck, T.L., Fairbanks, G. and Wallach, D.F.H. /1971/. Biochemistry 10, 2617.

Steck, T.L., Weinstein, R.S., Straus, J.H. and Wallach, D.F.H. /1970/. Science 168, 255.

Szász, I. and Gárdos, G. /1974/. FEBS Letters 44, 213.

Szelényi, J.G., Breuer, J.H., Győrffy, Gy., Hasitz, M., Horányi, M. and Hollán, S.R. /1972/. Haematologia 6, 327.

Szelényi, J.G. and Hollán, S.R. /1967/. Vox Sang. 12, 234.

Tillack, T.W., Marchesi, S.L., Marchesi, V.T. and Steers, E. Jr. /1970/. Biochim. biophys. Acta /Amst./ 200, 125.

Vanderkooi, G. /1972/. Ann. N.Y. Acad. Sci. 195, 6.

Verkileij, A.J., Zwaal, R.F.A., Roelofsen, B., Comfurius, P., Kastelijn, D. and van Deenen, L.L.M. /1973/. Biochim. biophys. Acta 323, 178.

Wallach, D.F.H. /1972/. In: The plasma membrane: dynamic perspectives, genetics and pathology. The English University Press LTD, London. Springer, New York, p. 352.

Wallach, D.F.H. and Zahler, P.H. /1966/. Proc. nat. Acad. Sci. /Wash./ 56, 1552.

Weed, R.I. /1970/. Seminars in Hemat. 7, 249.

Winterbourne, C.C. and Carrell, R.W. /1973/. Brit. J. Haemat. 25, 585.
Zwaal, R.F.A. and van Deenen, L.L.M. /1968/. Biochim. biophys. Acta /Amst./ 163, 44.
Zwaal, R.F.A., Roelofsen, B. and Colley, C.M. /1973/. Biochim. biophys. Acta /Amst./ 300, 159.

Mechanisms of Sugar and Amino Acid Transport

MODES OF COUPLING OF ENERGY WITH TRANSPORT

A. Kotyk

Laboratory for Cell Membrane Transport,
Institute of Microbiology, Czechoslovak Academy
of Sciences, Budějovická 270, 142 20 Praha, ČSSR

INTRODUCTION

Practically all major translocations across cell membranes, serving both nutrition and maintenance of cell homeostasis, proceed by some type of energy-coupled mechanism. Like the whole field of membranology, this particular topic has undergone in the past five years a tremendous expansion and, like any other area of intense investigation, it may be characterized by the observation of Blaise Pascal who, some 320 years ago, compared the total of understood natural phenomena to a sphere placed in the middle of the unknown. As the extent of our understanding increases so does the area of contact with the unknown. It may be a somewhat comforting thought to realize that the surface of the sphere in question increases at a lower rate than its volume. Whatever the obstacles and uncertainty prevailing in some areas of the field it may be useful even at this stage of development to summarize what is known about energy involvement in transport processes and to offer some general views on the subject.

PRINCIPAL WAYS OF COUPLING

No truly uphill transport can proceed in isolation from a source of energy other than the thermal movement of molecules or ions in solution. For a solute element to be moved against its chemical or electrochemical gradient a supply of energy must be present and made available in one of three general mechanisms.

1. Chemical modification of the solute. In some cases, such as the phosphotransferase system, this is not rigorously active transport since the solute in question is present in different chemical forms at the two membrane sides.

2. Chemical modification of the carrier, either by a specific substituent, such as a phosphoryl group, or by a more subtle conformational change, resulting in different affinity for solute or mobility in a complex sith the solute. This is the typical, primary active transport.

3. Modification of carrier affinity, transmembrane movement or, under some stringent conditions, simply of its effective concentration, by binding an activating ligand, a cation.

Since the effect is proportional to the difference in the "activating" ligand concentrations, a higher concentration at the *cis* side will result in a gradient of the ternary carrier-ion-solute complex and, eventually, in building up a higher concentration of the solute at the *trans* side. This has often been called secondary active transport although, according to a rigorous definition, it is not active transport even if it goes uphill.

It should be noted parenthetically that mechanisms belonging to all the above categories may actually operate without participation of metabolic energy; thus: (1) sucrase-isomaltase in the intestinal wall which splits sucrose into its component monosaccharides probably with concomitant transport across the membrane (Storelli et al., 1972); (2) the mediated diffusion of monosaccharides both in human erythrocytes (Krupka, 1971) and in baker's yeast (Kalsow and Doyle, 1973) probably involves a conformational change of the carrier; (3) under some conditions secondary active transport can be elicited by simply adding the driving ion to the appropriate side of the membrane.

Chemical modification of substrate

The now almost classical example is the phosphotransferase system, known to translocate various monosaccharides, disaccharides and polyols in a number of microbial species: *Escherichia coli*, *Salmonella typhimurium*, *Bacillus subtilis*, *Aerobacter aerogenes*, *Lactobacillus plantarum*, *Streptococcus lactis*, *Staphylococcus aureus*, *Rhodospirillum rubrum* and *Mycoplasma* species. The complete reaction sequence as described for the case of lactose in *S.aureus* (Simoni et al., 1973) is as follows:

$$3 \text{ enzyme I} + 3 \text{ phospho-enol-pyruvate} \xleftrightarrow{Mg^{2+}} 3 \text{ P-enzyme I} + 3 \text{ pyr}$$

$$3 \text{ P-enzyme I} + 3 \text{ HPr} \longleftrightarrow 3 \text{ HPr-P} + 3 \text{ enzyme I}$$

$$3 \text{ HPr-P} + \text{factor III}^{lac} \longleftrightarrow \text{factor III}^{lac}\text{-P}_3 + 3 \text{ HPr}$$

$$\text{factor III}^{lac}\text{-P}_3 + 3 \text{ lactose} \xleftrightarrow{EIIb^{lac}} 3 \text{ lactose-P} + \text{factor III}^{lac}$$

HPr is a small protein (8500-9600 daltons) in all the species examined and it apparently has an identical amino acid composition at least in the Gram-negative species (underlined in the list). Enzyme I is a constitutive "soluble" protein with molecular weight of about 80,000. Factor III^{lac} is also a soluble protein of molecular weight of about 35,000 and contains three identical subunits. Enzyme IIb^{lac} is a part of a large-molecular weight membrane-bound enzyme. The stoichiometry of transport by the system is always one sugar per one phosphate bond.

Another, considerably less advertised, system involving chemical transformation of substrate, is the γ-glutamyltranspeptidase apparently functioning in the transport of all amino acids, with the exception of proline, in kidney tubules (cf. Bodnaryk, 1972; Meister, 1973). The cycle of reactions may be depicted as shown in scheme II. The efficiency is low, three molecules of ATP being required for the transfer of one molecule of amino acid under continuous-cycle conditions.

A third system probably involving a chemical change of solute directly in transport is the translocation of adenine which is converted to adenosine phosphate in *Escherichia coli* vesicles (Hochstadt-Ozer and Stadtman, 1971) during the process.

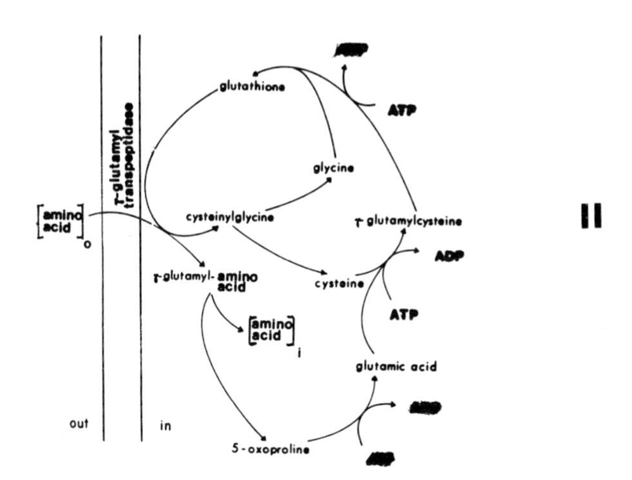

II

Although the chemical modification of the transported substrate in all the above cases is an essential condition for the substrate to be translocated, the actual molecular events are difficult to unravel. The possibility always exists that the substrate crosses the membrane by a process of mediated diffusion but at the inner face has no route for escaping from the carrier but along the active site of the transforming enzyme.

Chemical modification of carrier

The paradigm of this category is the family of membrane adenosinetriphosphatases. The Na,K-ATPase, in particular, has been studied a great deal and found to be the vehicle for carrying sodium ions out of and potassium ions into the cells of a variety of biological objects, with the exception of yeast and some plants (cf. Bonting, 1970). The point where energy enters the system is the phosphorylation by ATP of one of the enzyme forms at the γ-carboxyl group of L-glutamic acid (Kahlenberg et al., 1967). The efficiency of this system is relatively high, there being under optimal conditions 3 Na^+ and 2 K^+ ions transported per one molecule of ATP split. The general course of reactions is as follows:

$$\begin{aligned} E_1 + ATP &\xrightleftharpoons{Mg^{2+}, Na^+} E_1\sim P + ADP \\ E_1\sim P &\longrightarrow E_2\text{-}P \\ E_2\text{-}P + H_2O &\xrightleftharpoons{K^+} E_2 + \text{phosphate} \\ E_2 &\longleftrightarrow E_1 \end{aligned} \qquad \text{III}$$

In general it may be thought that enzyme form E_1 exists at the internal face of the membrane while the E_2 form appears at the outside. However, the detailed picture of the two enzyme activities involved (a phosphorylase and a phosphatase) and of the translocation of ions is extremely complicated (cf. Skou,1972). It is attractive to view the enzyme as an oligomeric protein with all its kinetic consequences (cf. Stein et al., 1973; Repke and Schön, 1973).

There is a group of ATPases activated by bivalent cations, especially by Ca^{2+} and Mg^{2+}, the activation being (not obligatorily) coupled to the translocation of these ions across the membrane (in the Na,K-ATPase, Mg^{2+} is also an activating ion but it is not translocated in a net fashion). The sequence of molecular events in these ATPases is only partly understood, most being known about the calcium pump in the sarcoplasmic reticulum where it resembles the Na,K-ATPase in the involvement of an acylphosphoprotein and in the translocation of 2 Ca^{2+} per ATP split (e.g. Martonosi et al., 1974).

A special type of magnesium-activated membrane ATPase has been found to translocate hydrogen ions from cells into the medium, the function of these ions being to drive various nonelectrolytes inward (see below). It is suspected to operate in various bacteria (cf. Harold, 1972), as well as yeasts (e.g. Seaston et al., 1973) and algae (Komor and Tanner, 1974). The mechanism of action of this ATPase is analogous to that of the Na,K-translocating one and the resulting H^+ gradient must not be confused with the general consequence of ATP splitting which, due to a change in the number of dissociable groups during the process, results in a pH lowering. Such a decrease of pH would be expected to occur intracellularly where ATP is split but in fact it is the outside of the cell which is acidified by this action. It is likely, moreover, that two H^+ are translocated by this ATPase per molecule ATP split (Greville, 1969).

Another type of carrier modification may be involved in the transport of some sugars (e.g. lactose), amino acids (e.g. proline, serine, tyrosine, lysine) and even ions (Rb^+) in bacterial vesicles (*Escherichia coli*, *Salmonella typhimurium*, *Pseudomonas aeruginosa*, *Bacillus*, *Micrococcus* and *Staphylococcus* species), as articulated most convincngly by Kaback's group (for the latest summary see Lombardi et al., 1974). The concept underlying this mechanism is that a part or a side branch of the electron transfer chain in these vesicles acts a a carrier, showing a high affinity for the transported solute when in the oxidized state and a low one when reduced.

lac pyr
$C_{ox}S_I$ ⇄ $C_{red}S_{II}$
S_I S_{II} **IV**
lac pyr
C_{ox_I} ⇄ $C_{red_{II}}$
cyt^{2+} cyt^{3+}

The system has several attractive features that are at present being employed to differentiate it from a proton-driven symport to be discussed below. Thus (1) oxamate causes a cessation of net flow of lactose or proline while it permits an exchange diffusion to take place; (2) the pH dependence of D-lactate-generated H^+ release and of D-lactate-generated Rb^+ uptake by valinomycin-treated vesicles are pronouncedly different; (3) there is no competition for H^+ symport when lactose and various amino acids are added in combination; (4) valinomycin-induced tyrosine uptake (due to K^+ efflux) in the absence of added substrate is strongly inhibited by cyanide and amytal although these are not recognized as proton conductors or uncouplers. There are also a number of data on the varying rates of uptake and accumulation ratios of different solutes which are being enlisted in support of the oxidation-reduction mechanism. However, it is the present author's impression that this is not the strongest point of the arguments as may be gathered from expressions (*8*) or (*9a*) in the subsequent section of the paper.

Modification of carrier properties by a second ligand

The vast amount of evidence attesting to the operation of ion-driven nonelectrolyte transports makes a comprehensive review to a gargantuan task (cf. the book by Heinz, 1970). In a brief outline, one may discern two principal coupling ions, H^+ in practically all unicellular organisms as well as mitochondria (and possibly plants), Na^+ in practically all animals (and possibly plants). The involvement of other ions, such as K^+ or anions, appears to be secondary in this context. To be of any use in driving the transport of a nonelectrolyte, the cation must be available at the positive face of the membrane (with respect to the membrane potential difference - in all interesting cells apparently the outer face) and/or form a gradient from the *cis* to the *trans* side of nonelectrolyte transport (usually again the outer face). The ion-motive force thus generated (proton-motive for H^+, sodium-motive for Na^+) is then composed of two terms, viz.

$$\text{i.m.f.} = \Delta\psi - 2.3RT/F \log \frac{[\text{ion}]_o}{[\text{ion}]_i} \qquad (1)$$

(The $\Delta\psi$ is taken to be negative in this connection.) Whereas this expression, in the form of

$$\text{p.m.f.} = \Delta\psi - Z\Delta\text{pH} \qquad (2)$$

is the cornerstone of Mitchell's (e.g. 1973) views on the role of H^+ symport both in mitochondria and across plasma membranes, it is rarely seen in the context of sodium-driven transports of sugars and amino acids in various mammalian cells where the argument is frequently about the adequacy or the inadequacy of ion gradients to drive the transport of a nonelectrolyte, the potential not being taken into consideration. On the other hand, the kinetics of sodium-driven transport has been worked out to a greater degree of sophistication (e.g. Schultz and Curran, 1970) than the kinetics of proton-driven transport. Perhaps a certain benefit may be gained from the perusal of the general kinetic scheme which applies both to proton-driven and to sodium-driven transports (symports in Mitchell's terminology). We are not concerned here with the mechanism by which the different concentrations of the driving ions are built up - let us simply take it for established that such mechanisms exist (they may be of ATPase, oxido-reductive or even photosynthetic character). If a carrier is capable of binding both the transported solute and the "activating" ion we can write the scheme as follows:

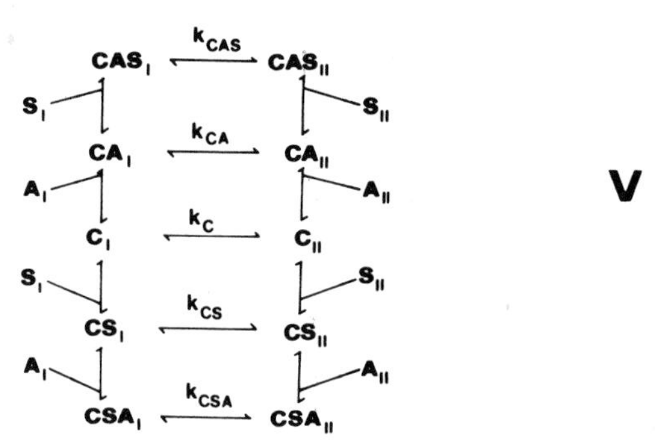

If the association-dissociation reactions at the two membrane surfaces are substantially more rapid than the translocation of the various carrier forms across the membrane the system can be treated as an equilibrium one (with $K_1 = C.A/CA$, $K_2 = CA.S/CAS$, $K_3 = C.S/CS$, $K_4 = CS.A/CSA$). If the corresponding dissociation constants are assumed to be identical at both membrane sides and if the translocation constants are equal in both directions we can write for the initial rate according to the upper part of the scheme ($C+A+S \longrightarrow CAS$) a Michaelis-Menten type expression

$$J_s = J_{max} \frac{S}{S + K_m} \tag{3}$$

with

$$J_{max} = 2C_t k_{CAS} \frac{k_C K_1 + k_{CA} A_{II}}{K_1(k_C + k_{CAS}) + A_{II}(k_{CA} + k_{CAS})} \tag{4a}$$

and

$$K_m = \frac{K_2[(K_1 + A_{II})(k_C K_1 + k_{CA} A_I) + (K_1 + A_I)(k_C K_1 + k_{CA} A_{II})]}{A_I[K_1(k_{CAS} + k_C) + A_{II}(k_{CAS} + k_{CA})]} \tag{4b}$$

with C_t being the total carrier concentration at one side of the membrane.

If the CA complex is immobile the values simplify thus:

$$J_{max} = 2C_t \frac{k_{CAS} k_C K_1}{K_1(k_C + k_{CAS}) + A_{II} k_{CAS}} \tag{5a}$$

$$K_m = \frac{k_C K_1 K_2 (2K_1 + A_I + A_{II})}{A_I[K_1(k_C + k_{CAS}) + k_{CAS}A_{II}]} \quad (5b)$$

Thus, in this system, J_{max} is independent of the ion concentration at the *cis* side while K_m is decreased by increasing A_I.

If the lower part of scheme V is now valid (C+S+A ⟶ CSA) an analogous set of expressions is obtained:

$$J_{max} = 2C_t k_C \frac{k_{CS}K_4 + k_{CSA}A_I}{K_4(k_C + k_{CS}) + A_I(k_C + k_{CSA})} \quad (6a)$$

$$K_m = \frac{2k_C K_3 K_4}{K_4(k_{CS} + k_C) + A_I(k_C + k_{CSA})} \quad (6b)$$

If now the CS complex is immobile we have

$$J_{max} = \frac{2C_t k_{CSA} k_C A_I}{k_C K_4 + A_I(k_C + k_{CSA})} \quad (7a)$$

$$K_m = \frac{2k_C K_3 K_4}{k_C K_4 + A_I(k_C + k_{CSA})} \quad (7b)$$

Thus, in this system, at least under special conditions (e.g. K_4 rather large), the maximum rate may depend positively on A_I while K_m will be unaffected.

One could vary the simplifying assumptions further (e.g. $k_{CS} = k_{CSA}$; $k_{CA} = k_{CSA}$; CAS = CSA; $k_C = k_{CS} = k_{CSA}$) to obtain a range of expressions, some of which would satisfy various special cases of ion-driven initial rates.

A characteristic feature of all active transports which deserves attention is the steady-state accumulation ratio, obtained most simply by setting $J_s = 0$ and extracting (if possible) the value of S_{II}/S_I. For all the above-treated cases of secondary coupled transport the value of S_{II}/S_I is related to A_I/A_{II} and in some restricted cases actually equal to it (cf. Kotyk, 1973). The only exception is the case that $K_2 = K_3; K_4 = \infty$ and all the translocation constants identical. Then $S_{II}/S_I = 1$. It would thus appear that the ratio is constant over the whole range of concentrations, something that is not compatible with experimental findings. It is generally found in all uphill transports that the ratio decreases with increasing S_I (Fig. 1) either toward unity or in such a way that S_{II} is practically constant at high S_I. To account for this behaviour in a primary active transport one has to postulate a slow change of the carrier forms (from a high- to a low-affinity one) or a limited supply of the chemical effector bringing about this alteration, respectively. In an ion-coupled transport, a similar "dissipation" of the gradient due to formation of the ternary complex CSA must be assumed. In the model of Kimmich (1973) the Na^+-activated intermediate required for transport would be present in limited supply.

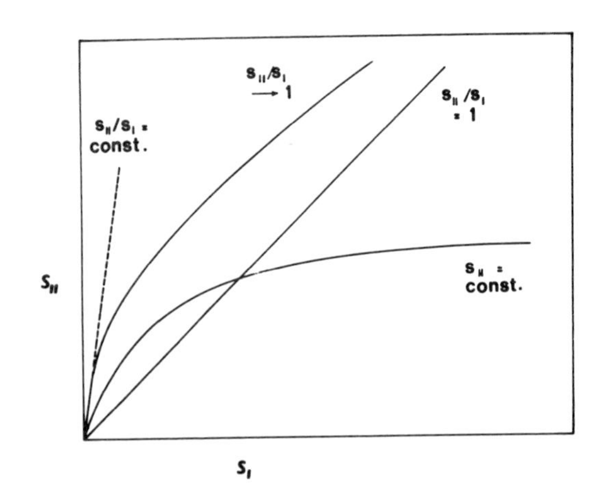

Fig. 1

Theoretical possibilities of dependence of steady-state intracellular on extracellular substrate concentration

For the (C+S+A) path of formation of the ternary complex the accumulation ratio is defined as follows:

$$S_{II}/S_I = \frac{g(d + 2e) + ceA_I}{g(d + 2e) + ceA_{II}} \qquad (8)$$

the various letters representing rate constants of the following model:

$$\begin{array}{ccc} C_I & \underset{f}{\overset{f}{\rightleftarrows}} & C_{II} \\ b \updownarrow as_I & & b \updownarrow as_{II} \\ CS_I & \underset{g}{\overset{g}{\rightleftarrows}} & CS_{II} \\ d \updownarrow cA_I & & d \updownarrow cA_{II} \\ CSA_I & \underset{e}{\overset{e}{\rightleftarrows}} & CSA_{II} \end{array} \qquad \textbf{VI}$$

It can be easily shown that if relatively little A_I is available (which is equivalent to an inadequacy of the ion pump) two cases may arise: (1) if CS is mobile, the ratio will tend to unity; (2) if CS is immobile, the ratio will depend strictly on A_I/A_{II} and may be less than unity as frequently observed experimentally. For $A_I = A_{II}$, obviously $S_I = S_{II}$.

There are two kinds of observations that apparently involve active ion-coupled transport and still are not in agreement with the simple derivations above: (1) Those of Harold's group referring to work in *Streptococcus faecalis* (Asghar et al., 1973) where in complete absence of an ion gradient a transient uphill transport is driven by the membrane potential alone; (2) those by Kimmich (1970) and others that uphill transport of sugars in the intestine and of amino acids in tumour cells can under some conditions proceed even against an equally directed sodium gradient.

If the effect of a superimposed membrane potential (inside negative) is embraced in the lower part of kinetic scheme V, its probable effect will be to increase the movement of CSA to the right so that we have $\overrightarrow{k}_{CSA} = k' > \overleftarrow{k}_{CSA} = k$. Without going through the rather laborious derivations the resulting formula is presented here:

$$s_{II}/s_{I} = \frac{k_{CS}K_{4} + k'A_{I}}{k_{CS}K_{4} + kA_{II}} \qquad (9a)$$

If CS is immobile, evidently

$$s_{II}/s_{I} = k'A_{I}/kA_{II} \qquad (9b)$$

Thus if $k/k' > A_{II}/A_{I}$ one may still expect uphill movement of S even if $A_{I}/A_{II} < 1$. Using the Ussing flux ratio formula we have

$$J_{CSA_{I}}/J_{CSA_{II}} = k'CSA_{I}/kCSA_{II} = CSA_{I}/CSA_{II}\ e^{-nF\psi/RT} \qquad (10)$$

so that, under surface-equilibrium conditions, a membrane potential (inside negative) of 60 mV at about 30°C (for a univalent cation) will bring about a flux ratio of the CSA complexes equal to 10, a potential of 120 mV a flux ratio of 100, etc. It thus appears that this approach, although only semiquantitative, predicts satisfactorily the above-mentioned observations both on bacteria and on intestine.

In spite of the amount of evidence suggesting that membrane ATPases are involved in the transport of both cations and indirectly nonelectrolytes, there remain some notable exceptions to the rule (ion transport in yeasts, fungi, algae; amino acid transport in some microorganisms) where both the mechanism of uptake and the source of energy remain unknown.

Likewise, the necessity for Na^{+} in some coupled transports may reflect another role, auxiliary perhaps to the gradient effect, such as sodium activation of a system essential in providing the energy for the uphill transport of nonelectrolytes.

AMOUNT OF ENERGY REQUIRED FOR TRANSPORT

It appears that several more or less intricate mechanisms have evolved to minimize the amount of energy required for maintaining an unequal distribution of a transported solute. The analogy of a Nernst formula (relating here to a nonelectrolyte for the sake of simplicity) would require that

$$\text{Energy/time} = J_s RT \ln(S_{II}/S_I) \quad \text{in cal.s}^{-1} \qquad (11)$$

J_s being (in mol.s^{-1}) in the simplest case a saturable function of S of the type $J_s = aS_I/(cS_I + dS_{II} + eS_I S_{II} + K)$. It is inherent in the expression that the energy consumption for an active transport will pass through a maximum as the concentration of S is raised, tending toward zero at very high concentrations when often $S_{II}/S_I = 1$.

The amount of energy calculated from actual fluxes and apparent concentrations for the various "active" transport systems is generally too large to be accommodated within the energy possibilities of a cell, particularly if an energy-depleted cell is used. This apparent inadequacy has led to views about a caloric catastrophe (e.g. Minkoff and Damadian, 1973) but one should be wary about making this type of conclusions for a number of reasons.

1. ATP, which is invariably taken as the sole source of energy for driving active transports, may in fact play a minor role in cases where another type of direct coupling is involved (phospho-enol-pyruvate, polyphosphate, oxidative chain).

2. The true activities both of the uphill-transported nonelectrolyte and of the driving ion may be quite remote from the calculated ones, two principal reasons being compartmentation within the cell and the presence of unstirred layers which, unless steady state obtains, tend to reduce the ratio of activities at the membrane surfaces as compared with those in the bulk solution.

3. A number of compounds and especially ions are believed to be transported by a process of exchange diffusion, the carrier being able to operate exclusively or preferentially when saturated with its substrate; depending on the actual concentrations at the two sides, it may well transport one species predominantly in one direction and another species in the opposite (an exchange of K^+ for H^+ of this type appears to be common in microorganisms).

4. The carrier may undergo such affinity changes on going from one membrane side to the other as to be unable to transport back the substrate which it readily transports forward. A typical example is the transport of amino acids in yeasts and fungi (Fig. 2).

5. Adsorption-like processes may account for a substantial part of seemingly uphill transport, particularly at low concentrations of substrate (Fig. 3).

The fact that many cells can perform uphill transport with

minimum expenditure of energy is documented by the example of the strictly aerobic yeast-like organism *Rhodotorula glutinis*. Even under conditions when no metabolism is detectable and no oxygen consumption can be measured, very substantial active transport persists, ceasing only when the last traces of oxygen are removed (Fig. 4).

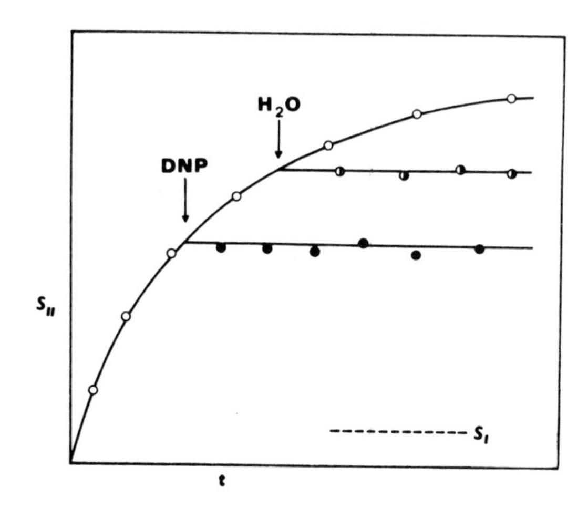

Fig. 2

Lack of efflux of accumulated amino acids in baker´s yeast; 1 mM α-aminoisobutyric acid; DNP - 2,4-dinitrophenol

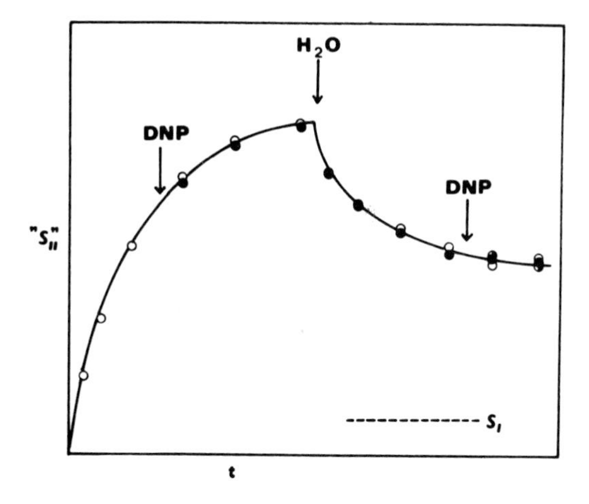

Fig. 3

Intracellular binding of 0.1 mM labelled D-arabinose by baker´s yeast. DNP - 2,4-dinitrophenol

TOWARD A UNIVERSAL ACTIVE TRANSPORT MECHANISM ?

It is an indisputable fact that some widely distributed patterns of active transport have emerged during the past few years, to wit: phosphotransferase in bacteria, Na,K- and Ca-adenosinetriphosphatases in animal cells and ion-driven nonelectrolyte transports in various organisms. Still, none of these mechanisms is ubiquitous and many of them are accompanied even in the archetypal cells by others, perhaps of a less general distribution but of a unique importance for a given type of substrate. Moreover, our present knowledge covers only some of the possible substrates and but a minute fraction of bacterial, plant and animal species.

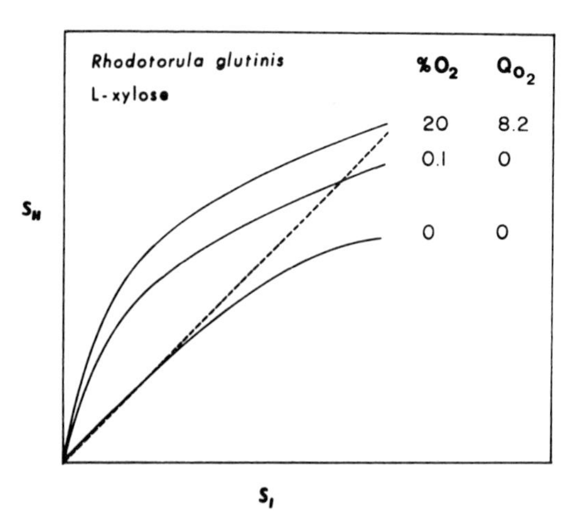

Fig. 4
Uptake of L-xylose and endogenous metabolism in *Rhodotorula glutinis*, the final steady-state levels being shown

It is perhaps appropriate to conclude this brief review with a warning to those who succumb to the temptation to extrapolate some species- or substrate-specific finding to other substrates in other species. Nature would be deprived of its pleasing variety if this were permissible.

REFERENCES

Asghar, S.S., Levin, E. and Harold, F. (1973). J. Biol. Chem. 248, 5225.

Bodnaryk, R.P. (1972). Can. J. Biochem. 50, 524.

Bonting, S.L. (1970). In: Membranes and Ion Transport, ed. E.E. Bittar (Wiley Interscience, London), vol. 1, p. 257.

Greville, G.D. (1969). In: Current Topics in Bioenergetics, ed. D.R. Sanadi (Academic Press, New York), vol. 3, p. 1.

Harold, F.M. (1972). Bacteriol. Rev. 36, 172.

Heinz, E. (ed.)(1972). Na^+-Linked Transports of Organic Solutes (Springer-Verlag, Berlin-Heidelberg-New York).

Hochstadt-Ozer, J. and Stadtman, E.R. (1971). J. Biol. Chem. 246, 5304.

Kahlenberg, A., Galsworthy, P.R. and Hokin, L.E. (1967). Science 157, 434.

Kalsow, C.M. and Doyle, R.J. (1973). J. Bacteriol. 113, 612.

Kimmich, G.A. (1970). Biochemistry 9, 3669.

Kimmich, G.A. (1973). Biochim. Biophys. Acta 300, 31.

Komor, E. and Tanner, W. (1974). Eur. J. Biochem. 44, 219.

Kotyk, A. (1973). Biochim. Biophys. Acta 300, 183.

Krupka, R.M. (1971). Biochemistry 10, 1143.

Lombardi, F.J., Reeves, J.P., Short, S.A. and Kaback, H.R. (1974). Ann. N. Y. Acad. Sci. 227, 312.

Martonosi, A., Lagwinska, E. and Oliver, M. (1974). Ann. N. Y. Acad. Sci. 227, 549.

Meister, A. (1973). Science 180, 33.

Minkoff, L. and Damadian, R. (1973). Biophys. J. 13, 167.

Mitchell, P. (1973). Bioenergetics 4, 266.

Repke, K.R.H. and Schön, R. (1973). Acta Biol. Med. Germ. 31, K 19.

Schultz, S.G. and Curran, P.F. (1970). Physiol. Rev. 50, 637.

Seaston, A., Inkson, C. and Eddy, A.A. (1973). Biochem. J. 134, 1031.

Simoni, R.D. (1972). In: Membrane Molecular Biology, eds. F.C. Fox and A. Keith (Sinauer Assoc., Stamford), p. 289.

Skou, J.C. (1972). Bioenergetics 4, 203.

Stein, W.D., Lieb, W.R., Karlish, S.J.D. and Eilam, Y. (1973). Proc. Nat. Acad. Sci. USA 70, 275.

Storelli, C., Vögeli, H. and Semenza, G. (1972). FEBS Lett. 24, 287.

ACTIVE TRANSPORT OF ELECTROLYTES AND NONELECTROLYTES IN THE INTESTINAL EPITHELIUM

T. Z. CSÁKY

Department of Pharmacology, University of Kentucky
College of Medicine, Lexington, Kentucky, U.S.A.

The notion that the ionic composition of the immediate environment may profoundly influence the permeability of the intestinal epithelium to non-electrolytes developed around the turn of the century. Understandably of the univalent anions sodium and potassium were of primary interest. Interestingly at first the effect of potassium was studied more systematically than that of sodium. Gardner and Burget (1955) found that potassium chloride stimulated glucose transport in the intestine, while Budolfsen (1955) found the opposite. In a well controlled experiment conducted by Riklis and Quastel (1958) in the surviving isolated guinea pig intestine the mucosal-to-serosal flux of glucose, placed into the lumen in an initial concentration of 14 mM, was found to be definitely stimulated by increasing the potassium concentration in the medium. This observation was confirmed and clarified in rats *in vivo* by Csáky and Ho (1966) who established that only the carrier-mediated downhill transport of glucose but not that of 3-0-methylglucose was stimulated by high potassium. Moreover the effect of potassium is connected to an increased intracellular metabolism of glucose thereby creating a large lumen-to-cell concentration gradient.

The possible effect of sodium upon intestinal glucose absorption was suggested as early as in 1902 by Reid (1902), later by Clark and McKay (1942). Also Riklis and Quastel (1958) in the study referred to above found that if the potassium concentration in the medium was very high, with a concomitant necessary decrease of the sodium content, the glucose transport was diminished. They speculated that perhaps also sodium is essential for the intestinal glucose transport. The first systemic study of the effect of sodium upon the active glucose transport was performed by Csáky and Thale (1959). They found that in the isolated small intestine of the toad: (a) sodium was absolutely essential for the active 3-0-methylglucose transport; (b) replacement of sodium with potassium, lithium, magnesium or choline completely abolished the transport of the sugar-ether against a gradient; (c) sodium had to be present only in the lumen, the ionic composition of the serosal bathing surface was not influential for the active sugar transport; and (d) bathing the mucosa with a sodium-free isosmotic medium did not permanently damage the intestinal epithelium: re-introduction of sodium restored the active transport of 3-0-methylglucose.

In connection with their report Csáky and Thale (1959) proposed that perhaps the sodium-potassium-magnesium stimulated ATPase (the "pump ATPase") is involved in the transport of glucose and the lack of sodium inhibits the function of this enzyme causing the inhibition of the active sugar transport. The correctness of this assumption can be tested by various manipulations which specifically inhibit this enzyme, or the sodium pump in general, causing the inhibition of the sugar transport as well. In the following some experiments are summarized which prove the validity of this assumption.

1. It is generally accepted that cardioactive steroids (digitalis) are specific inhibitors of the membrane bound pump ATPase. Ouabain and thevetin were found to be strong inhibitors of the uphill sugar transport in the isolated intestine of the frog (Csáky et al., 1961). The anatomical localization of the digitalis action on the intestinal epithelium is probably on the baso-lateral membrane because this was proven to be the site of the sodium pump (Schultz and Zalusky, 1964). The barrier through which the uphill transport of glucose occurs is most likely the brush border (Csáky and Fernald, 1961). If ouabain, which does not pass readily across cell membranes because of its high polarity, is placed on the side of the brush border which is involved in the transport of sugars, no inhibition of the sugar transport occurs. On the other hand if the glucoside is placed on the serosal membrane, where the sodium pump is localized, the inhibition of the active transport of 3-O-methylglucose is

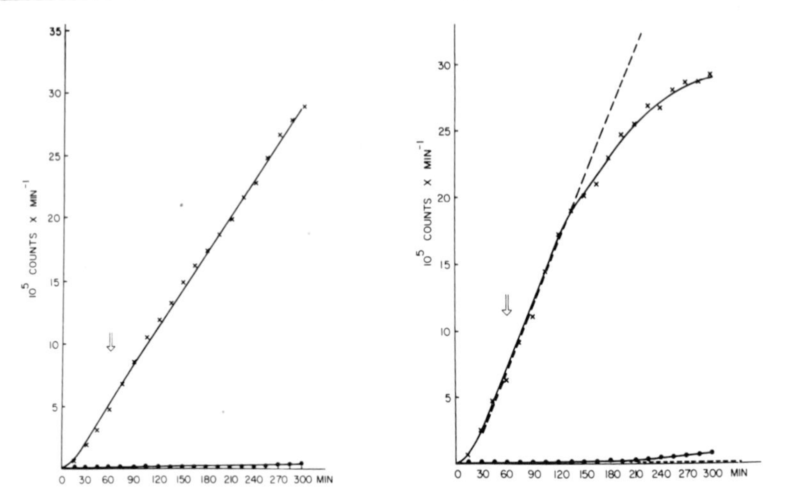

Fig. 1. Effect of ouabain on the mucosal-to-serosal (x______x) and serosal-to-mucosal (.______.) flux of 3-O-$C^{14}H_3$-glucose. At the arrow ouabain was added into the mucosal (A) or serosal (B) compartment. Note the lack of effect of ouabain when placed on the mucosal side; if added to the serosal compartment it depresses the mucosal-to-serosal and enhances the serosal-to-mucosal flux causing a flux ratio ∿ 1.0, thus abolishing the active transport. (From Csáky and Hara, 1965).

rather clear as shown in Fig. 1 (Csáky and Hara, 1965). It is known that potassium antagonizes the inhibitory action of ouabain upon the sodium pump. Should the inhibitory effect of ouabain upon the sugar transport be mediated through the sodium transport then this inhibition should be antagonized by potassium. The data summarized in Table 1 indicate that if the serosal facing membrane is bathed with a potassium rich medium, the inhibitory action of ouabain is strongly antagonized: about hundred

Table 1.

Inhibition of mucosal-serosal flux of 3-0-methylglucose by ouabain

Steroid in Compartment	Concn.	Principal Cation in Serosal Compartment	
		Na^+	K^+
Ouabain, mucosal	10^{-5} M	0 (6)	
	10^{-4} M	0 (2)	
Ouabain, serosal	10^{-7} M	0 (2)	0 (2)
	10^{-6} M	+ (3)	0 (2)
	10^{-5} M	++ (6)	0 (2)
	10^{-4} M	++ (2)	+ (4)

0 = No inhibition; + = slight inhibition; ++ = marked inhibition. Numbers in parentheses indicate number of experiments performed in that category. (From Csáky and Hara, 1965).

times more ouabain is needed to produce the same inhibition as without the enrichment of potassium. This clearly indicates that the primary inhibitory action of ouabain is on the sodium pump.

2. If sodium is needed for the functioning of the pump ATPase there should be a minimal amount of sodium which has to be present intracellularly. Consequently one can assume that if sodium on the mucosal side is replaced by cations which rapidly penetrate into the cell and thus replace the intracellular sodium the inhibition due to the lack of sodium will be more quick and complete. On the other hand if substances are utilized which do not penetrate very readily, the inhibition will probably be slower and not complete. Table 2 indicates that if the replacement is carried

Table 2.

Uphill absorption of glucose (μmole/hr g dry gut) from various perfusing solutions*

	Na_2SO_4	Li_2SO_4	K_2SO_4	$MgSO_4$	Mannitol
	520	55	80	116	385
	680	50	99	165	340
	710	65	92	206	380
Avg.	637	57	90	162	368
% Inhibition		91	86	75	42

*From Csáky, 1963.

out with lithium or potassium, both of which rapidly penetrate the cell, the inhibition due to the lack of sodium in the bathing medium is almost complete; whereas magnesium which penetrates slowly and mannitol which penetrates even more slowly produce considerably less degree of inhibition (Csáky, 1963).

3. If the strong inhibitory action of potassium and lithium is due to the intracellular replacement of sodium then returning sodium into the medium should have no immediate effect on the resumption of the active transport as a time lag is needed for the intracellular exchange of the lithium or potassium with sodium. Figure 2 indicates that after perfusing the gut with a lithium or potassium containing solution which is then

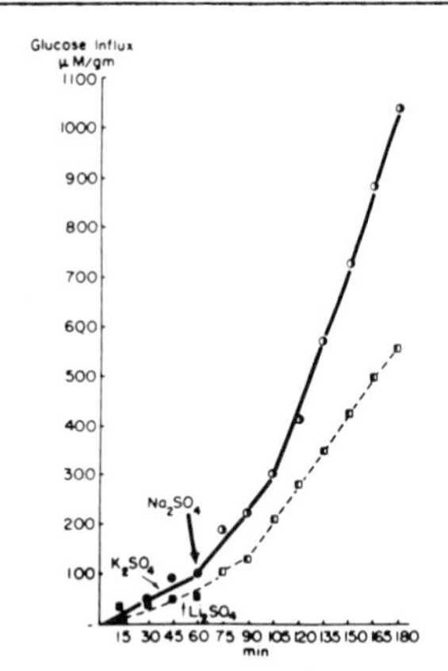

Fig. 2. Active glucose absorption (µmole/g dry gut tissue) from the small intestine perfused with an isosmotic solution of K_2SO_4 (solid line) or Li_2SO_4 (interrupted line). At the arrow the mineral composition of the perfusing solution was changed to Na_2SO_4. Note the gradual increase of the glucose uptake. (From Csáky and Zollicoffer, 1960).

replaced with a sodium containing medium the resumption of the active transport of glucose is gradual, about 45-60 minutes before it is complete. It is likely that this much time is needed to rebuild the intracellular sodium concentration (Csáky and Zollicoffer, 1960).

These experiments were conducted _in vivo_ perfusing the intestine of an anesthetized rat. The question can be legitimately raised: could the relatively low inhibition obtained by replacing sodium with magnesium or mannitol be due to the leakage of sodium from the blood into the unstirred layer adjacent to the brush border from which it is again rapidly absorbed thus maintaining a higher functional luminal concentration of the univalent cation. In order to examine this question the following experiment was performed: the intestine was perfused with a solution containing potassium which markedly decreased the rate of sugar transport. After a time the perfusing fluid was changed to isosmotic magnesium sulfate. Should magnesium have a specific action on the transport, the rate should have immediately increased; however, this was not the case: the rate of sugar transport stayed at the same low level as obtained when potassium was in the perfusing fluid. (Figure 3).

4. It is known that high proton concentration usually strongly inhibits the functioning of the sodium pump. It was found that if the hydrogen ion concentration in the intestinal lumen is increased the active sugar transport diminishes very rapidly (Csáky, 1971). In this

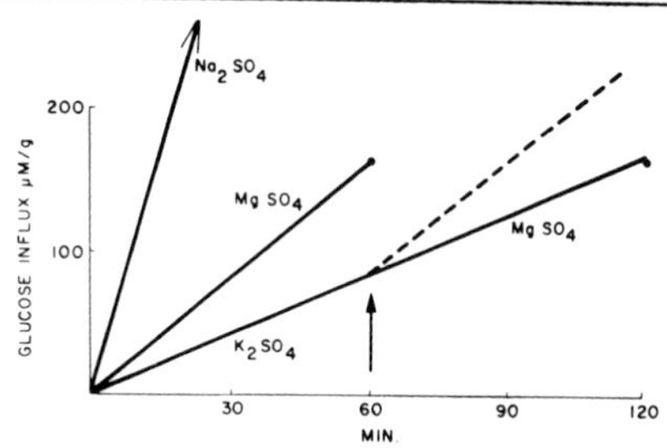

Fig. 3. Active absorption of glucose from the small intestine of the rat perfused in situ with an isosmotic solution of Na_2SO_4, $MgSO_4$ or K_2SO_4. At 60 min. the K_2SO_4 was replaced by $MgSO_4$. The absorption rate did not increase to that expected (and indicated by broken line) when $MgSO_4$ is in the initial perfusing mineral.

regard it is rather significant that the high proton concentration affects only the active sugar transport and not the carrier mediated passive diffusion, i.e. it has no effect upon the sugar carrier. From previous studies conducted in this laboratory it was known, namely, that sodium is needed only for the active uphill glucose transport but not the carrier mediated downhill process. Furthermore it was found earlier that if the initial concentration of glucose in the lumen is approximately 100 mM or higher then the transport proceeds via mediated diffusion which completely overshadows the active transport; mediated diffusion is completely independent from the presence of sodium (Csáky and Ho, 1966). With this in mind the effect of a pH 4.0 on the sugar transport in the intestine was examined at varying initial sugar concentrations. As Figure 4 indicates if the initial sugar concentration in the gut was less than approximately

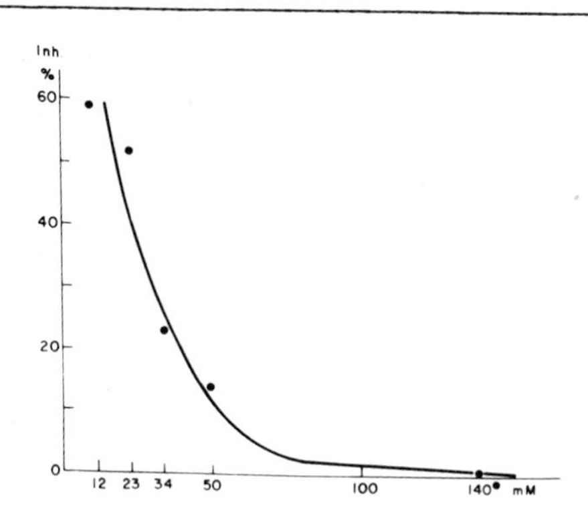

Fig. 4. Inhibition of the intestinal glucose absorption by pH 4.0 at varying initial glucose concentrations.

100 mM the inhibition of the sugar transport due to the low pH increases with the decrease of the sugar concentration. However above approximately 100 mM the high proton concentration has no effect upon the intestinal sugar transport.

5. Another interesting observation, which indicates that the sodium

potassium stimulated ATPase is most likely involved in the active transport of nonelectrolytes, was made in two kinds of frogs: the North American green frog, Rana pipiens, and the tropical bullfrog, Rana catesbeiana. The former is an animal which over the thousands of years developed a seasonal biological clock adapting to winter and summer. The bullfrog, being a tropical animal, was not exposed to seasonal temperature variations therefore its biological clock is not tuned to winter and summer. It was consistently observed in our laboratory that the green frog gut does not transport actively sugars and amino acids during the winter months. Figure 5 indicates that the depression of the transporting ability of the

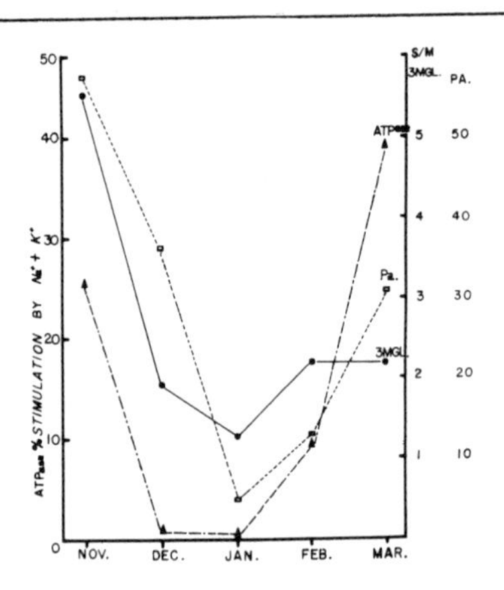

Fig. 5. Accumulation of 3-0-methylglucose (3MGl) and 1-phenylalanine (PA) in the serosal compartment in the isolated intestine of Rana pipiens from November through March. The decrease of the active transport is accompanied by decrease of the Na-K stimulated ATPase.

green frog gut during the winter runs parallel with the decrease of the sodium potassium ATPase activity. It is interesting that no such seasonal variation in either active transport or ATPase activity was observed in the bullfrog.

The above experiments amply justify the assumption that not only is the sodium-potassium-stimulated ATPase, in other words the sodium pump, involved in the active sugar transport in the gut but one could assume that there is only one pump in the epithelium, namely the sodium pump and all other uphill transports are energized through this single pump. This thesis is gaining general acceptance. However, there is considerable controversy as to how is the sodium pump coupled with the active transport of sugars or amino acids. Without going into the details of this exciting problem, which will probably be amply covered by other participants in this symposium, only one of the theories will be mentioned which presently enjoys popularity. This is the ionic gradient hypothesis (Crane, 1965). According to this if the sodium concentration is higher in the lumen than within the epithelial cell, the sodium will flow downhill from the lumen into the cell. This downhill movement of sodium provides energy for the uphill movement of the nonelectrolytes. Without going into the various agruments pro and con about this theory, and without elaborating on other hypotheses, it is interesting to call to mind one set of our own experiments to which reference was made above. (Fig. 2). The gut was perfused with a solution containing lithium or potassium as the only cation: the active glucose transport diminished to almost zero. When in this state the sodium was estimated in the cell it was decreased. When such a preparation was suddenly exposed to a sodium containing Ringer's in the lumen, clearly the sodium gradient between the lumen and the intracellular space will be now larger than normally particularly in the initial phase. This should result in a larger influx of sodium, according to the Fick's law. Should such an influx provide the energy for the active sugar transport, the latter should immediately pick up and, if anything, should be larger than normal. As could be seen on Fig. 2, this is not the case: the active sugar transport reaches full capacity considerably slower. It is hard to reconcile this finding with the sodium gradient hypothesis.

A few years ago this author ventured to suggest that it was perhaps the flow of water produced by the function of the sodium pump which may be the connecting link between the concentrative transport of sugars and amino acids and sodium. This suggestion is still only a working hypothesis as the methodology to examine it is not yet on hand. However there are a few points which are worth recalling:

(a) The epithelial cell is not a bag therefore one cannot speak of the uniform intracellular space; instead this space most likely is divided into compartments by various membranes. One of these membranous structures which is known is the Golgi apparatus which divides the sub-brush border area from the rest of the cells. One could assume that there are other membranes which produce compartmentalization.

(b) The function of the sodium pump is connected with a flow of water.

(c) By continuously withdrawing water from a multiple compartmental system with varying membrane permeabilities, one could create a model which could produce a concentrative increase of a given solute in at least one compartment. In fact such a model was produced not by withdrawing water but by adding a solute to one compartment (Ussing, 1969).

In summary: there is little doubt that sodium is essential for the uphill transport of glucose, amino acids and other nonelectrolytes in the intestinal epithelium. Furthermore, it is most likely that all uphill transports are energized by one single transducer, the sodium pump or the sodium-potassium-magnesium stimulated ATPase. The exact linking of these ATPase to the sugar transport is not clearly understood at the moment but various possibilities are worth exploring. Among these the possibility exists that, the cell is being a multiple compartment system, the flow of water may create a concentration increase of solutes in some compartments.

ACKNOWLEDGEMENTS

The research reported here was supported by grants from the U.S. Public Health Service.

REFERENCES

Budolfsen, S. E. (1955). Acta Physiol. Scand. 33, 132.
Clark, W. G. and McKay, E. M. (1942). Amer. J. Physiol. 137, 104.
Crane, R. K. (1965). Fed. Proc. 24, 1000.
Csáky, T. Z. and Thale, M. (1960). J. Physiol. 151, 59.
Csáky, T. Z. and Zollicoffer, L. (1960). Am. J. Physiol. 198, 1056.
Csáky, T. Z. and Fernald, G. W. (1961). Nature 191, 709.
Csáky, T. Z., Hartzog, H. G. III and Fernald, G. W. (1961). Amer. J. Physiol. 200, 459.
Csáky, T. Z. (1963). Fed. Proc. 22, 3.
Csáky, T. Z. and Hara, Y. (1965). Amer. J. Physiol. 209, 467.

Csáky, T. Z. and Ho, P. M. (1966). J. Gen. Physiol. 50, 113.
Csáky, T. Z. (1971) in "Intestinal Transport of Electrolytes, Amino Acids and Sugars" (Charles C. Thomas, Springfield) p. 188.
Gardner, J. W. and Burget, G. C. (1955). Amer. J. Physiol. 121, 475.
Reid, W. (1902). J. Physiol. 28, 241.
Riklis, E. and Quastel, J. H. (1958). Can. J. Biochem. Physiol. 36, 347.
Schultz, S. G. and Zalusky, R. (1964). J. Gen. Physiol. 47, 567.
Ussing, H. H. (1969). Quart. Rev. Biophys. 1, 365.

ENERGETIC COUPLING OF AMINO ACID TRANSPORT

E. HEINZ, C. PIETRZYK, P. GECK and B. PFEIFFER

Gustav-Embden-Zentrum der Biologischen Chemie
Universität Frankfurt/Main, FRG

INTRODUCTION

Active transport of amino acids and sugars depends on metabolism for supply of energy. It is therefore coupled ultimately to either fermentation or respiration, or both. This coupling, however, is only indirect, it rather involves several intermediate steps of energy transfer. The immediate source of energy for transport, i.e. the exergonic reaction, or process, to which the transport, in particular the translocation step is coupled directly, is not precisely known for many systems. For microorganisms two processes appear to have emerged to serve such function for most transport systems: the hydrolysis of energy rich phosphates (ATP, PEP) and the dissipation of a proton electrochemical potential gradient across the membrane (Fig.1, left) views that other processes function in this way,

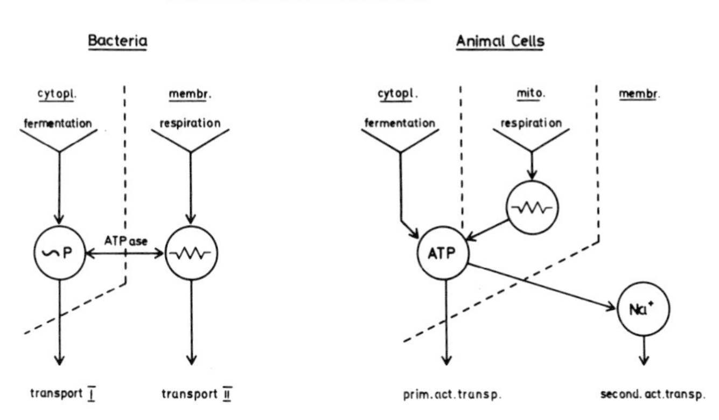

Fig. 1. Major pathways of energy supply transfer between metabolism and active transport in bacteria (left) and animal cells (right).

such as redox reactions, are controversial. The so-called "energized state of the membrane", invoked by several authors as the immediate driving force for certain transport processes, is probably closely related to, or even identical with the above mentioned proton gradient.

In animal cells the linkage between metabolism and active transport involves pathways different from those of microorganisms, but the immediate sources of energy to be considered here may be analogous: namely an energy rich phosphate, most likely ATP, and the electrochemical potential gradient[1] of an electrolyte ion, most likely of Na^+ (Fig. 1, right). The direct involvement of other such gradients, as of H^+ and K^+, has also been considered but is still uncertain. Whereas in microorganisms the proton gradient (energized state of the membrane) is under normal conditions directly linked to the respiratory chain, this is probably not so with animal cells: in which the Na^+ gradient requires ATP for its generation and maintenance, so that transport should be coupled to ATP hydrolysis in any case, either directly or indirectly via a Na^+ gradient. For this reason one has come to distinguish between primary transport on the one hand, presuming a direct (chemiosmotic) coupling to ATP hydrolysis, and secondary active transport on the other hand, presuming a direct (osmoosmotic) coupling to an ion flux. This distinction, based on the priority within a sequence of coupled processes, is not justified for microbiological systems, in which both energy sources, ATP and ion gradient, appear to be on equal footing, as seen on Fig.1.

In earlier times, most systems of active transport in animal cells were assumed by most authors to be directly linked to ATP hydrolysis. That active transport of amino acid and sugars in these cells depends on the presence of Na^+ ions has been known for a long time, but only in more recent years has the gradient of this ion be invoked to function as the only energy source for this transport (gradient hypothesis). Meanwhile sufficient evidence has accumulated to show beyond doubt that the mentioned active transport processes are intimately linked to parallel movements of Na^+ (cotransport) and that energy from the Na^+ gradient can be utilized in this transport. Serious doubts, on the other hand, have persisted as to whether the ion gradients are the only source of energy for transport processes, whether not a major, if not the major, part of the energy is supplied directly from ATP hydrolysis, via direct (chemiosmotic) coupling (Heinz, 1972, Jacquez, 1972). There are indeed several arguments which, while not ruling out electrolyte gradients as an energy source, favor a direct coupling to ATP hydrolysis, or which, at least, are not easily compatible with the gradient hypothesis. The most important ones are the following: 1. the electrochemical potential gradient of Na ions appears to be grossly inadequate to account for the amino acid accumulation observed in Ehrlich cells. Only the combination of both Na and inverse K gradient may just meet the energy requirement, but only if the efficiency of energy transfer is about hundred percent (Jacquez and Schafer, 1969). 2. Active transport of amino acids, though at a strongly reduced rate, has been shown to occur while the electrochemical potential gradient of Na^+ and K^+ ions were inverted (Schafer and Heinz, 1971).

[1] in the following simply called "gradient"

RESULTS AND DISCUSSIONS

To elucidate the above problems we have recently attempted to revise the actual amount of energy present in the electrolyte gradients and, furthermore, to investigate whether the coupling between ion flows and amino acid transport is tight enough to meet the requirements of energy transfer. Finally we have compared the effectiveness of this coupling with that between amino acid transport and ATP hydrolysis.

As to the energy present in electrolyte gradients, all previous estimates were based on at least two assumptions, first, that the activity coefficient of the ions is the same inside and outside the cell, and, second, that the ions as well as the transported substrate are evenly distributed over the whole cell water, with and without metabolic inhibition (Jacquez, 1972). The activity coefficients of the ions involved inside the Ehrlich cells are not precisely known, but from measurements with other cells it has been inferred that differences between activity coefficients outside and inside the cell may indeed be too small to account approximately for the above mentioned discrepancies (Pfister, 1970). As to the second assumption, however, it has been found for liver cells, that Na^+ ions are highly accumulated in the nuclei (Siebert et al., 1965). We therefore studied the Ehrlich cells with respect to this distribution by similar methods, breaking the cells in nonaqueous solvents and fractionating the various cellular components according to their density (Pietrzyk and Heinz, 1974). Assuming that DNA occurs predominantly in the nucleus, this nucleic acid was used to monitor the content of nuclear material. For each fraction the contents of Na^+, K^+, Cl^- and amino acid were then plotted versus the DNA content, and from the slope of the curve obtained the amount of ion sequestered in the nuclei was estimated. Contamination with extracellular material appeared to interfer only in the low DNA fractions, whereas with higher DNA fractions the correlation between ion and DNA content appears to be fairly linear. From the results it appears that only Na^+ and Cl^- ions are strongly sequestered in the nuclei whereas both K^+ ions and amino acid seem to be distributed rather evenly over the whole cell material. Upon lowering the extracellular Na^+ content, however, e.g. upon reversing the ion gradients, the nucleic sequestration of Na^+ largely disappears. This phenomenon is also known for other cells, where it is interpreted to indicate a direct connection between nucleus and extracellular medium (Siebert et al., 1965). With due consideration of possible sources of error, we then corrected the cytoplasmic ion concentration for the sequestered part, so that the effective ion gradients, namely those between medium and cytoplasm, could be estimated. Clearly these gradients for Na^+ and Cl^- came out much higher than was previously assumed as is illustrated on Fig.2. The resulting increments of driving forces are likely to be fully effective since the amino acid appears to distribute itself rather rapidly and evenly all over the whole cell. The sum of the corrected driving forces, as shown on this Figure, especially if a (possibly indirect) contribution of the K^+ ion gradient is taken into account, are well above those required for the amino acid accumulation observed. The question now arises whether the efficiency of energy transfer be high enough to permit appropriate utilization of the energy inherent in the electrolyte ion gradients. This has been studied by a different set of experiments.

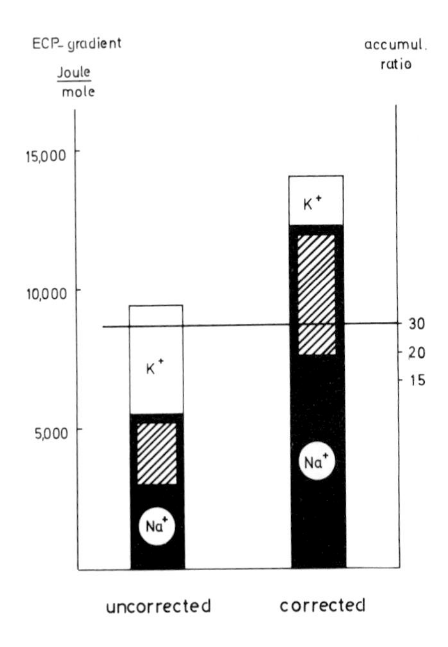

Fig. 2. Possible driving force for amino acid transport in Ehrlich Ascites cells, as derived from ionic electrochemical potential differences. The solid bars represent the chemical potential difference of Na^+ ions only. The cross hatched bars give the contribution of the electrical potential difference as derived from the Cl^- ion distribution. The white parts refer to the contribution of the electrochemical potential difference of K^+ ions. The left bar is based on raw intracellular ion concentrations. The right bar is based on the same values after correction for nuclear sequestration of Na^+ and Cl^- ions.

The coupling between two processes can be described in terms of non-equilibrium thermodynamics, to the extent that this is applicable, by the following set of equations

$$J_1 = L_{11}X_1 + L_{12}X_2 \tag{1}$$

$$J_2 = L_{12}X_1 + L_{22}X_2 \tag{2}$$

J stands for flow (or reaction rate), X for the (negative) electrochemical potential difference or for the affinity of a chemical reaction, respectively. The subscript 1 refers to the transported species, subscript 2 to the species of the driving process, i.e. Na^+ or ATP, respectively. L_{11} and L_{22} are the straight rate coefficients, and L_{12} is the cross coefficient responsible for the coupling. A quantitative approach of the tightness of coupling between J_1 and J_2 is based on the "degree of coupling" (q), defined by the equation

$$q = \frac{L_{12}}{\sqrt{L_{11} \cdot L_{22}}} \tag{3}$$

q is experimentally accessible on the basis of the following relationship

$$q^2 = \left(\frac{\partial J_1}{\partial J_2}\right)_{X_1} \cdot \left(\frac{\partial J_2}{\partial J_1}\right)_{X_2} \tag{4}$$

q^2 is thus equal to the dependence of transport on the rate of Na^+ flux or ATP hydrolysis, respectively, times the inverse relationship, i.e. the dependence of Na^+ influx or ATP hydrolysis, respectively, on amino acid influx (Kedem and Caplan, 1965). This approach has been carried out for both Na movement and ATP hydrolysis (Geck et al., 1972, 1974). A possible interrelationship between Na^+ flux and ATP hydrolysis, which might influence the q values, has been considered irrelevant as far as direct coupling is concerned. The results of our experiments are summarized on Table 1.

COUPLING OF AIB TRANSPORT TO Na-INFLUX AND ATP HYDROLYSIS

		J_i	
		$i = Na^+$ (Na-INFLUX)	i = ATP (ATP HYDROLYSIS)
DEGREE OF COUPLING	q	0.5 - 0.6	not detectable
EFFICACY OF ACCUMULATION	$-\frac{X_A}{X_i}$	≈ 0.6	0
EFFICIENCY OF ENERGY TRANSFER	η	$\approx \frac{\nu_A}{\nu_{Na}}$ 0.6	0

It is seen that q for the coupling between amino acid transport and Na^+ influx is seizable ranging between 0.50 and 0.65, whereas any coupling between amino acid transport is directly coupled to Na^+ influx rather than to ATP hydrolysis. Furthermore, we tried to find out whether the efficiency of this coupling is tight enough to account for the efficiency requirement of this coupling. What counts in the present context is the "intrinsic efficiency", i.e. the efficiency of energy transferred to amino acid transport from the coupled fraction of Na^+ influx only, as there is evidence that a considerable part of Na^+ influx is not directly coupled to the transport under investigation. This "intrinsic efficiency" depends on the stoichiometric coupling coefficient (r_ν) between amino acid influx and sodium influx, which is not precisely known. It has frequently been claimed that this ratio is unity but there are too many uncertainty factors involved to rely on this figure. It is, instead, possible to determine the efficacy of accumulation (ε_A), defined as the ratio of the maximum chemical potential of amino acid accumulation possibly from a given driving force, in this case electrochemical potential difference of Na^+ ion (Heinz and Geck, 1974). ε_A is experimentally approachable according to the relationship

$$\varepsilon_A = -\left(\frac{X_1}{X_2}\right)_{J_1=0} = \left(\frac{\partial J_2}{\partial J_1}\right)_{X_2} \tag{5}$$

From experimental values this efficacy of accumulation has been found to be about 0.6, owing to considerable scattering of the data , the true value may be appreciably higher or lower than this (Geck et al., 1972). The normal distribution of Na^+ and Cl^- ions, corrected for nucleus sequestration, would give us a driving force of about 12 200 j/mole. Considering an efficacy of accumulation of about 0.6, the experiments on q would predict a maximum accumulation ratio of about 15 for AIB. This value is lower than the highest accumulation ratios found for AIB. The electrical potential, however, may be higher than that derived from the chloride distribution. It might in reality lie between the value derived from the Cl^- distribution and that derived from the K^+ distribution. If the electrochemical potential difference of K^+ ion thus contributes to the driving force, whether by direct coupling of K^+ efflux to amino acid transport or, more likely, by increasing the electrochemical p.d. of the Na^+ ion, one might predict a maximum accumulation ratio for amino acid of about 25. This value is already in the range of the highest accumulation ratios ever observed, so that the gradient hypothesis can hardly be rejected on energetic grounds only.

Even if the efficacy of accumulation as determined in our experiments appears to fall somewhat short of what is required, the difference is very small and may be accounted for by some technical imperfection of the experimental approach. Anyway, the coupling of Na^+ ion flux and amino acid transport appears to be much tighter than that between ATP hydrolysis and this transport which under the same condition cannot be detected at all. The experimental results are therefore compatible with the view that, at least under normal conditions, amino acid transport mainly, if not exclusively, is driven by the electrolyte gradients, whereas there is no evidence of a direct linkage between amino acid transport and ATP hydrolysis. Hence the first of the above mentioned arguments against the gradient hypothesis, based on the apparent insufficiency of energy available from electrolyte gradients seems to be invalidated, at l ast for normal conditions.

There remains the other argument against the gradient hypothesis, namely that active transport of amino acid may take place with inverted gradients of Na^+ and K^+. This argument is not invalidated by our experiments. As already mentioned, the sequestration of Na^+ ions by nucleic material vanishes if the Na^+ concentration in the medium is lowered, as is the case in experiments in which the electrolyte gradients are inverted. Accordingly, no correction for sequestered ions is indicated so that the driving forces, this time opposing amino acid uptake, are probably correct as previously estimated. Under such conditions we still get some active amino acid uptake (Fig. 3).

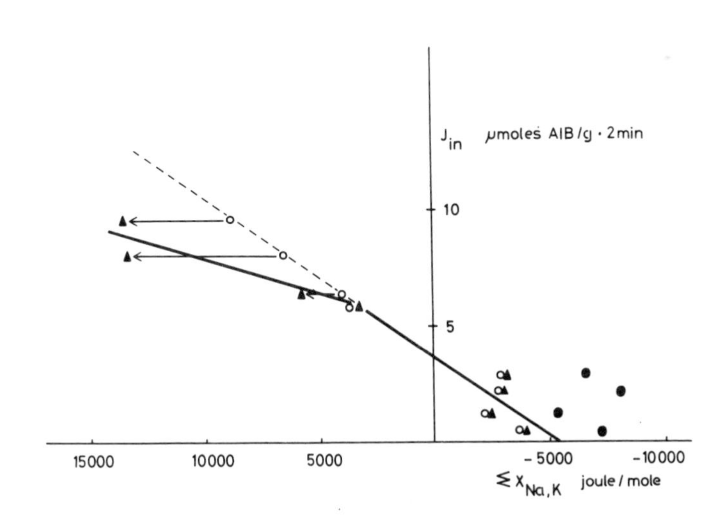

Fig. 3. The net influx of 2-aminoisobutyrate as a function of the ionic driving forces. The final results of 8 typical experiments with varying Na^+ and K^+ gradients are summarized. Abscicca: $X_{Na^+} - X_{K^+}$ the sum of the ionic driving forces of Na^+ and K^+, in J/mole. Ordinate: Influx of 2-aminoisobutyrate (J_{AIB}) in µmoles/g per 2 min. **O** , uncorrected driving forces; ▲ , corresponding driving forces corrected for nuclear sequestration of the alkali ion. The corresponding values are connected by a horizontal arrow. Since in the experiments with inverted gradients (right side of the ordinate) the intracellular concentrations of Na^+ and K^+ changed considerably towards normalization of the gradients during the 2-min incubation, the initial values are also given (●). Hence the intermediate values are somewhere in between, in any case more towards the right from the final (corrected) values. Each pair of values represents a whole experiment.

This observation cannot easily be reconciled with the gradient hypothesis, but does not necessarily indicate an additional driving force other than the ion gradients. The maintenance of an electrical potential difference of about 40 mV, outside positive, might under certain conditions suffice to account for this seemingly paradoxical amino acid transport. There are various possibilities to explain such a potential difference on the basis of a certain pattern of permeabilities and pump activities for Na^+ and K^+ under the condition of abnormal distribution of these ions. These possibilities have not been tested yet. Furthermore, a small uptake of amino acid by tertiary active transport, i.e. via forced exchange with endogenous amino acids, cannot be fully excluded.

SUMMARY

A major argument against the "gradient hypothesis", which postulates that the total energy for the active transport of amino acids derives from the electrochemical potential gradients of Na^+ and K^+, is based on the apparent energetic insufficiency of these gradients. In order to test the energy available from the ion gradients for amino acid transport in Ehrlich cells, the magnitude of these gradients has been revised, taking into account the sequestration of Na^+ and Cl^- in the nucleus, and the fraction of this energy source utilized for amino acid transport has been estimated on the basis of the tightness of coupling between the influxes of Na^+ and aminoisobutyrate. It was found that under normal conditions the electrochemical gradient of Na-ions is much higher than has been assumed previously and that the coupling of Na-ion flux to amino acid transport is tight enough to ensure an almost adequate transfer of energy to account for the highest accumulation ratios measured of this amino acid. By contrast, no direct coupling between amino acid transport and ATP hydrolysis could be detected. Hence under normal conditions the gradient hypothesis for the transport system investigated cannot be rejected on energetic grounds. The fact, however, that amino acid mac be taken up in spite of inverted gradients of Na^+ and K^+ is not explained by the present results but cannot be taken as evidence against the gradient hypothesis without further experimental studies.

ACKNOWLEDGEMENTS

The work described in this paper has been supported by a grant of the Deutsche Forschungsgemeinschaft (HE 102-12/13) and to a smaller part by Verband der Chemischen Industrie e.V. The authors thank Mrs. E. Kemsley for preparing the manuscript.

REFERENCES

Geck, P., Heinz, E. and Pfeiffer, B. (1972).
Biochim. Biophys. Acta 288, 486.

Geck, P., Heinz, E. and Pfeiffer, B. (1974).
Biochim. Biophys. Acta 339, 419.

Heinz, E. (1972). In: Metabolic Pathways Vol. VI, ed. L. Hokin (Academic Press, New York and London) p.455.

Heinz, E. (1974). In: Current topics in membranes and transport, Vol. 5, eds. F. Bronner, A. Kleinzeller (Academic Press, New York and London) p.137.

Heinz, E. and Geck, P. (1974).
Biochim. Biophys. Acta 339, 426.

Jacquez, J.A. and Schafer, J.A. (1969).
Biochim. Biophys. Acta 193, 368.

Jacquez, J.A. (1972). in: Na-linked transport of organic solutes, ed. E. Heinz (Springer Verlag Berlin, Heidelberg, New York) p.4.

Kedem, O. and Caplan, S.R. (1965)
Trans. Faraday Soc. 61, 1897.

Pfister, H. (1970). Z. Naturforsch. B.25, 1130.

Pietrzyk, C. and Heinz, E. (1974).
Biochim. Biophys. Acta 352, 397.

Schafer, J.A. and Heinz, E. (1971).
Biochim. Biophys. Acta 249, 15.

Siebert, G., Langendorf, J., Hannover, R., Nitz-Litzow, D., Pressman, B.C. and Moore, C. (1965).
Hoppe-Seylers Z. Physiol. Chem. 343, 101.

Energy coupling to Na^+-dependent transport systems: Evidence for an energy input in addition to transmembrane ion gradients.

G.A. Kimmich and J. Randles

Dept. Rad.Biol.& Biophys., Univ. of Rochester, Rochester, N.Y. 14642

INTRODUCTION:

A considerable controversy has developed over the past several years concerning the nature of energy coupling events serving in support of Na^+-dependent transport systems for various amino acids and monosaccharide sugars. An important aspect of this controversy relates to whether the Na^+-dependent systems are entirely coupled to transmembrane gradients of monovalent ions, or are to a degree coupled to a second energy source acting in conjunction with the ion gradients. The latter possibility might represent direct coupling to cellular metabolic activity in the sense that ATP expenditure is required for energization of specific transport components; or a more indirect relationship might be involved in which ATP is necessary, but in a role not involving immediate interaction between the nucleotide and elements of the entry system for the organic solutes.

RESULTS AND DISCUSSION:

Concepts important to an understanding of the manner in which monovalent ion gradients might energize solute transport systems can be summarized schematically as shown in Figure 1. A solute carrier is envisioned which has binding sites for both solute and a monovalent ion.

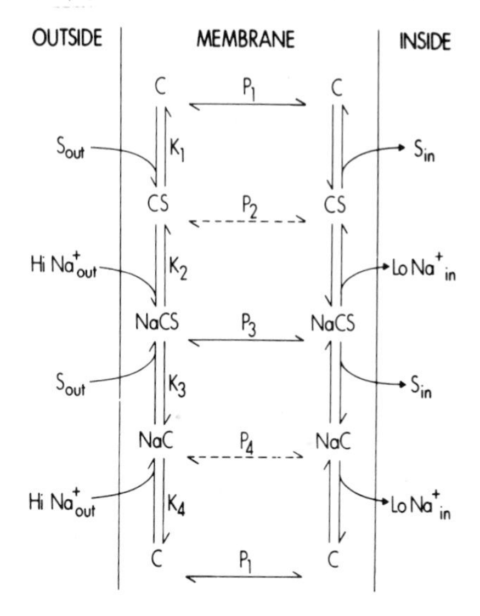

Fig. 1: Schematic model for Na^+-dependent transport systems. C = carrier, S = substrate, K = dissociation constants, P = permeability coefficient.

Kinetic evidence drawn from a variety of systems suggests at least two possible roles for monovalent ion in altering carrier characteristics; change in carrier affinity for its substrate (Curran et al., 1967) or a change in mobility of the ternary carrier complex compared to the ion-free binary complex (Goldner et al., 1969). Evidence has been provided which is consistent with a role for Na^+ in enhancing carrier affinity or mobility, while K^+ may act in a converse sense to decrease affinity or mobility (Crane et al., 1965). Any capability for generating a transmembrane gradient of solute is regarded as being dependent upon and a consequence of transmembrane gradients for the monovalent cations. In the high $[Na^+]$-low $[K^+]$ environment typical of extracellular fluids a Na^+-solute carrier ternary complex is generated due to favorable mass action ratios. After traversing the membrane, the low $[Na^+]$-high $[K^+]$ environment characteristic of intracellular milieu favors dissociation of Na^+ (and/or K^+ binding) and allows a carrier form with poor substrate affinity (or poor membrane mobility) to be generated. Carrier-mediated net entry of solute will continue until the degree of saturation of the carrier with solute at the two membrane surfaces is equal (assuming no exit or entry by other routes), at which point solute influx and efflux rates equalize and a steady state solute distribution ratio is maintained. Before the steady state is attained, the cell solute concentration must be greater than extracellular in order to compensate for the poor affinity for solute (or membrane mobility) of those Na^+-free carrier forms which predominate at the inner surface of the membrane.

It is important to recognize that no direct input of metabolic energy in the form of ATP is necessary at the locus of the solute carriers if concepts implicit in the ion gradient hypothesis are correct. Expenditure of ATP would only be required in order to satisfy the energy requirements of the monovalent ion transport systems which in turn maintains the assymetric distribution of Na^+ and K^+ essential for active accumulation of solute. The monovalent ion pump sites need not even be located at the same cell surface as the entry systems for organic solutes, and indeed some evidence indicates a complete physical separation at opposite poles of the cell for the two systems (Schultz et al., 1966; Fujita et al., 1971). It should also be recognized that conceptually the ion gradient hypothesis predicts functional symmetry for the organic solute transport components. If the monovalent ion gradients could be reversed in polarity from the usual physiological sense, the ion-dependent carriers should produce an active extrusion of organic solute from cell to extracellular fluid.

The latter aspect has been examined experimentally in a number of laboratories under various conditions, and the results have only heightened the controversey regarding the sufficiency of the ion gradient hypothesis as an adequate mechanistic basis for the Na^+-dependent transport systems. In those situations where a metabolic inhibitor was employed to aid in elevating cellular Na^+ concentrations, subsequent incubation in low Na^+ (or Na^+-free) medium led to solute extrusion (Hajjar et al., 1970; Goldner et al., 1972) but questions remain regarding whether the observed extrusion was against a concentration gradient. In each case intact tissue was employed and exact definition of cellular solute concentrations are difficult to achieve in this situation. If the epithelial cells were loaded to a distribution ratio greater than unity during the preincubation the observed solute efflux might represent only transition to a new steady state of solute accumulation (in response to decreased influx rates induced by lower $[Na^+]_{out}$) rather than active efflux driven by a reversed ion gradient. An especially alarming aspect is the fact that no re-entry of solute was observed even after 30 minutes incubation during which time the imposed reversed Na^+ gradient was significantly dissipated. If active extrusion had indeed been induced, re-entry to a distribution ratio of unity would be predicted as the unfavorable Na^+ gradient disappears.

The situation is still more complex when one considers experiments in which reversed Na^+ (or Na^+ and K^+) gradients are imposed on cells which are not simultaneously treated with a metabolic inhibitor. Experiments by Jacquez and Schafer (1969), Potashner and Johnstone (1971), Schafer and Heinz (1971), as well as our own (Kimmich, 1970; Tucker and Kimmich, 1973) have indicated that active accumulation of sugars or amino acids can occur even when Na^+ gradients of reversed polarity have been imposed. An example of this type of experiment performed with isolated intestinal epithelial cells is shown in Figure 2. The cells were preincubated at 0° in the

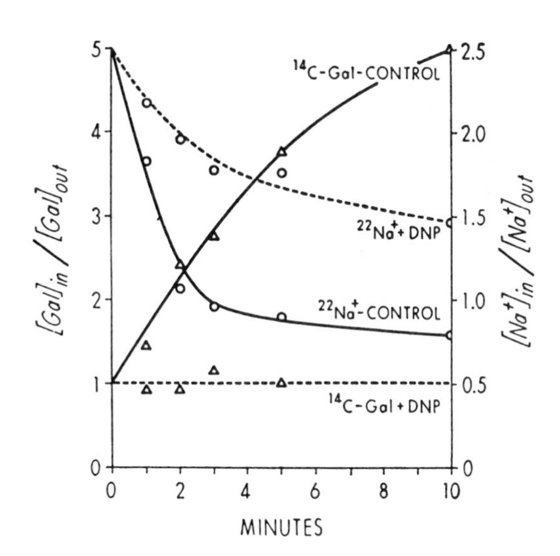

Fig. 2: Fluxes of ^{22}Na and ^{14}C-galactose in isolated chick intestinal cells with a reversed gradient of Na^+ imposed (from Kimmich, 1970).

presence of 80 mM $^{22}Na^+$ and 1 mM ^{14}C-galactose to allow a degree of Na^+ and galactose loading. At time 0 the cells were diluted into Na^+-free medium containing 1 mM ^{14}C galactose at the same specific activity as used during the pre-incubation. Extracellular $[Na^+]$ was decreased to 20 mM by the dilution procedure. Cellular $[Na^+]$ at the time of dilution was approximately 40 mM. Despite the reversed Na^+ gradient note that the cells begin to accumulate from the start of the incubation. High concentrations of ouabain do not prevent loss of cell $^{22}Na^+$ since cellular Na^+ concentrations exceed those in the incubating media and outward diffusion of $^{22}Na^+$ can still occur. The steady state level of $^{22}Na^+$ reached by the ouabanized cells should indicate the point at which cellular activity of Na^+ is equal to that in the bathing medium assuming ouabain is fully effective in preventing active Na^+ extrusion. Note that approximately 1.5 minutes are required before the nonouabanized cells reduce their $^{22}Na^+$ content to that level. The cells are able to establish a 2-fold concentration gradient of galactose during that time interval, i.e., while the Na^+ gradient is reversed. That active accumulation of galactose is indeed occurring during the early part of the experiment can be demonstrated by adding a metabolic inhibitor after a short incubation interval has elapsed. When this is done, galactose leaks out of the cells by diffusion to exactly the level observed at the beginning of the experiment, indicating that a concentration gradient had been established in the early stage of incubation (see Figure 3). If inhibitor is included at the start of incubation no net flux of galactose is observed in accord with the cells having been preloaded with galactose to a distribution ratio of unity during the preincubation.

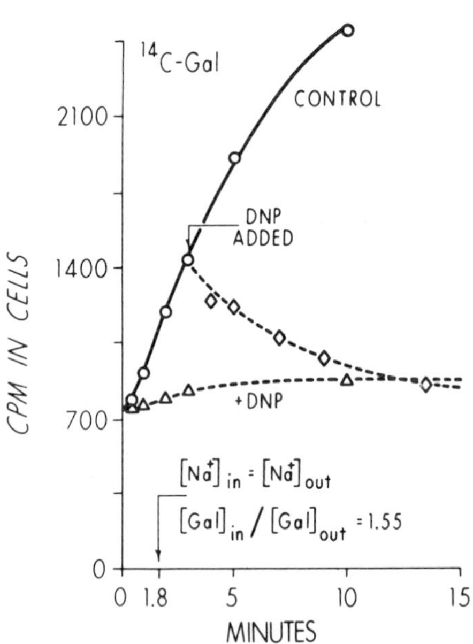

Fig. 3: Effect of DNP on ^{14}C galactose accumulation by isolated intestinal cells. The initial Na^+-gradient imposed was reversed in polarity as described for Figure 2. $^{22}Na^+$ flux data indicated the reversed gradient was maintained for 1.8 minutes as indicated (from Kimmich, 1970).

Experiments of this sort performed with non-inhibited ascites and intestinal cells have been important in establishing the possibility that an energy source other than transmembrane gradients of monovalent ions may be important in supporting Na^+-dependent transport systems for various organic solutes. The possibility is further indicated by the fact that ATP-depleted ascites cells maintain amino acid gradients only 30% as great as metabolically competent cells even though monovalent ion gradients equal to the physiological situation are imposed (Eddy, 1968). These experiments suggested that a form of energy coupling may exist in addition to that supplied by monovalent ion gradients, and which might involve direct expenditure of ATP. At the same time, any direct coupling model must account for the well-defined characteristics of Na^+-dependent transport systems, particularly those aspects which relate to monovalent ions and a general relationship between ion and solute transport capability. In an effort to devise a transport model consistent with known properties of Na^+-dependent transport, we suggested that the chemical reactions associated with monovalent ion transport might constitute a general energy transducing system acting in support of a variety of membrane-associated energy-dependent events (Kimmich, 1970; Kimmich and Randles, 1973). In that regard we envisioned the possibility that the phosphorylated intermediates which have been described for Na^+-K^+ ATPase might energize adjacent membrane components serving active accumulation systems for sugars and amino acids as illustrated in Figure 4. Na^+ might

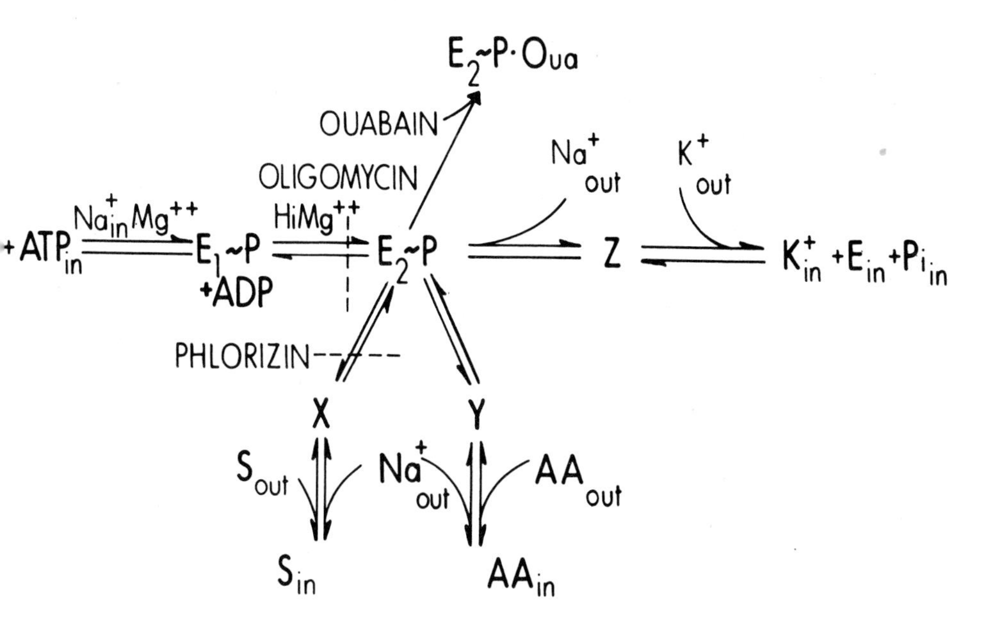

Fig. 4: Schematic representation indicating a possible role for Na^+-K^+ ATPase in direct energization of Na^+-dependent transport systems for sugars and amino acids (from Kimmich and Randles, 1973).

then play a dual role, by meeting a direct requirement of the solute carrier, as well as being necessary for generating intermediates of high chemical potential important for chemical energization of transport.

While a degree of direct chemical coupling between transport systems for monovalent ions and organic solutes is consistent with a wide variety of experimental observations certain facts are extremely difficult to reconcile with coupling via Na^+-K^+ ATPase. In particular, there is an increasing body of evidence which indicates a complete physical separation between cellular transport sites for monovalent ions and organic solutes. While sugar and amino acid transport systems are believed to be localized at brush border surfaces, Na^+-K^+ ATPase is thought by most investigators to be primarily or exclusively confined to the basolateral membranes (Fujita et al., 1971). If complete physical separation is indeed the case, any direct coupling model involving certain common steps for the two systems must be incorrect. It should also be recalled that ouabain is not inhibitory to Na^+-dependent sugar or amino acid transport if added at the mucosal surface of intact intestinal tissue, while it is fully inhibitory when added at the serosal surface (Csaky and Hara, 1965). If any Na^+-K^+ ATPase activity exists at the mucosal boundary, one would expect mucosal ouabain to be an effective inhibitor and to concomitantly inhibit any transport system to which it is directly coupled.

The above facts cast serious doubt on the liklihood of direct energy coupling between Na^+-K^+ ATPase and Na^+-dependent transport systems for organic solutes. They also raised the question as to whether our initial observations concerning sugar accumulation while reversed Na^+ gradients are imposed were valid. A frequent objection voiced against the use of isolated cells is the fact that the opportunity for discerning functional polarity is lost for free cells in contrast to the situation for intact tissue. The observed active accumulation might represent transport events occurring at other cell surfaces rather than defining transport mediated by brush border components. We doubt the validity of this objection for several reasons. Figure 5 illustrates the unidirectional

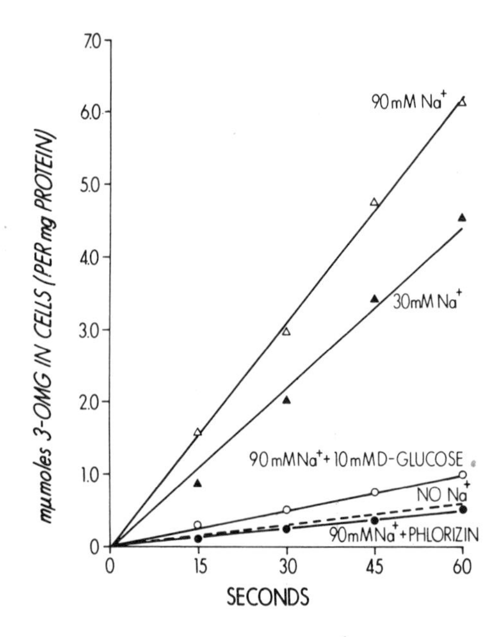

Fig. 5: Effect of phlorizin, glucose, and Na^+ on the unidirectional influx of 3-OMG into isolated chick intestinal cells.

influx of 3-OMG into isolated intestinal cells in a variety of situations. Note that more than 90% of the 3-OMG influx is Na^+-dependent and inhibited by the presence of 200 µM phlorizin. Neither Na^+ dependence nor phlorizin sensitivity has ever been reported as characteristic of sugar influx across baso-lateral membranes of intact intestinal tissue (Bihler, 1973; Kinter and Wilson, 1965). Furthermore, the percentage inhibition induced by phlorizin or absence of Na^+ is very similar to that reported for intact tissue preparations in which unidirectional influx across only the brush border boundary was studied (Goldner et al., 1972). It seems likely, therefore, that sugar fluxes measured with isolated cells represent primarily transport events localized in the brush border region of the cell. Transport kinetic parameters established with the aid of isolated cells are calculated only after correcting for non-phloridzin sensitive (or non-Na^+-dependent) entry.

A potentially more serious drawback regarding interpretation of experiments in which cellular Na^+ gradients have been manipulated is possible ambiguity originating from non-uniform distribution of Na^+ within the cells. It must be recognized that measured values for cellular [Na^+] refer only to average concentrations. Any tendency for certain cellular organelles (e.g., nuclei) to preferentially load with Na^+ might allow elevated cellular Na^+ concentrations to be produced in a situation in which cytoplasmic concentrations have not been significantly changed. It has also been suggested that microenvironments might exist near membrane surfaces in which the Na^+ concentration does not accurately reflect Na^+ concentration in the bulk cytoplasmic compartiment. Reversed Na^+ gradients may be more apparent than real in terms of those membrane surface environments important to the functional operation of an ion dependent carrier.

At first consideration, the prospect of dealing with objections based on the possible existence of membrane microenvironments seems remote. We have, however, made an approach to the problem through a consideration of mechanistic concepts believed to be the basis of mutual inhibitory interaction between transport systems for sugars and amino acids. In terms consistent with the ion gradient hypothesis, interaction between Na^+-dependent transport systems has been attributed to partial dissipation of cellular Na^+ gradients during transport of one substrate, with consequent inhibitory effects on other Na^+-dependent transport systems deriving energy from the same gradient (Chez et al., 1966; Semenza, 1971). For those systems in which Na^+ acts to alter carrier affinity for substrate, increments in cellular Na^+ would tend to enhance efflux of substrate without affecting influx, and thus lead to a decreased steady state concentration of substrate maintained by the cell. If these concepts are accurate, one would logically expect a correlation between the rate of transport of a given substrate and its ability to act as inhibitor of a second Na^+-dependent transport system. That is, the greater the rate of transport of a particular substrate, the greater the rate of Na^+ entry (by co-transport), and more significant expected disturbances in cellular Na^+ concentrations with resultant effects on other transport systems. We have not found the expected correlation (Kimmich and Randles, 1973).

Figures 6 and 7 show the effect of 10 mM 3-OMG on the accumulation of 1 mM valine and converse effect of 10 mM valine on uptake of 1 mM 3-OMG. Note that as much as 75 % of the active valine uptake can be inhibited by 3-OMG. On the other hand, active accumulation of 3-OMG is only 25% inhibited by 10 mM valine. In light of this data, one might predict that 3-OMG is transported at a much higher rate than valine; can

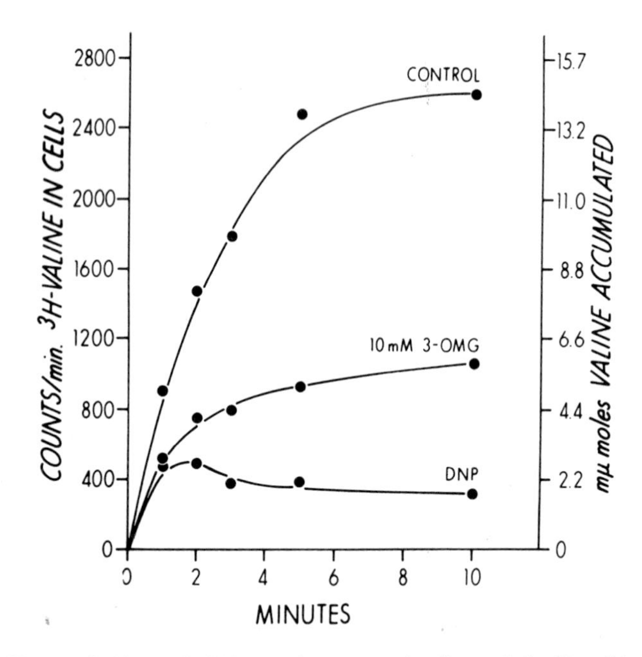

Fig. 6: Effect of 10 mM 3-OMG on the accumulation of 1 mM valine by isolated chick intestinal cells.

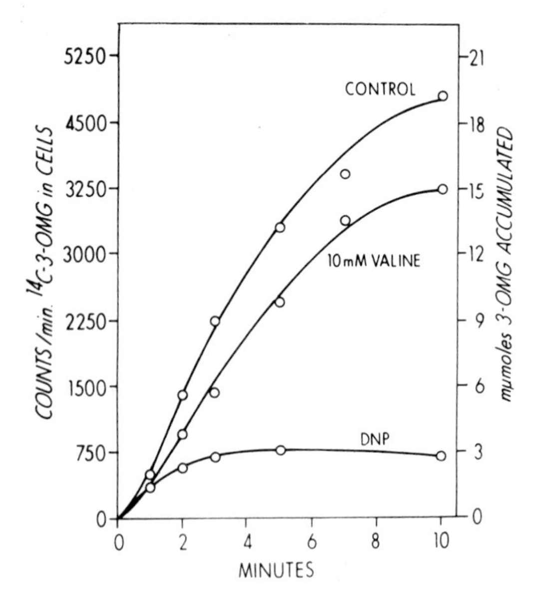

Fig. 7: Effect of 10 mM valine on the accumulation of 1 mM 3-OMG by isolated chick intestinal cells.

disturb cellular Na^+ gradients more significantly; and is therefore the better inhibitor of other events which depend on maintenance of a Na^+ gradient. Instead we have found that 10 mM valine is transported at nearly twice the rate of 10 mM 3-OMG. Even after corrections for non-carrier mediated (diffusional) entry, and unequal coupling coefficients (Na:substrate entry ratios) it can be calculated that valine would be expected to disturb cellular Na^+ gradients to a greater degree than 3-OMG (see Table I).

TABLE I: Comparison of unidirectional flux rates for 10 mM valine and 10 mM 3-OMG into isolated chick intestinal cells.

Substrate	Entry rate (nmoles·min^{-1}·mg prot.$^{-1}$) Total	Passive	Carrier	Coupling Coefficient*	Calculated Na^+ Entry Rate
10 mM valine	19.0	3.0	16.0	0.8	12.8
10 mM 3-OMG	12.5	3.0	9.5	1.0	9.5

*Values reported for rabbit ileum at 80 mM Na^+ (Goldner et al., 1969; Curran et al., 1967).

Initially we concluded that the lack of correlation between transport rate and inhibitory effectiveness for a particular substrate indicated that interaction between Na^+-dependent transport systems is not mediated by partial dissipation of cellular Na^+ gradients. The same conclusion is reached whether one considers the entire cytoplasmic compartment or localized microenvironments near membrane surfaces. In order for two carriers to interact in response to changes in cellular Na^+ gradients they must be sensing a common intracellular environment, whether that is a macro- or microenvironment. If delivery of Na^+ on the sugar carrier to that environment is able to inhibit 75% of the active accumulation of valine, it is difficult to imagine that an even greater rate of delivery of Na^+ on the amino acid carrier would have less severe effects on the sugar carrier.

Semenza (1971) has pointed out, however, that non-symmetrical interaction might occur between two solute transport systems if Na^+ interacts with each in a different manner. For instance, for those systems in which Na^+ alters carrier mobility, an increase in cellular $[Na^+]$ will tend to decrease unidirectional influx of substrate as well as to accelerate efflux rate. A transport system of this type would be predicted to have greater sensitivity to changes in cellular Na^+ than one in which only carrier affinity for substrate is affected. For this reason, it was important to determine the nature of Na^+ interaction with the transport systems for 3-OMG and valine, in order to learn if the interaction assymetry might be explained on this basis. The data indicate significant differences for the two transport systems, but the observed relationships cannot satisfactorily explain the nature of interaction between the two transport systems. As shown in Figures 8 and 9, Na^+ alters the KT (carrier affinity) for valine with no change in V_{MAX} (carrier mobility). In contrast carrier mobility (V_{MAX}) of the 3-OMG carrier is altered by Na^+, with little or no change in K_T. These observations suggest that one might expect the 3-OMG carrier to be more susceptible than the valine carrier to interaction with other Na^+-dependent transport systems, if interaction is mediated via partial

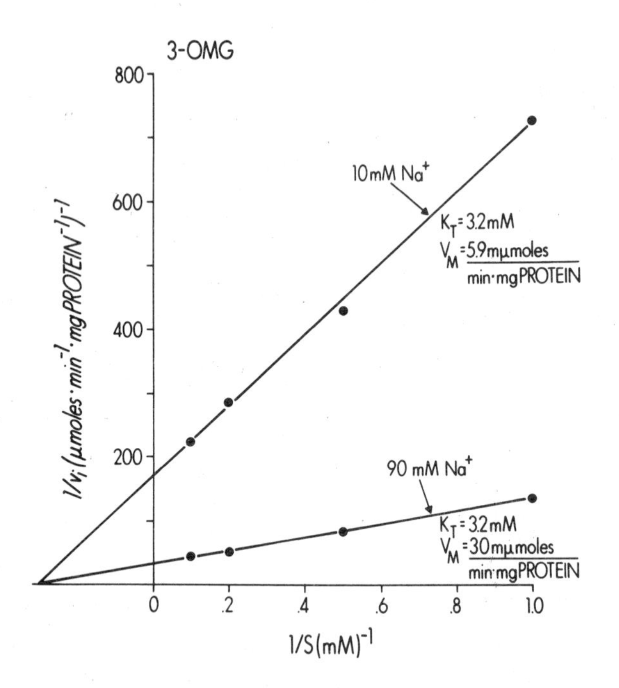

Fig. 8: Lineweaver-Burk plot illustrating the effect of Na^+ concentration on the unidirectional influx of 3-OMG into isolated chick intestinal cells.

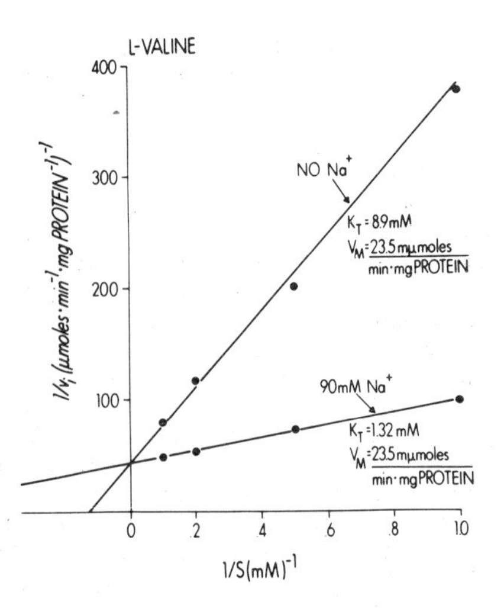

Fig. 9: Lineweaver-Burk plot illustrating the effect of Na^+ concentration on the unidirectional influx of L-valine into isolated chick intestinal cells.

dissipation of the Na^+ gradient. Since the observed interaction exhibits converse characteristics we continue to believe that interaction between transport systems must be ascribed to reasons other than competition for energy inherent in trans-membrane Na^+ gradients.

If Na^+ gradient dissipation is not the underlying cause for transport interaction, what is the basis? Steric or allosteric interaction between solutes competing for carriers in close apposition at the outer membrane surface, seems unlikely in view of data indicating that certain competitive inhibitors of the sugar transport system (e.g., phlorizin) are completely without inhibitory effect on the accumulation of amino acids. Furthermore, transport interaction disappears in cells depleted of energy reserves by preincubation with a metabolic inhibitor (Frizzell and Schultz, 1971), suggesting that competition for some form of energy must be involved.

Considering these facts we decided to explore the effect of metabolic inhibitors on the unidirectional influx of 3-OMG and valine for possible indications of direct integration between metabolic activity and Na^+-dependent transport capability. The data are presented in Table II. Note

TABLE II: Effect of rotenone or DNP on unidirectional influx rates for 3-OMG into isolated chick intestinal cells.

		Unidirectional Influx Rate* (Pre-Incubation Conditions)			
Inhibitor	Substrate	$-Na^+$	% Change	$+Na^+$	% Change
none	valine	1.4	-	2.35	-
20 μM Rotenone	valine	1.4	0%	1.80	-23%
200 μM DNP	valine	0.5	-64%	1.70	-28%
none	3-OMG	2.12	-	2.44	-
20 μM Rotenone	3-OMG	2.27	+7%	1.27	-52%
200 μM DNP	3-OMG	1.17	-45%	1.28	-47%
hypotonic incubation	3-OMG	-	-	2.6	+8%

*Influx rate was measured over a 1-min. interval at a Na^+ concentration of 80 mM, regardless of preincubation conditions.

that rotenone has a marked effect on unidirectional influx of both solutes when preincubation is performed in the presence of Na^+, but loses its effectiveness if Na^+ is absent during preincubation. These observations are similar to those reported for various metabolic inhibitors for intact intestinal tissue (Chez et al., 1967) obtained from rabbit. Lack of an inhibitory effect in that instance was taken as an indication that ATP does not play a role in carrier energization. The effect observed after preincubation in the presence of Na^+ was attributed to nonspecific changes in transport capability due to cell swelling induced by Na^+ entry. We feel the latter explanation is inadequate in light of the fact that influx is not altered in cells swollen by preincubation in hypotonic media, and an inhibitory effect of DNP is noted even when preincubation is carried out in the absence of Na^+ (Table III). It seemed possible that there might be differences in cellular ATP levels after preincubation in the two situations. One might expect extracellular Na^+ levels to influence the rate of dissipation of ATP in the presence of inhibitor by altering energy turnover via Na^+-K^+ ATPase. Lack of Na^+ might lead to a rather slow decrease in cellular ATP due to limited expenditure by the ion pump. Data presented in Table III suggest that

TABLE III: Influence of extracellular Na^+ concentration on residual ATP remaining after preincubation of isolated chick intestinal cells for 10 minutes with various inhibitors

Inhibitor	ATP (nμmoles/mg protein) 0 Na^+	80 mM Na^+
none	3.9	3.1
20 μM Rotenone	0.40	0.25
200 μM DNP	0.38	0.23

this might indeed by the case. Note that there is 50% more residual ATP after preincubation in the absence of Na^+ than when it is present.

While the above data indicate a relationship between residual ATP levels and unidirectional influx rates for organic solutes the significance of this relationship is not yet clear. For reasons discussed earlier direct interaction between ATP and transport components seems unlikely. At the same time, differences in the chemical gradient for Na^+ would not explain an effect on unidirectional influx if the role of Na^+ is to modify carrier affinity for substrate (Semenza, 1971) as is the case for valine. Nor would the effect of DNP on unidirectional influx be explained for that case where preincubation was performed in the absence of Na^+.

Perhaps the significant factor to consider is the difference in mode of action between rotenone and DNP. While both inhibitors cause depletion of cellular ATP, it is possible that DNP exerts further effects on membrane potential due to its ability to act as a lipophilic H^+ carrier. Murer and Hopfer (1973) and Gibb and Eddy (1972) have recently demonstrated that membrane potentials may be an important, but thus far relatively overlooked determinant of Na^+-dependent transport capability. In those situations where reversed Na^+ gradients were imposed on metabolically competent cells it now seems likely that membrane potentials of normal polarity might have been maintained, and have accounted for the observed active solute accumulation. The role for Na^+ in such a situation might be to create a carrier substrate complex of appropriate charge which can respond to the membrane potential. Full transport inhibition would require discharge of both the chemical gradients of monovalent ions as well as the capability for maintaining a membrane potential. In the present situation, residual ATP remaining after preincubation with rotenone in the absence of Na^+ could have supported modest monovalent ion transport capability and a small membrane potential. When the ATP is more fully discharged by preincubation in the presence of Na^+, the membrane potential would be less significant, and a decrease in solute influx might be expected. DNP could function as a more potent inhibitor due to its effects on both ATP levels and membrane potential.

While this interpretation must be treated as speculative, the concept of membrane potentials as an important determinant of transport capability by Na^+-dependent transport systems should be considered further. It offers some common ground for resolving present controversey regarding the sufficiency of chemical gradients of monovalent ions as the driving force for solute accumulation. It seems clear that active accumulation of solute has been demonstrated in a number of situations in which reversed gradients of monovalent ions had been imposed. It seems equally clear that any mechanism involving a direct interaction between ATP and Na^+-dependent transport components is unlikely in light of evidence summarized earlier. Inadequate consideration of the membrane potential as a driving force for transport may have been a significant factor in the origin of different mechanistic models which ultimately will prove to have been based on consideration of different aspects of a single general mechanism.

This work was supported by grants #AM70166 and AM15365 from the National Institutes of Health, Division of Arthritis and Metabolic Diseases, and under contract with the Atomic Energy Commission at the University of Rochester Atomic Energy Project, and has been assigned Report No. UR-3490-595.

REFERENCES:

Bihler, I. (1973) Biochim. Biophys. Acta 298, 429.

Chez, R.A., Palmer, R.R., Schultz, S.G. and Curran, P.F. (1967) J. Gen. Physiol. 50, 2357.

Chez, R.A., Schultz, S.G. and Curran, P.F. (1966) Science 153, 1012

Crane, R.K., Forstner, G. and Eicholz, A. (1965) Biochim. Biophys. Acta 109, 467.

Csaky, T.Z. and Hara, Y. (1965) Am. J. Physiol. 209, 467.

Curran, P.F., Schultz, S.G., Chez, R.A. and Fuisz, R.E. (1967) J. Gen. Physiol. 50, 1261.

Eddy, A.A. (1968) Biochem. J. 108, 489.

Frizzell, R.A. and Schultz, S.G. (1971) Biochim. Biophys. Acta 233, 485.

Fujita, M., Matsui, H., Nagano, K., and Nagano, M. (1971) Biochim. Biophys. Acta 233, 404.

Gibb, L.E. and Eddy, A.A. (1972) Biochem. J. 129, 979.

Goldner, A.M., Hajjar, J.J. and Curran, P.F. (1972) J. Mem. Biol. 10, 267.

Goldner, A.M., Schultz, S.G. and Curran, P.F. (1969) J. Gen. Physiol. 53, 362.

Hajjar, J.J., Lamont, A.S. and Curran, P.F. (1970) J. Gen. Physiol., 55, 277.

Jacquez, J.A. and Schafer, J.A. (1969): Biochim. Biophys. Acta 193, 368.

Kimmich, G.A. (1970) Biochemistry 9, 3669.

Kimmich, G.A. and Randles, J. (1973) J. Mem. Biol. 12, 47.

Kinter, W.B. and Wilson, T.H. (1965) J. Cell Biol. 25, 19.

Murer, H. and Hopfer, U. (1974) Proc. Nat. Acad. Sci. 71, 484.

Potashner, S.J. and Johnstone, R.M. (1971) Biochim. Biophys. Acta 203, 445.

Schafer, J.A. and Heinz, E. (1971) Biochim. Biophys. Acta 249, 15.

Schultz, S.G., Fuisz, R.E. and Curran, P.F. (1966) J. Gen. Physiol. 49, 849.

Semenza, G. (1971) Biochim. Biophys. Acta 241, 637.

Tucker, A.M. and Kimmich, G.A. (1973) J. Mem. Biol. 12, 1.

THE SUCRASE-ISOMALTASE COMPLEX (SI) FROM SMALL INTESTINE: A POSSIBLE HYDROLYTIC MECHANISM AND INDICATIONS ON ITS ROLE IN THE MEMBRANE TRANSPORT OF SOME SUGARS.

G. SEMENZA, A. COGOLI, A. QUARONI and H.VOEGELI

Laboratorium für Biochemie der ETH, 8006 Zürich

In the small intestine sugars can cross the brush border membrane through various mechanisms: (i) Two (Honegger and Semenza, 1973) Na-dependent and concentrative carriers for free aldoses of the glucose-galactose type (for reviews see Crane, 1965; Schultz and Curran, 1970; Kimmich, 1973) (the transport system(s) for aldoses in the baso-lateral membrane is (are) Na-independent and are not concentrative (Bihler and Cybulsky, 1973; Murer et al, 1974). (ii) One Na-independent and non-concentrative carrier for fructose (Schultz and Strecker, 1970; Honegger and Semenza, 1973; Nelson and Hopfer, 1974). (iii) One or more disaccharidase-dependent systems for the sugars provided as hydrolysbar disaccharides (Crane et al, 1970; Malathi et al, 1973; Ramaswamy et al, 1974). The latter systems are not accessible to the corresponding free monosaccharides to any large extent; are little or not Na-dependent; are less sensitive to phlorizin; and are also little inhibited by Tris (tris-hydroxymethyl-amino methane), although the corresponding hydrolytic activities are strongly inhibited by it.

In 1972 we have reported (Storelli et al, 1972) the first successful reconstitution of a transport sytem using an isolated membrane protein (the sucrase-isomaltase,SI,complex from rabbit small intestine, Cogoli et al, 1972) and an artificial membrane, i.e., a planar black lipid membrane (BLM). The protein was mixed with lipids prior to the formation of the BLM. The artificial membranes so obtained showed a permeability for glucose and fructose, when provided as ^{14}C-sucrose, which was some three orders of magnitude larger than that of corresponding BLM not containing the sucrase-isomaltase complex.

The purpose of our present report is to describe a mechanism for the hydrolysis by sucrase and isomaltase; to present new data on the reconstituted system, this time using liposomes; to discuss some possible mechanisms for the coupling of hydrolysis and transport in this system.

THE HYDROLYTIC MECHANISMS:

Any proposed mechanism for the hydrolytic activity of sucrase and isomaltase must account for the following:

(i) The kinetic mechanism is a rapid equilibrium PingPong Bi Bi for transglucosidase (Ordered Uni Bi for hydrolase) activity (Semenza and Balthazar, 1974). Its identification was based on the following: glucose is a competitive, fructose a poor non-competitive inhibitor; in the presence mannose, hydrolysis and transglucosidase activity have the same apparent K_m for sucrose; Na, while increasing the V_{max} does not change the apparent K_m.

(ii) The glucose is liberated as α-pyranose, as identified by both GLC and enzymatic procedures (Semenza et al. 1967).

(iii) Both sucrase and isomaltase split the bond between C_1 of the glucosyl moiety and the glucosydic oxygen, (Stefani, Janett, Semenza; in preparation). This was identified by carrying out the enzymic hydrolysis in $H_2{}^{18}O$; stopping the reaction by freezing; making the TMS-derivatives and studying them by GLC and mass spectrometry. Some results are summarized in Table I, which clearly shows during hydrolysis the water ^{18}O is incorporated into the glucose moiety alone, and that no exchange of ^{18}O is catalyzed by the enzyme.

In view of the suggested possible participation of C_2 in the hydrolysis by β-galactosidase (Wallenfels and Weil, 1972), it was of interest to localise the site of incorporation of ^{18}O during hydrolysis.

Table I

Incorporation of ^{18}O from $H_2{}^{18}O$ during the hydrolysis of sucrose (by sucrase) and of palatinose (by isomaltase) during 3 min. incubation at 37°C. The content in ^{18}O in the monosaccharides (expressed as % enrichment of the original $H_2{}^{18}O$) was calculated by comparing the fragments of mass 435 and 437 (i.e., F + 2) which arise from the molecular ion of perTMS-glucose (- $\dot{C}H_3$, then - TMSiOH, see DeJongh et al., 1969) and of perTMS-fructose (- $\dot{C}H_2OTMSi$ at C_6, see Curtius et al., 1968). Other fructose fragments also did not contain significant amounts of ^{18}O. Averages of two agreeing experiments.

Substrate	Sucrase-Isomaltase	^{18}O content of glucose	^{18}O content of fructose
Sucrose 29.2 mM	1.7 mg/ml	66%	6.3%
Palatinose 36.6 mM	45 mg/ml	60%	7.7%
Glucose 50 mM	1 mg/ml	approximately 20 % in each	
Glucose 50 mM + fructose 50 mM	1 mg/ml	approximately 20 % in each	
id.	2.9 mg/ml	approximately 20 % in each	
Glucose 50 mM	-	approximately 20 % in each	
Fructose 50 mM	-	approximately 20 % in each	

From A. Stefani, M. Janett & G. Semenza

This was done by comparing the mass spectra of the TMS derivatives of α-1-^{18}O-glucose, plain α-glucose, and of the α-glucose arising from sucrose and sucrase in the presence of $H_2{}^{18}O$ (Table II). The identification of the individual peaks rested on the work of De Jongh et al, (1969). Clearly the mass spectra of the α-^{18}O-glucose liberated during hydrolysis of sucrose corresponds to that of α-1-^{18}O-glucose: ^{18}O is present only in those peaks which contain C_1. We conclude that ^{18}O is incorporated in C_1 alone.

Table II

Distribution of ^{18}O in the fragments of α-D-glucopyranose obtained from the hydrolysis of sucrose by the sucrase-isomaltase complex in the presence of $H_2{}^{18}O$. (Mass spectra of persilylated derivatives).

Fragment	Mass	Main contrib. from	Intensity ratio F/(F+2)		
			α-D-Glu	Sample	α-1-^{18}O-Glu
1	393	C-1 thru C-4	4.6	1.0	0.3
2	319	C-2 thru C-5	2.6	2.7	3.2
3	217	C-2 thru C-4	6.8	6.9	4.5
4	204	C-2 + C-3 and C-3 + C-4	10.5	8.0	5.3
5	191	C-1 (by 90%)	11.5	1.4	0.4

Structure of the fragments (based on the work of DeJongh et al., J. Amer. Chem. Soc., 91 (1969) 1728-1733):

(1) $(CH_3)_2Si{=}CH(TMSiO)$–CH–CH–$CH{=}\overset{+}{O}TMSi$ (OTMSi)

(2) $\cdot CH{=}C{-}CH{-}CH^{+}$ (TMSiO, OTMSi, OTMSi)

(3) $CH{=}CH{-}CH^{+}$ (TMSiO, OTMSi)

(4) $TMSiO{-}\overset{+}{C}H{-}\overset{\cdot}{C}H{-}OTMSi$

(5) $TMSiO{-}CH{=}\overset{+}{O}TMSi$

From A. Stefani, M. Janett & G. Semenza

(iv) The secondary deuterium effect, i.e., the decrease in reaction velocity due to the substitution of the C_1 hydrogen in the glucosyl moiety by a deuterium, was studied using p-Cl-phenyl-α-glucoside. The calculated k_H/k_D via the formation of a carbonium-oxonium ion is 1.38 (Richards, 1970) or 1.14 (Halevi, 1963). Acid-catalysed hydrolysis of phenyl-β-glucopyranoside (which takes place via carbonium-oxonium ion, Capon 1969) has a k_H/k_D of 1.13 (Dahlqvist et al, 1968), (as compared to the k_H/k_D of the alkali-catalysed reaction, which is 1.03, Dahlqvist et al, 1968).

The k_H/k_D for the hydrolysis of p-Cl-phenyl-α-glucoside by sucrase and isomaltase was found to be between 1.13 and 1.20, which is strongly indicative of the formation of a carbonium-oxonium ion in the rate-limiting step of these enzymes (Cogoli and Semenza, in preparation).

(v) Whereas the classical Hammett function has rarely given satisfactory correlations in enzymology with either the reaction rate or the K_m-values, Hansch has shown that a much better correlation is found if a hydrophobicity parameter π is introduced into Hammett equation (Hansch et al, 1968).

Table III shows the Hammett-Hansch functions obtained using five p-substituted phenyl-α-glucopyranosides. The ρ values obtained for sucrase and isomaltase are very small and fall within the range for acid-catalysed hydrolysis of glucosides (Nath and Rydon, 1954) and are quite different from the values for alkali-catalysed hydrolysis of the same glycosides, which range between 2.8 and 4 (Hall et al. 1961). This observation again points to the formation of a carbonium-oxonium ion in the rate limiting step of sucrase and isomaltase.

Table III

Hammett-Hansch relationships in sucrase and isomaltase

$$\log k_{cat} = k' \pi + \rho' \sigma + c'$$

$$\log V/K_s = k'' \pi + \rho'' \sigma + c''$$

	R	s
Sucrase subunit		
$\log k_{cat} = 0.254\,\pi + 0.089\,\sigma + 2.33$	0.89	0.20
$\log (V/K_m \times 10) = -0.195\,\pi + 0.423\,\sigma + 0.70$	0.68	0.22
Isomaltase subunit		
$\log k_{cat} = -0.014\,\pi - 0.385\,\sigma + 2.65$	0.81	0.22
$\log (V/K_m \times 10) = -0.037\,\pi + 0.737\,\sigma + 1.03$	0.95	0.16

R: multiple correlation coefficient
s: standard deviation

From Cogoli and Semenza

(vi) Conduritol-B-epoxide was introduced by Legler (1966) as an active site directed inhibitor of β-glycosidases. It also reacts with both sucrase and isomaltase, the loss of enzymic activity being prevented by the presence of substrates and competitive inhibitors. One single molecule of conduritol-B-epoxide binds covalently to each of the two subunits of the sucrase isomaltase complex. From the specifity pattern of protection we have deduced that this compound reacts with the glucosyl, rather than with the "aglycone", subsite of the active sites. (Quaroni, Gershon and Semenza, 1974).

The pH-dependence of the reaction and the sensitivity to NH_2OH

of the conduritol-active site bond (Table IV) strongly indicate that this active site directed inhibitor forms an ester bond with a carboxylate group of the active site.

Table IV Hydroxylamine sensitivity of the bond formed between ^{3}H-conduritol-B-epoxide and the active sites in the sucrase-isomaltase complex.
Sucrase-isomaltase complex which had bound covalently one mole of inhibitor per active site was treated as listed below and dialysed exhaustively. (From Quaroni et al, 1974)

	Radioactivity lost
NH_2OH (0.5 M), Na carbonate buffer (50 mM), pH 9.0............................	90± 4
Na carbonate buffer alone................. .	5±10

(vii) Double logarithmic Dixon plots for both sucrase and isomaltase activity point to the partecipation in the catalysis of a group having the same pK'_a as that which reacts with conduritol-B-epoxide, and being reactive in the deprotonated form. We conclude that, in all likelyhood, the carboxylate group which reacts with conduritol-B-epoxide plays a role in the catalysis also. Inspection of the models of conduritol-B-epoxide (whose configuration approaches the half chair) and of the carbonium-oxonium ion arising from the glucosyl moiety during catalysis, immediately suggests that a carboxylate group of the active site positioned trans to the epoxide ring, would also be in the proper position to stabalize the carbonium oxonium ion.

These observations have led us to suggest the mechanism for sucrase and isomaltase (Cogoli and Semenza, in preparation), which is depicted in Fig. 1,B. The rate limiting step (IV-V), which was identified by kinetic analysis as including the liberation of the "aglycone" (i.e., of fructose) is suggested to go through the formation of a carbonium-oxonium ion at the glycosyl moiety (secondary deuterium effect, the ρ-values in the Hammett-Hansch function). The carbonium-oxonium is first stabilized by a carboxylate group (as suggested by the work on conduritol-B-epoxide inactivation) and finally by HO^- from the water (see the results with $H_2{}^{18}O$), which reaches the carbonium-oxonium from the side opposite to that of the COO^- group, that is, from the same side as fructose has left (in neither hydrolytic nor transglucosidase activity there is inversion at the C_1 of glucose). Little information is presently available as to the nature of the acid which protonates the glucosidic O in step III-IV (perhaps a COOH with a high pK'_a ?), and on whether a partial transition from chair to half chair configuration takes place during the binding of the substrate prior to the formation of the carbonium-oxonium i.e., in step II-III (δ-glucono-lactone is only a slightly better inhibition than glucose).

CHARACTERISTICS OF THE RECONSTITUTED SUCRASE-DEPENDENT TRANSPORT SYSTEM.

As reported previously (Storelli et al, 1972), we could reconstitute this system by drawing BLM with lipids containing

Fig. 1.
The mechanism of sucrase and isomaltase

A: The kinetic mechanism of sucrase (from Semenza and Balthazar, 1974).

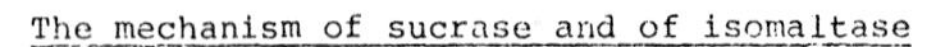

A: The kinetic mechanism B: The suggested chemical mechanism

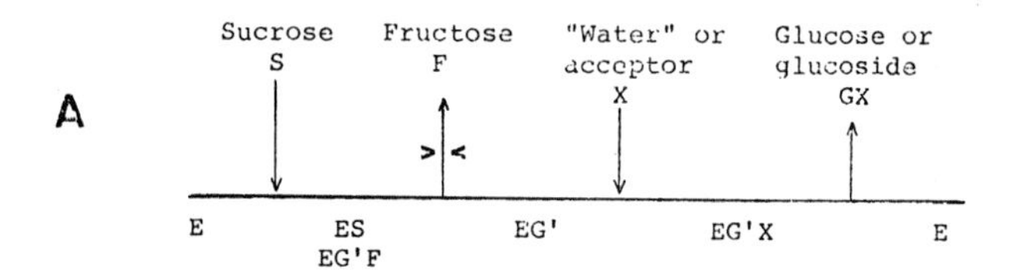

the sucrase-isomaltase complex. Recently we have achieved a similar reconstitution using liposomes made from a similar lipid-sucrase complex mixture: egg lecithin + phosphatidic acid in decane were mixed with a small volume of the sucrase-isomaltase complex in water, and the solvent was removed. This lipid-protein mixture was used to make multilamellar liposomes in a suitable swelling medium. Unilamellar liposomes were made by sonicating the multilamellar liposomes.

Fig. 2 shows the efflux of ^{14}C (from sucrose) and ^{3}H (from mannitol) from multilamellar liposomes at room temperature. It is clear that the presence of the sucrase-isomaltase complex in the lipids produces liposomes which are considerably more permeable to sucrose than to mannitol (in multilamellar liposomes the increase in permeability coefficient is considerably larger than the increase in efflux rate observed, see Chowhan et al, 1972). The ^{14}C released from liposomes-SI loaded with sucrose was identified as glucose + fructose by paper chromatography and autoradiography.

HOW DOES THE RECONSTITUTED SYSTEM COMPARE WITH THE ORIGINAL ONE?

During the last couple of years more characteristics of the sucrase-dependent sugar transport system have become known, both in the original brush border membrane (Malathi et al, 1973 and Ramaswamy et al, 1974) and in artificial lipid membranes,(BLM and/or liposomes), so that a more accurate comparison can now be made between original and reconstituted systems. They seem to be fundamentically identical, on the basis of the following criteria:

(i) substrate specificity: The sucrase-dependent sugar transport system is little or not accessible to free glucose

Fig. 1. (cont'd.)
B: The suggested chemical mechanism (from Cogoli and Semenza).

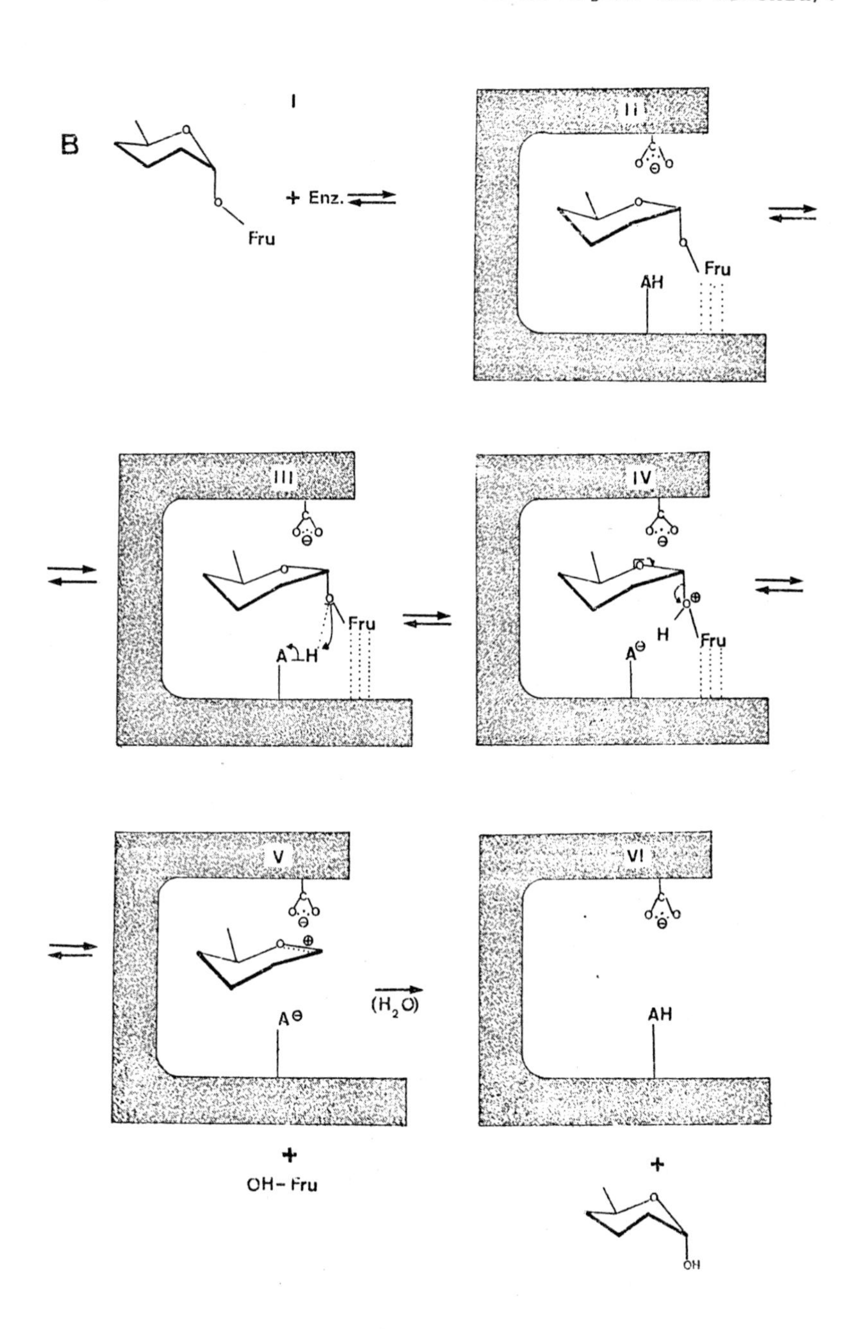

Fig. 2. Efflux of ^{14}C-sucrose and of ^{3}H-mannitol from multilamellar liposomes.

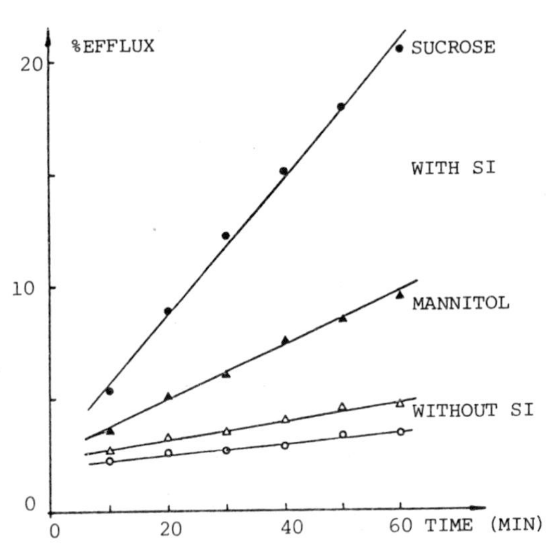

Efflux of ^{14}C-sucrose (o,•) and of ^{3}H-mannitol (Δ,▲) from multilamellar liposomes made of lipids alone (empty symbols) or of lipids and of SI-complex (full symbols). The radioactivity compounds arising from ^{14}C-sucrose and appearing in the outer fluid were identified as fructose, glucose and sucrose.

(Crane et al, 1970 and Malathi et al, 1973); BLM (Storelli et al, 1972) and liposomes (Vögeli et al, in preparation) made of lipids containing the SI-complex show a small, if any, increase in permeability for free glucose.
(ii) Na dependence: The original system is little Na-dependent, the effect of Na^+ being totally accounted for by the well known Na-activation of sucrase's hydrolytic activity (Semenza & Balthazar, 1974). The reconstituted system is also little or not Na-dependent in both BLM (Storelli et al, 1972) and in liposomes (Fig. 3).
(iii) Tris,a strong inhibitor of both membrane-bound and solubilized sucrase (Semenza & Balthazar, 1974; ref. therein) inhibits little the sucrase-dependent transport in the original membrane, and does not inhibit the reconstituted liposomes-SI system, (Fig. 3).

There are clearly quantitative differences between the reconstituted and the original systems, e.g. the little sensitivity of the original system vs. the total insensitivity of the liposomes-SI system towards Tris. Unless this is due to species differences, it can be given the following tentative explanations. We suggest (a) that in the original brush border membrane most of the SI-molecules are located (either constantly or exchangeably) in such a manner as to protrude from the outer surface of the brush border membrane, and/or that (b) the sucrase molecules acting as translocators may be not accessible to Tris in intact cells (see Ramaswamy et al, 1974).

Fig. 3. Lack of Na-activation and of Tris inhibition of sucrase-dependent sugar efflux.

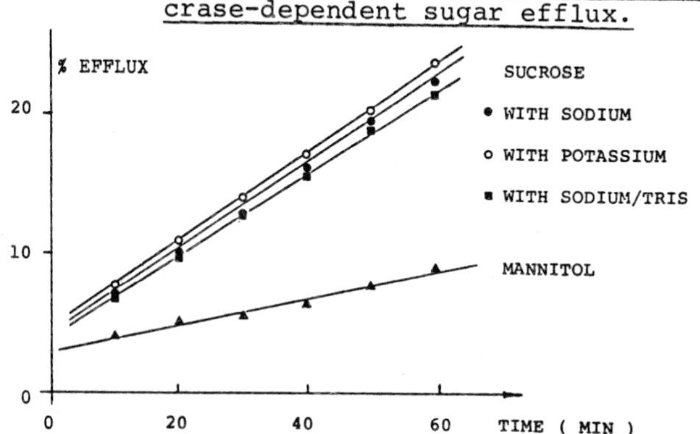

Efflux of ^{14}C-sucrose and of ^{3}H-mannitol (▲) from multilamellar liposomes in sodium (●) or potassium (○) phosphate buffers (100 mM, pH 4,8), or in sodium buffer in the presence of 37,5 mM Tris (■). The efflux of mannitol was independent of the cations present in the medium.

Explanation (a), but also (b) would agree with the fact that all the kinetic parameters of sucrase hydrolytic activity investigated (K_m, energy of activation, pH-activity curves, Na-activation, Tris inhibition) are indistinguishable between solubilised and brush border bound sucrase. In particular, the three latter parameters should differ, since, due to negative fixed charges at the membrane, the pH in the microenvironment is calculated to be more acidic by one pH unit than that of the bulk phase (Hogben et al, 1959).(Tris inhibition is also pH-dependent, Semenza and Balthazar, 1974). The glucose and fructose liberated by these sucrase molecules may enter the brush border membrane by the carrier systems for free glucose and for free fructose. On the other hand, a few percent of natural SI and much of the SI associated with the liposomes (which contain phosphatidic acid) would sit deeper in the lipid membrane, so as to be exposed to the microenvironment of prevealing negative fixed charges. These SI molecules only would be effective in carrying out the hydrolysis — coupled transport, would be little or not sensitive towards Tris (the action of which is pH-dependent) and would show a pH-activity curve apparently displaced towards alkaline values.

POSSIBLE MECHANISMS OF SUCRASE-DEPENDENT SUGAR TRANSPORT.

As mentioned above, conduritol-B-epoxide is an active site directed irreversible inhibitor of both sucrase and isomaltase, which reacts with a carboxylate group in the glucosyl subsite. A single conduritol epoxide molecule is incorporated during inactivation (Quaroni et al, 1974). It is thus possible to obtain a totally inactivated sucrase-isomaltase complex with minimal and selective chemical modification and reasonably presumed to have undergone no major changes in the secondary or tertiary structure. Inclusion of this inactivated complex into liposomes resulted in no increased permeability for ^{14}C-sucrose (Fig. 4). This observation, of course, provides and excellent "blank", i.e. it shows that the increase permeability for ^{14}C-sucrose brought about by native SI is specific and is not mimicked by other proteins. It also shows that integrity of sucrase activity and/or binding ability at the active site is necessary prerequisite for sucrase-dependent transport. Conduritol-inactivated SI did not show any detectable binding capacity for sucrase in equilibrium dialysis. Although the latter observation is restricted to ligand binding having a dissociation constant better than 10^{-4}M, it provides no evidence for a hypothetical second binding site for sucrose in sucrase-isomaltase.

The observations reported here and those reported previously allow a selection among unlikely and likely mechanisms for sucrase-dependent sugar transport. (i) Does the SI-complex bring about the increased sucrose permeability of BLM at the torus or across the lipid bilayer?. The second alternative seems to be the right one, since we failed to detect any effect in bulk membranes, and since liposomes-SI also show increased permeability. (ii) Does the SI-complex, once bound to ^{14}C-sucrose, change its ability to interact with lipids, so that it leaves the lipid bilayer alltogether? This possibility also seems very unlikely, since the amounts of SI present in the system (and thus the more so those presumed to be in the lipid membrane) are in catalytic, rather than stoichiometric relation with the ^{14}C-sucrose (or monosaccharides) passing the membrane. (iii) Is the increased permeability accounted for by the flip flop rate of "non specific" proteins in lipid bilayers? An estimation of the "transport turnover number" of the SI-complex in BLM indicates that it must be in the same order of magnitude as its hydrolytic turnover number (3000, see Kolínská and Semenza, 1967). Although few data are available in the literature on the flip flop rate of membrane proteins, data on erythrocyte membrane indicate that this rate must be negligeably small, at least for the glycoproteins considered (Nicolson and Singer, 1972; Bretscher, 1973). (The sucrase-isomaltase complex contains some 15% carbohydrates, Cogoli et al, 1972).(iv) A simple increase in ^{14}C-sucrose permeability due to splitting into sugars of smaller size occurring in the bulk phase can easily ruled out, because the permeability of glucose is not increased to similar extent; because the SI-complex, when added to the water compartments of performed BLM or to the swelling medium of liposomes, is ineffective; and because Tris, which inhibits sucrase in solution, is ineffective in inhibiting the sucrase-dependent increase in sucrose permeability (Fig.3). (v) The active site of the sucrase involved in transport could be totally and constantly confined to the one side of the membrane. If hydrolysis is vectorial (as it has been suggested for other carbohydrases, e.g., Robyt and French, 1970), and if the pro-

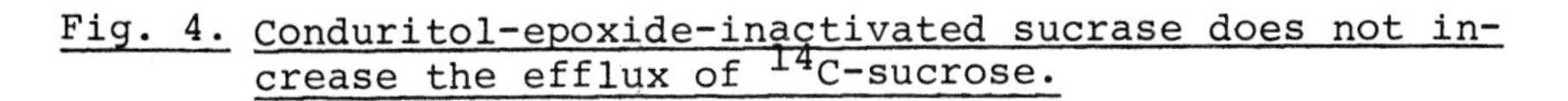

Fig. 4. Conduritol-epoxide-inactivated sucrase does not increase the efflux of ^{14}C-sucrose.

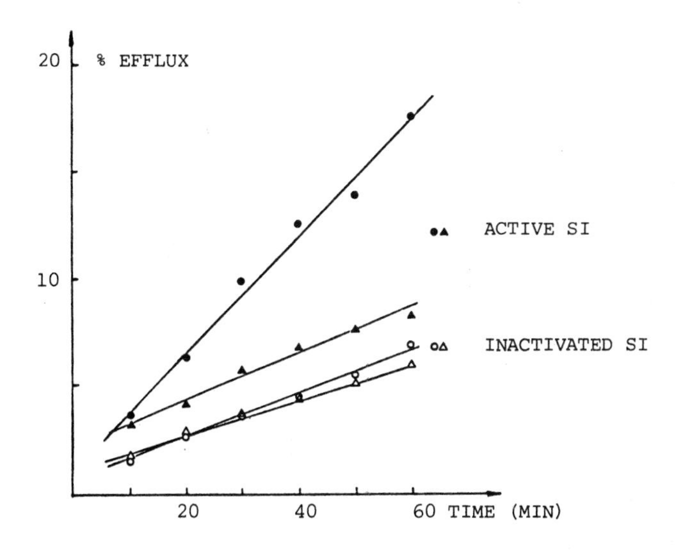

Efflux of ^{14}C-sucrose (o,●) and of ^{3}H-mannitol (△,▲) from multilamellar liposomes made of lipids and native sucrase-isomaltase complex (full symbols) or of lipids and conduritol-B-epoxide-inactivated sucrase isomaltase complex (empty symbols).

ducts are liberated in a microenvironment at the surface of (or perhaps even within) the lipid bilayer, from which diffusion into the bulk should be slow, a local hyperconcentration of glucose and fructose may ensure, providing the concentration "head" for an apparently increased passive diffusion. The free energy of sucrose hydrolysis ($-\Delta G^{o} \approx 5$ kcal/mol) may be sufficient to sustain this local hyperconcentration.

(vi) The active site of the sucrase involved in transport may have access, either all the time or alternatively, to both sides of the lipid membrane providing either a kind of static pore endowed with enzymic activity (which should be very specific, however, because the increase in electrical conductance of BLM is negligeable) or a dynamic translocator (e.g., with alternative gate opening or with actual vectorial movement). Transport would have to take place before the hydrolytic liberation of the first product (fructose), because the two monosaccharides appear in the trans compartment in equimolar amounts.

Clearly, it is not possible at the present time to make a choice among the possibilities mentioned under (v) and (vi). To this goal work is presently in progress.

Our work was rendered possible by the financial support of the SNSF, Berne, and of Nestlé, Vevey.

SUMMARY

The hydrolytic mechanism of the sucrase-isomaltase complex from the brush border membrane of rabbit small intestine was shown to go, in all likelyhood, through the formation of a carbonium-oxonium ion.

The presence of this protein into lipid membranes (either BLM or liposomes) increases their permeability to ^{14}C-sucrose specifically, and with the same characteristics as the original sucrase-dependent sugar transport system.

REFERENCES:

Bihler, J., Cybulsky, R. (1973) Biochim. Biophys. Acta, 298, 429-437.

Bretscher, M., (1973), Science, 181, 622-629.

Capon, B. (1969) Chem. Rev., 69, 407-498.

Chowhan, Z.T., Yotsuyanagi, T., Higuchi, W.I. (1972) Biochim. Biophys. Acta, 266, 320-342.

Cogoli, A., Mosimann, H., Vock, C., von Balthazar, A.K., Semenza, G., (1972) Eur.J.Biochem., 30, 7-14.

Crane, R.K., (1965) Fed.Proc., 24, 1000-1005.

Crane, R.K., Malathi, P., Caspary, W.F., Ramaswamy, K. (1970) Fed. Proc., 29, 595.

Dahlqvist, F.W., Rand-Meir, T., Raftery, M.A., (1968) Proc. Nat.Acad.Sci., 61, 1194-1198.

De Jongh, D.C., Radford, T., Hribar, J.D., Hanessian, S., Bieber, M., Dawson, G., Sweeley, C.C., (1969), J.Amer. Chem. Soc., 91, 1728-1733.

Halevi, E.A., (1963) Progr. Phys. Org. Chem., 1, 109.

Hall, A.N., Hollingshead, S., Rydon, H.N. (1961) J. Chem. Soc. 4290-4295.

Hansch, C., Deutsch, E.W., Smith, R.N., (1965) J. Am. Chem. Soc., 87, 2738-2742.

Hogben, C.A.M., Tocco, D.J., Brodie, B.B., Schanker, L.S., (1959) J. Pharmac, Exp. Ther., 125, 275-282.

Honegger, P., Semenza, G., (1973) Biochim. Biophys. Acta., 318, 390-410.

Kimmich, G.A., (1973) Biochim.Biophys. Acta, 300, 31-78.

Kolínská, J. and Semenza, G. (1967), Biochim.Biophys.Acta, 146, 181-195.

Legler, G.,(1966), Z.Physiol.Chem., 345, 197-214.

Malathi, P., Ramaswamy, K., Caspary, W.F., Crane, R.K., (1973) Biochim.Biophys. Acta, 307, 613-626.

Murer, H., Hopfer, U., Kinne-Saffran, E., Kinne, R., (1974) Biochim.Biophys.Acta, 345, 170-179.

Nath, R.L., Rydon, H.N., (1954) Biochem. J., 57, 1-10

Nelson, K.,Hopfer, U., (1974), Biochim.Biophys.Acta, in press.

Nicolson, G., and Singer, S.J., (1972) Ann.NY. Acad.Sci. 195, 368-375.

Quaroni, A., Gershon, E., Semenza, G., (1974) J. Biol.Chem. in press.

Ramaswamy, K., Malathi, P., Caspary, W.F., Crane, R.K., (1974) Biochim.Biophys.Acta, 345, 39-48.

Richards, J.H.,(1970)in: The Enzymes (Boyer, E.D., Ed.) Vol.II Acad.Press, 329-333.

Robyt, J.F., French, D., (1970)Arch.Biochem.Biophys. 138, 662-670.

Schultz, S.G., Strecker, C.K., (1970) Biochim.Biophys.Acta. 211, 586-588.

Schultz, S.G., Curran, P.F., (1970) Phyiol.Rev. 50, 637-718.

Semenza, G.,von Balthazar, A.K., (1974) Eur.J.Biochem. 41, 149-162.

Semenza, G., Curtius, C.H., Kolínská, J., Müller, M., (1967) Biochim.Biophys.Acta. 146, 196-204.

Storelli, C., Vögeli, H., Semenza, G., (1972) FEBS-Letters, 24 287-292.

Wallenfels, K., Weil, R., (1972) in: The Enzymes (Boyer E.D., ed.) Vol.VII. Acad. Press, 655-663.

HEXOSE-PROTON COTRANSPORT OF CHLORELLA VULGARIS

W. Tanner and E. Komor

Fachbereich Biologie, Universität Regensburg

INTRODUCTION

Chlorella vulgaris possesses an inducible, non-phosphorylating hexose uptake system (Tanner, 1969; Komor and Tanner, 1971), which is responsible for accumulation of sugar analogues up to 1500 fold (Komor et al., 1973 a). The energy for this transport can be supplied by respiration and under anaerobic conditions by cyclic photophosphorylation (Komor et al., 1973 b) or - last and least - by fermentation.

Recently it has been observed that this sugar transport is coupled to a stoichiometric influx of protons (Komor, 1973; Komor and Tanner, 1974a) according to Mitchell's theory (1963). Proton coupled sugar uptake has also been observed for bacteria (West and Mitchell, 1972) and fungi (Seaston et al., 1973; Slayman and Slayman,1974).

In this paper we shall first summarize what is known about the sugar-proton cotransport of Chlorella and subsequently data will be presented which show how the transport system is affected by changing the proton concentration. Mainly these latter results led us to postulate the existence of a carrier molecule in a protonated and unprotonated state with differing properties (Komor and Tanner, 1974 b).

RESULTS AND DISCUSSION

I) The cotransport of protons with hexoses

When 6-deoxyglucose is added to a suspension of induced Chlorella cells, an immediate transient pH shift to more alkaline values occurs in the suspending medium. This pH shift cannot be observed with non-induced cells. Furthermore only sugars which are transported give the pH shift (like glucose, 6-deoxyglucose, 1-deoxyglucose), sugars not transported like α-methylglucoside do not exert any alkalization (Komor,1973; Komor and Tanner,1974 a).

The velocity of proton uptake is dependent on the concentration of sugar added. A saturation behaviour is observed with a K_m-value identical with the K_m-value for sugar uptake. This is shown in table 1 for three sugars with largely differing affinities to the transport system.

Table 1.

K_m-values for sugar uptake and sugar-induced pH shift

sugar	K_m for sugar uptake µM	K_m for pH shift µM
Glucose	10 - 20	8
6-Deoxyglucose	200 - 300	300
3-0-Methylglucose	1000 -2000	1900

When the initial rate of proton uptake after the addition of sugar is compared with the initial uptake rate of the sugar, a fixed stoichiometry of one proton per sugar molecule is observed. This is true for glucose as well as for 6-deoxyglucose and it also holds for different energy sources, i.e. respiratory, photosynthetic or fermentative energy supply (table 2).

Table 2.

Stoichiometric uptake of protons and extra oxygen per sugar molecule transported. (From Komor and Tanner, 1974 a; Decker and Tanner, 1972)

Sugar	Condition	Protons taken up / Sugar taken up	Extra O_2 taken up / Sugar taken up
Glucose	O_2, dark	0.98	0.41
	N_2, light	0.80	-
	N_2, dark	0.87	-
6-Deoxyglucose	O_2, dark	1.06	0.20
	N_2, light	1.00	-
	N_2, dark	1.00	-

Under aerobic conditions in the dark each molecule of 6-deoxyglucose taken up induces the additional consumption of 0.2 molecules of O_2 (corresponding to 1.2 ATP or an equivalent per sugar) and in the case of glucose the extra uptake is 0.4 O_2 per sugar (corresponding to 2.5ATP) From the metabolic fate of glucose within the cells it is known that 1.5 ATP per glucose are required for assimilation reactions (Decker and Tanner, 1972). Thus for 6-deoxyglucose as well as for glucose the same amount of energy, i.e. close to 1 ATP or an equivalent is required for the actual transport work. The proton stoichiometry reported above indicates, therefore, that proton uptake is related only to sugar transport through the membrane and not to subsequent metabolic reactions.

II) What is the cotransported proton good for?

According to Mitchell (1973) the total protonmotive force (ΔP) is given by

$$\Delta P = \Delta \Psi - Z \Delta pH$$

where $\Delta \Psi$ is the membrane potential in mV, Δ pH is the transmembrane pH gradient and -Z is a factor converting pH units into electrical units,

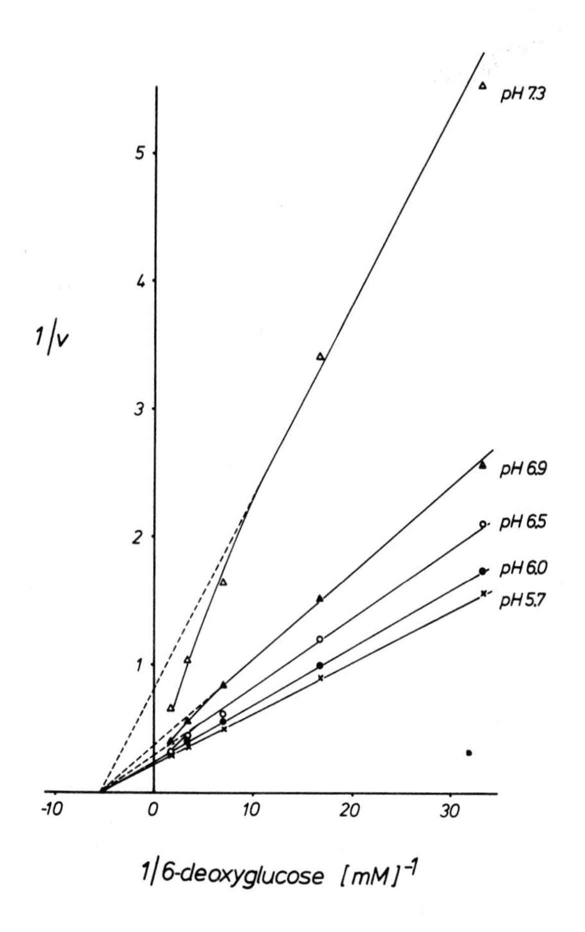

Fig. 1. Lineweaver-Burk plot of 6-deoxyglucose uptake at various pH-values; v is expressed as μmole/min/ml·p.c. (From Komor and Tanner, 1974 b)

which at 25° is 59 mV/pH unit. This protonmotive force, generated by some energy dissipating reaction, is thought to drive endergonic reactions, here the active transport of hexoses. To learn more about the problem how this is achieved, the uptake characteristics were studied at varying proton concentrations.

From fig. 1 it is evident that at all pH-values the apparent K_m measured at low sugar concentration is about 3×10^{-4} M, whereas the intercept at

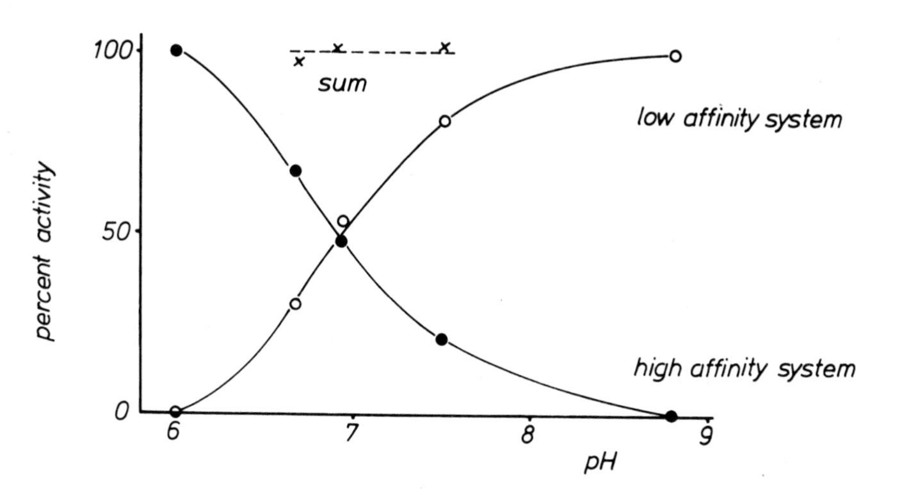

Fig. 2. Relative amount of high- and low affinity carrier at different pH values.
As amount of high affinity carrier was taken the extrapolated maximal velocity of 6dG uptake at low sugar concentration as in fig. 1. As amount of low affinity carrier was taken the difference in influx velocity, which was produced by elevating the 6dG concentration from 10 mM to 200 mM. (From Komor and Tanner, 1974 b).

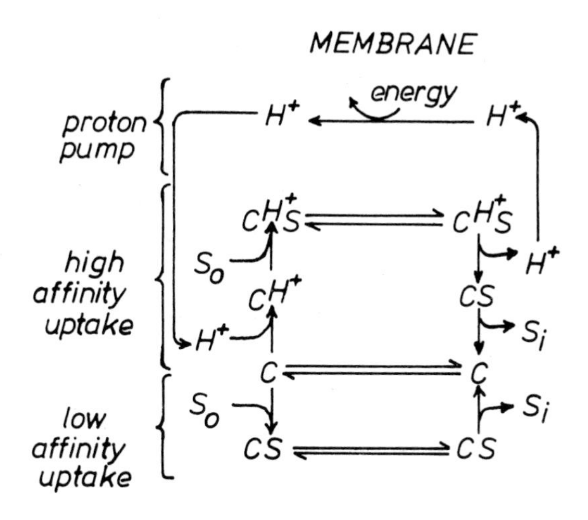

Fig. 3. Model for hexose uptake by Chlorella.

the $1/_v$ axis is quite different for each pH. The apparent maximal velocity V_{max} for sugar influx thus is decreased with increasing pH-values. But at relatively high sugar concentrations a deviation from the straight lines in the Lineweaver-Burk plots has been observed at the more alkaline pH-values. When the sugar concentration dependence of influx velocity is measured with hundred to thousand times higher 6-deoxyglucose concentrations an additional saturation function with a large K_m-value (20-50 mM) is observed at alkaline pH; this saturable component is missing however at an acidic pH like 6.3 (Komor and Tanner, 1974 b). Thus at acid pH values and at very alkaline pH values the uptake of 6-deoxyglucose corresponds to simple Michaelis-Menten kinetics at all substrate concentrations, whereas at slightly alkaline pH values the observed kinetics are similar to those expected for the simultaneous action of two uptake systems with largely differing K_m-values for 6-deoxyglucose.

Both uptake systems are strictly pH dependent whereby a surprising correlation exists: the high affinity system decreases with increasing pH in the same way as the low affinity system increases (fig. 2), so that at pH 6.9 half of each is present. Both uptake activities, expressed in percent of maximal uptake activity at optimal conditions, are summing up to one hundred percent at all pH-values tested (fig. 2; for experimental details see Komor and Tanner, 1974 b). Since a coincidence of two independent uptake systems with opposite but otherwise identical pH dependences is unlikely, the simplest explanation of these results seems to be the following one (see also Fig. 3): at low pH the carrier mainly exists in a protonated form which shows the high affinity towards the substrate. With increasing pH values more and more carrier C remains unprotonated; it represents the low affinity system. At all pH values the sum of both has to be the same.

If the model in fig. 3 were correct two predictions should hold: (1) The affinity of 6-deoxyglucose in the cells towards the carrier should be

Table 3

K_m-values for 6-deoxyglucose influx and efflux
(For experimental details see Komor et al., 1973).

pH of the Medium	K_m-value (mM) for		Experimental Conditions
6.5	influx	0.2	Net influx
	efflux	21	Net efflux.
	efflux	32	Determined from the saturation behaviour of transstimulated influx
9.0	influx	20-50	Net influx

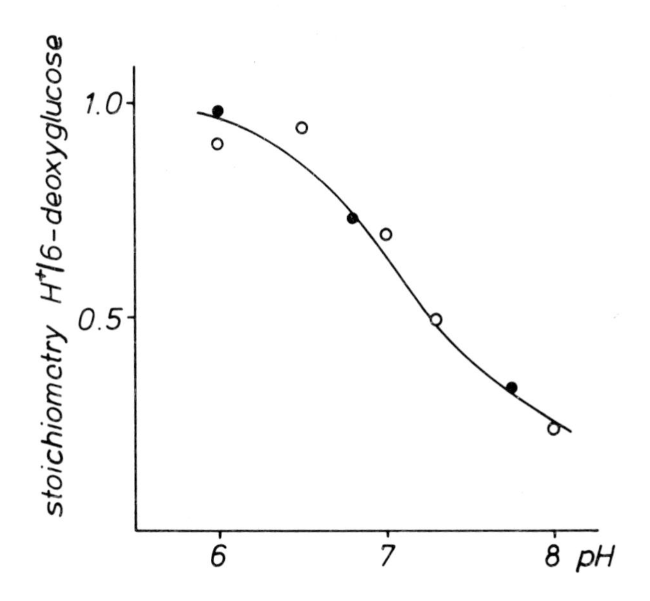

Fig. 4. Stoichiometry of proton-6dG cotransport at different pH values (from Komor and Tanner, 1974 b)

comparable to the affinity at the outside under alkaline conditions. This has indeed been observed (table 3). (2) With increasing pH the stoichiometry of H^+ cotransported per sugar should decrease. This also has been found (fig. 4): at the high sugar concentration used, 3 out of 4 sugar molecules are translocated without protons at pH 8.0. Sugar is still accumulated at this pH and only at pH values > 9.0 the low affinity system works solely; no active transport, only facililated diffusion is then observed (Komor and Tanner, 1974 b).

The question arises, which chemical entity gets protonated in this uptake system. From the pK of 6.9 (fig. 2) it seemed possible that a histidyl residue of the carrier is responsible for the transition from high to low affinity uptake. Since diethylpyrocarbonate, a specific modifying agent for histidin (Ovadi et al., 1967), clearly inhibits hexose uptake under conditions where the respiration rate of the cells is not yet significantly affected, and since the presence of glucose completely prevents this inhibition (table 4), the involvement of a histidin does not seem unlikely.

Table 4

Evidence for the involvement of a histidyl-group in proton-sugar cotransport

Treatment	Inhibition of Uptake of 6dG	Inhibition of Endogenous Respiration
(a) 40 μM Diethylpyro-carbonate	31 %	
$0^{o}C$; 10 min; pH 6.0		4 %
(b) as (a), but Glucose present (6×10^{-3} M)	0 %	

The data presented so far suggest that a transmembrane pH gradient exists in Chlorella (table 3) and that this part of the protonmotive force causes a difference in binding of the sugar to the carrier on either side of the membrane. Could this account, however, for the accumulation factor of more than 1000 fold? The Δ pH inside/outside has been found to be only close to one (inside more alkaline) at an outside pH of 6.0

and no pH difference has been observed at an outside pH of 7.6 (Komor and Tanner, 1974 b). Determinations of an "inside pH" are of course fairly problematic in a eucaryotic cell. Nevertheless the facts that active transport is still observed at an outside pH far above 7.6 and that the existence of an inside pH larger than 9 seems highly unlikely, suggest that $\Delta\Psi$ also contributes to accumulation. This is supported by the following two observations: (1) a diffusion potential set up by 0.2 M K^+ or Na^+ in the outside inhibits sugar uptake in a transient manner (Komor and Tanner, 1974 a) and (2) FCCP inhibits all net and steady state fluxes in either direction (Komor et al., 1972). This observation had already previously been interpreted to mean that a translocational step requires energy. We now consider the membrane potential, which most likely is greatly reduced in the presence of FCCP, to be responsible for a high rate of the translocational step from the outside to the inside.

SUMMARY

The hexose proton cotransport system of Chlorella can exist in a protonated (high affinity) and unprotonated (low affinity) form. This change in affinity is in part responsible for hexose accumulation as long as a transmembrane pH gradient exists. The positively charged sugar-carrier-proton complex will, however, also be affected in its rate of translocation and thus in the degree it accumulates sugar by the membrane potential. The positively charged group of the carrier possibly is a protonated histidin.

ACKNOWLEDGEMENT

This work has been supported by the Deutsche Forschungsgemeinschaft.

REFERENCES

Decker, M. and Tanner, W. (1972). Biochim. Biophys. Acta 266, 661

Komor, E. (1973). FEBS Lett. 38, 16

Komor, E. and Tanner, W. (1971). Biochim. Biophys. Acta 241, 170

Komor, E. and Tanner, W. (1974 a). Eur. J. Biochem. 44, 219

Komor, E. and Tanner, W. (1974 b). J. Gen. Physiol., in press

Komor, E., Haaß, D. and Tanner, W. (1972). Biochim. Biophys. Acta 266, 649

Komor, E., Haaß, D., Komor, B. and Tanner, W. (1973 a). Eur. J. Biochem. 39, 193

Komor, E., Loos, E. and Tanner, W. (1973 b). J. Membrane Biol. 12, 89

Mitchell, P. (1963). Biochem. Soc. Symp. 22, 142

Mitchell, P. (1973). J. Bioenergetics 4, 63

Ovadi, J., Libor, S. and Elödi, P. (1967). Acta Biochim. et Biophys. Acad. Sci. Hung. 2, 455

Seaston, A., Inkson, C. and Eddy, A. A. (1973). Biochem. J. 134, 1031

Slayman, C. L. and Slayman, C. W. (1974). Proc. Nat. Acad. Sci. US 71, 1935

West, J. C. and Mitchell, P. (1972). J. Bioenergetics 3, 445

Tanner, W. (1969). Biochem. Biophys. Res. Commun. 36, 278

Ion Transport and Related Membrane Phenomena

EFFECTS OF PEPTIDE PV AND RELATED COMPOUNDS ON THE IONIC PERMEABILITY OF LIPID BILAYER MEMBRANES

D. C. Tosteson

Department of Physiology and Pharmacology
Duke University Medical Center
Durham, North Carolina 27710
U.S.A.

We have recently reported the synthesis (Gisin and Merrifield, 1972) and some of the properties (Ting-Beall et al., 1974) of a cyclic dodecapeptide analogue (PV) of valinomycin (val). In this paper, we summarize some aspects of the action of PV and related compounds on the ionic permeability of lipid bilayer membranes. The primary structures of the compounds are as follows:

Val	cyclo (Dval - Llac - Lval - Dhyv)$_3$
PV	cyclo (Dval - Lpro - Lval - Dpro)$_3$
PVPA	cyclo (Dval - Lpro - Lala - Dpro)$_3$
PVAV	cyclo (Dval - Lala - Lval - Dpro)$_3$

PV, PVPA, and PVAV differ from val by substitution of amino acids for the hydroxy acids lactate (lac) and hydroxyvalerate (hyv). In metal complexes of val (Pinkerton et al., 1969), it is known that the six valine carbonyl oxygens in ester linkage interact with the ion while the six hydroxy acid carbonyl oxygens in amide linkage are internally hydrogen-bonded to one another. In PV, ester bonds in val are replaced by amide bonds between the carbonyl carbons of valine and the amino nitrogen atoms of proline. Thus, the carbonyl oxygens which interact with the metal ion in complexes of PV are in imide rather than in ester bonds. Unlike val, PV is an optically inactive meso-form. The related compounds which we have synthesized were made optically active by introducing 3 L-alanine residues in replacement for either 3 L-valine residues (PVPA) or 3 L-proline residues (PVAV).

PV is an extremely effective complexing agent for monovalent cations. In two-phase extraction systems such as $H_2O:CHCl_3$, the apparent association constants of PV in the organic phase for K trinitro-cresolate (TNC) is about 10^4 times greater than the apparent association constant of val for KTNC (Tosteson, 1973). The selectivity of PV for K^+ over Na^+ in such systems is about 10 while the comparable value for val is about 10^3. The association constants of PV for K^+ in water and methanol measured from salt induced changes in the UV absorption spectrum are about 10^2 and 10^8 respectively, both at least 2 orders of magnitude greater than the comparable values for val (Grell, 1974). The proton nmr spectra of the complexes of PV with Li^+, Na^+, K^+, Rb^+, Cs^+ and Tl^+ in $CDCl_3$ show the chemical shifts and coupling constants to be about the same for all of these compounds (Davis and Tosteson, 1974). A comparison of these results with proton nmr spectra of val and with molecular models leads to the conculsion that the solution conformation of these molecules is similar to the A-2 "bracelet" conformation of val (Ivanov et al. 1969). The solution conformations of the Na^+ and K^+ complexes, as judged from nmr spectra, are much more similar for PV than for val, consistent with the lower selectivity of the former compound. The proton nmr spectra of the Li^+, Na^+ and K^+ complexes of PV in methanol is similar to that observed in $CHCl_3$. Furthermore, the rate coefficient for dissociation of the Na^+ PV and K^+ PV complexes in methanol are in the range of 1 sec^{-1} markedly slower than the values of 2×10^6 sec^{-1} and 1.3×10^3 sec^{-1} reported for the dissociation rate coefficients of Na^+ val and K^+ val in this solvent (Grell et al. 1972). The solution conformation of free monomeric PV in $CDCl_3$ as judged from proton nmr spectra and models is quite similar to that of the cation complexes. However, the monomer appears to undergo an aggregation reaction $3\ PV \rightleftarrows PV_3$ to form a trimer. The value of the association constant is 1.4 (±0.5) x 10^6 M^{-2} while the value of the rate coefficient for dissociation is about 1 sec^{-1} (Davis and Tosteson, 1974). These solution properties of PV are useful in interpreting the effects of the compound on the ionic permeability of bilayers.

Fig. 3a shows the relation between the current (Im) and the electrical potential difference (V_m) across a bilayer membrane made of sheep red cell lipids dissolved in decane and exposed to PV on only one side. Three striking features of this curve are (1) that current passes more readily from than to the side of the membrane exposed to PV, i.e. rectification, (2) that the slope decreases monotonically with increasing V_m, i.e. saturation, and (3) that V_m is not equal to zero when I_m=0 but rather that the side of the membrane exposed to PV is some 90 mV negative to the side not exposed to the carrier. The relation between the magnitude of this zero-current potential (V_m^o) and the concentration of PV present on one side of a bilayer membrane is shown in Fig. 1. Note that the magnitude of V_m becomes relatively independent of PV concentration and of the cation present in the bathing solutions (Na^+ vs K^+) when the PV concentration is greater than 10^{-5} M. We have devoted considerable effort to understand the origin of this zero-current potential difference because it develops upon addition of an uncharged molecule (PV) to one side of a bilayer separating identical salt solutions, i.e. a system in which the equilibrium potentials for all ions are zero. No such potential develops when val is added to one side of a bilayer separating identical KCl solutions (Andreoli et al.,1967). We interpreted V_m^o to be the equilibrum potential for the K^+PV or Na^+PV ions formed when the peptide makes a complex with the respective monovalent cations (Ting-Beall et al., 1974). We argued that the value of this equilibrium potential depends on the ratio of the effective permeability of the membrane to that of the unstirred layers for the carrier. When V_m^o=0, as is the case with unilateral val, the ratio must be high, while when V_m^o is relatively high as in the case with PV, this ratio must be low. From the value of V_m^o, a value for the permeability coefficient of the membrane to carrier (${}^mP_s$) may be computed if the permeability of unstirred layers, ${}^uP_s$, is known. For PV, we computed a value of ${}^mP_s$ of 1.3×10^{-5} cm sec^{-1} from values of V_m^o of -90 mV and ${}^uP_s$ of 4.3×10^{-4} cm sec^{-1}.

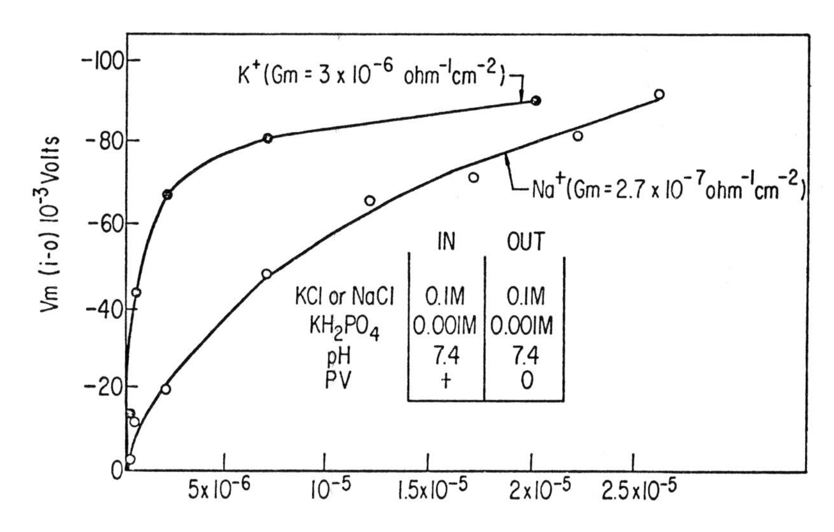

Figure 1. Effect of PV on electrical potential difference (V_m) across lipid bilayers. PV was added only to one side of the membrane. The aqueous solutions contained 0.1 M KCl (•) or 0.1 M NaCl (o).

Subsequently, we have estimated the permeability coefficient of bilayers to val and PV by measuring directly both the flux and the concentration difference for the compounds across the membrane system. The val measurements were made with ^{14}C acetyl lysine val synthesized by B. F. Gisin while the PV measurements were made using a potentiometric assay for the carrier (Benos and Tosteson, 1974). The results, shown in Table I, indicate that the measured permeability coefficient for val (4.5×10^{-4} cm sec^{-1}) agrees well with the expected permeability coefficient of an unstirred layer about 10^{-2} cm thick while the value for PV (4.5×10^{-5} cm sec^{-1}) agrees fairly well with the value calculated from the magnitude of V_m^o with unilateral PV (-90 mV).

TABLE 1

MEASURED PERMEABILITY OF BILAYERS TO VALINOMYCIN AND PEPTIDE PV

SHEEP RBC LIPIDS

0.1M KCl, 0.001M K_2HPO_4, 1% Ethanol, pH 7.4, T=23°C

COMPOUND	PERMEABILITY COEFFICIENT ($cm \cdot sec^{-1}$)
VALINOMYCIN	4.5 ($\pm$ 1.5) x 10^{-4} (n=10)
PEPTIDE PV	4.5 ($\pm$ 0.9) x 10^{-5} (n=17)

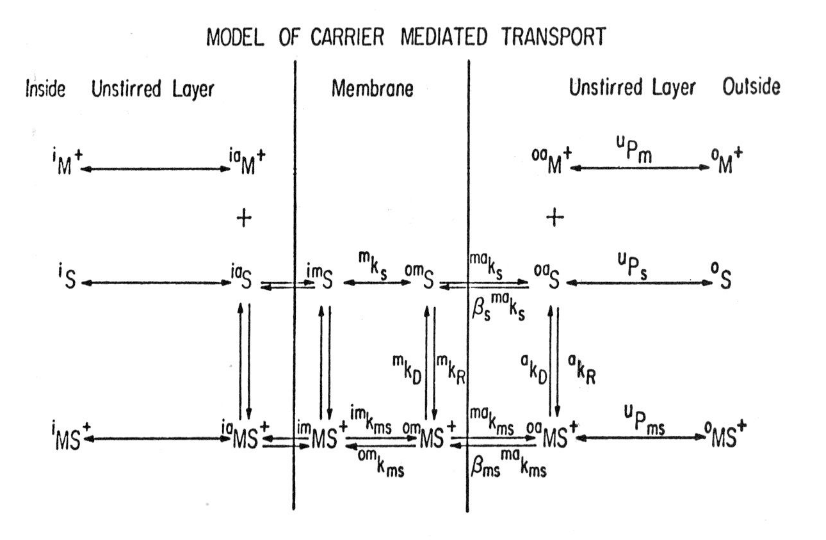

Figure 2. Proposed model for carrier-mediated ion transport. M^+, S, and MS^+ indicate concentrations of free metal ion, free carrier, and cation-carrier complex, respectively. The superscripts i, ia, im, om, oa, and o represent the inside bulk aqueous solution, the aqueous phase adjacent to the inner membrane surface, the inner membrane surface, the outer membrane surface, the aqueous phase adjacent to the outer membrane surface and the outer bulk aqueous solution, respectively. The symbols ${}^uP_m$, ${}^uP_s$, and ${}^uP_{ms}$ denote the permeability coefficients of the unstirred layers to free metal, free carrier, and complex, respectively. The symbols ${}^mk_s$, ${}^{im}k_s$, and ${}^{om}k_{ms}$ indicate rate coefficients for translocation of free carrier and complex across the membrane interior, ${}^{ma}k_s$, and ${}^{ma}k_{ms}$ are rate coefficients for exit of free carrier and complex from the membrane surface. β_s and β_{ms} are the partition coefficients of free carrier and complex (e.g. ${}^{om}S/{}^{oa}S$ where ${}^{om}S$ is the surface concentration in mol·cm^{-2} and ${}^{oa}S$ is the concentration in the adjacent aqueous phase in mol·cm^{-3}. Thus β has the unit centimeter.) k_D and k_R are the rate coefficients for dissociation and formation of the cation-carrier complex where the superscripts m and a indicate membrane surface and aqueous phase, respectively.

There are several possible explanations for the low permeability of bilayers to PV as compared with val. We have considered these in terms of the model shown in Fig. 2. This minimal model includes the possibilities of reactions between carrier (S) and metal ions (M^+) in the aqueous phases and on the membrane surface as well as translocation steps across the unstirred layers, the water-membrane interface, and membrane interior. We assumed the reaction on the membrane surface to be heterogeneous involving free and complexed carrier molecules adsorbed to the membrane surface but metal ions in the adjacent aqueous phase. We have solved the system of eight linear equations which describe the steady state behavior of this model using an IBM 370 Series 165 digital computer. The solutions permits the computation of steady state concentrations of S and MS^+ on the two membrane surfaces and in the two aqueous phases adjacent to the membrane for known values of the concentrations of these compounds as well as free metal ions in the two bulk aqueous solutions (${}^{o}M^+$, ${}^{i}M^+$, ${}^{o}MS^+$, ${}^{i}MS^+$, ${}^{o}S$,${}^{i}S$) and of all rate coefficients and of the membrane potential (Vm). The steady state value of I_m for each value of V_m was also computed. Fig. 3 displays the results of some computations of this kind. For comparison, Fig. 3a shows the experimentally observed relation between I_m and V_m across a sheep red cell lipid bilayer exposed to PV (5×10^{-6}M) on one side only. Figs. 3b, c and d show computed values of I_m as a function of V_m for the model shown in Fig. 2 with parameters chosen as indicated in the legend to Fig.3. In all cases, including the experiment shown in Fig. 3a, the concentrations of salt were identical (1M) in the two bulk aqueous phases while the concentrations of total carrier (free plus complex) were 5×10^{-6} M in the inside and zero in the outside bulk aqueous phases. For all three cases of the model, the parameters chosen to characterize the aqueous unstirred layer as well as the water-membrane interface were identical. They provided that near equilibrium conditions prevailed at the interface. The association constant of carrier for metal ions (k_R/k_D) was 50 in the aqueous phase and 400 on the membrane surface. These values were chosen to simulate the measurements of the association constants of PV for monovalent cations in water and methanol described above.

In model I, shown in Fig. 3a, the values of all membrane rate coefficients except for ${}^{m}k_D$ were chosen to be similar to those reported for valinomycin (Laüger, 1972). ${}^{m}k_D$ was reduced to obtain a value for membrane association constant (${}^{m}k_R/{}^{m}k_D$) of 400. Note that the computed I-V curve is different from the measured curve for PV (Fig. 3a) in at least three important respects. First, the value of V_m^o is zero in the model rather than the observed value of -90 mV. Second, the values of I_m in the model are about two orders of magnitude greater than the observed values. Third, the model does not while the experimental system does show rectification. Therefore, this model in which the value of the association constant in the aqueous phase is so high that almost all of the carrier is present in complex (MS^+) rather than free (S) form, fails to simulate the distinctive features of the action of PV on the electrical properties of lipid bilayers.

In model II, shown in fig. 3c, all parameters were chosen to be identical to those of model I except for the value of ${}^{m}k_S$ which was reduced from 2×10^4 to 20. This alteration produced a computed I-V curve which is much more like the observed curve. In model II, the rate limiting step in movement of carrier through the membrane at zero current is the translocation of the free form. This leads to a substantial difference in the

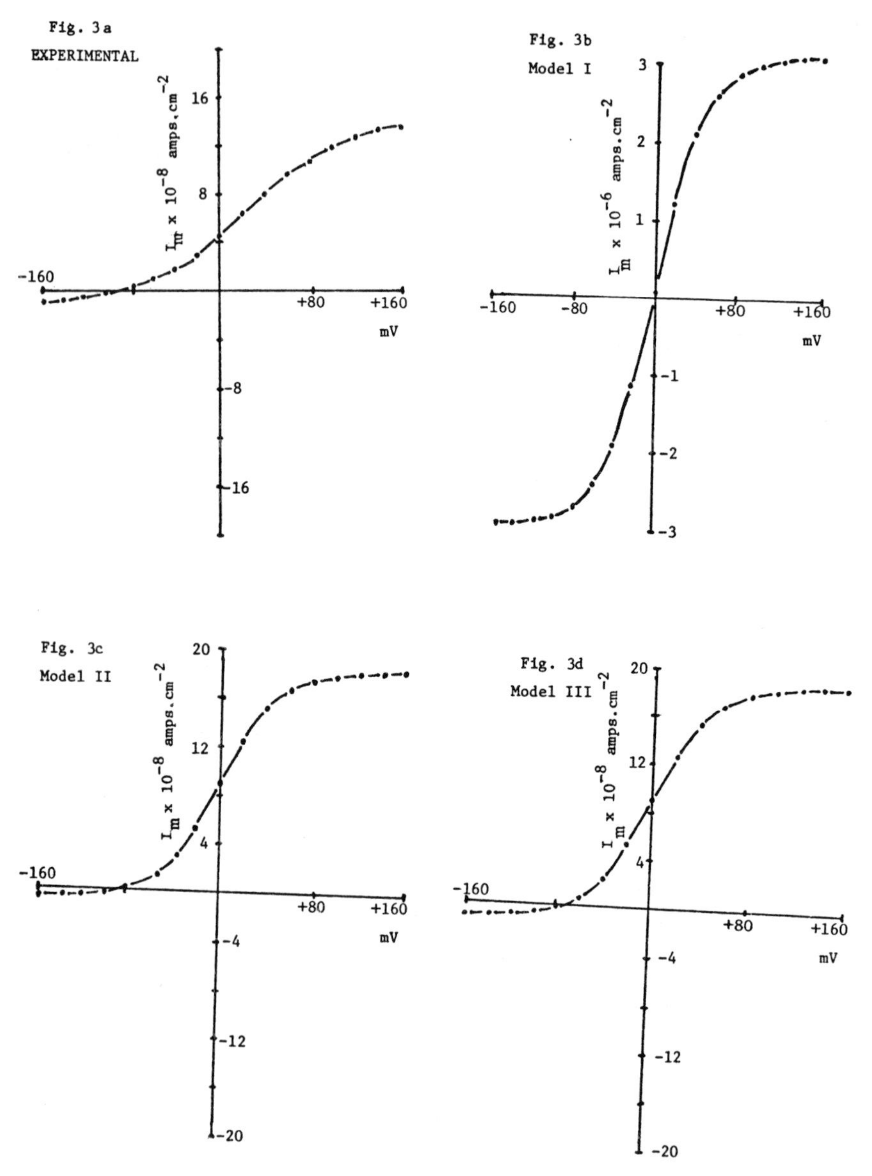

Figure 3. This figure shows the relation between membrane current (I_m) and membrane voltage (V_m) for sheep red cell lipid bilayers exposed to unilateral PV (5×10^{-6}M) (3a), and for three particular solutions of a membrane model system (3b, c, d). The equations describing the model are listed below. The symbols are defined in Fig. 2. Values for the parameters are noted in the following table.

Steady-State Rate Equations for the Carrier-Mediated Ion Transport Scheme Depicted in Figure 2:

$$\frac{d\,{}^{im}S}{dt} = 0 = {}^{m}k_D\,{}^{im}MS^+ - M^+\,{}^{im}S\,{}^{m}k_R - {}^{m}k_S\,{}^{im}S + {}^{m}k_S\,{}^{om}S + {}^{ma}k_S\beta_S\,{}^{ia}S - {}^{ma}k_S\,{}^{im}S$$

$$\frac{d\,{}^{im}MS^+}{dt} = 0 = M^+\,{}^{im}S\,{}^{m}k_R - {}^{m}k_D\,{}^{im}MS^+ + {}^{ma}k_{MS}\beta_{MS}\,{}^{ia}MS^+ - {}^{ma}k_{MS}\,{}^{im}MS^+ + {}^{m}k_{MS}\,{}^{om}MS^+\exp(-FV_m/2RT) - {}^{m}k_{MS}\,{}^{im}MS^+\exp(+FV_m/2RT)$$

$$\frac{d\,{}^{om}S}{dt} = 0 = {}^{m}k_D\,{}^{om}MS^+ - M^+\,{}^{om}S\,{}^{m}k_R - {}^{m}k_S\,{}^{om}S + {}^{m}k_S\,{}^{im}S + {}^{ma}k_S\beta_S\,{}^{oa}S - {}^{ma}k_S\,{}^{om}S$$

$$\frac{d\,{}^{om}MS^+}{dt} = 0 = M^+\,{}^{om}S\,{}^{m}k_R - {}^{m}k_D\,{}^{om}MS^+ + {}^{ma}k_{MS}\beta_{MS}\,{}^{oa}MS^+ - {}^{ma}k_{MS}\,{}^{om}MS^+ + {}^{m}k_{MS}\,{}^{im}MS^+\exp(+FV_m/2RT) - {}^{m}k_{MS}\,{}^{om}MS^+\exp(-FV_m/2RT)$$

$$\frac{d\,{}^{ia}S}{dt} = 0 = u_P\,{}^{i}S - u_P\,{}^{ia}S + {}^{ma}k_S\,{}^{im}S - \beta_S\,{}^{ma}k_S\,{}^{ia}S + {}^{a}k_D\,{}^{ia}MS^+ - M^+\,{}^{a}k_R\,{}^{ia}S$$

$$\frac{d\,{}^{ia}MS^+}{dt} = 0 = u_P\,{}^{i}MS - u_P\,{}^{ia}MS^+ + {}^{ma}k_{MS}\,{}^{im}MS^+ - \beta_{MS}\,{}^{ma}k_{MS}\,{}^{ia}MS^+ + M^+\,{}^{a}k_R\,{}^{ia}S - {}^{a}k_D\,{}^{ia}MS^+$$

$$\frac{d\,{}^{oa}S}{dt} = 0 = u_P\,{}^{o}S - u_P\,{}^{oa}S + {}^{ma}k_S\,{}^{om}S - \beta_S\,{}^{ma}k_S\,{}^{oa}S + {}^{a}k_D\,{}^{oa}MS^+ - M^+\,{}^{a}k_R\,{}^{oa}S$$

$$\frac{d\,{}^{oa}MS^+}{dt} = 0 = u_P\,{}^{o}MS^+ - u_P\,{}^{oa}MS^+ + {}^{ma}k_{MS}\,{}^{om}MS^+ - \beta_{MS}\,{}^{ma}k_{MS}\,{}^{oa}MS^+ + M^+\,{}^{a}k_R\,{}^{oa}S - {}^{a}k_D\,{}^{oa}MS^+$$

PARAMETER	MODEL I	MODEL II	MODEL III
${}^{m}k_D$ (sec^{-1})	1.25×10^{2}	1.25×10^{2}	1.00×10^{-1}
${}^{m}k_R$ (M^{-1}.sec^{-1})	5.00×10^{4}	5.00×10^{4}	4.00×10^{1}
${}^{m}k_S$ (sec^{-1})	2.00×10^{4}	2.04×10^{1}	1.00×10^{4}
${}^{m}k_{MS}$ (sec^{-1})	1.00×10^{4}	1.00×10^{4}	1.00×10^{4}
β_S (cm)	2.80×10^{-5}	2.80×10^{-5}	2.80×10^{-5}
β_{MS} (cm)	2.24×10^{-4}	2.24×10^{-4}	2.24×10^{-4}
${}^{ma}k_S$ (sec^{-1})	1.00×10^{6}	1.00×10^{6}	1.00×10^{6}
${}^{ma}k_{MS}$ (sec^{-1})	1.00×10^{6}	1.00×10^{6}	1.00×10^{6}
${}^{a}k_D$ (sec^{-1})	2.00×10^{-5}	2.00×10^{-5}	2.00×10^{-5}
${}^{a}k_R$ (M^{-1}.sec^{-1})	1.00×10^{-3}	1.00×10^{-3}	1.00×10^{-3}
u_P (cm.sec^{-1})	3.90×10^{-4}	3.90×10^{-4}	3.90×10^{-4}
${}^{i}M = {}^{o}M = {}^{ia}M = {}^{oa}M$	1 Molar	1 Molar	1 Molar
${}^{i}MS^+ + {}^{i}S$ (Molar)	5×10^{-6}	5×10^{-6}	5×10^{-6}
${}^{o}MS^+ + {}^{o}S$ (Molar)	0	0	0

concentrations of ^{im}S and ^{om}S. Because the reaction rates on the membrane surfaces are relatively fast, the reactions come to equilibrium and $(^{im}S/^{om}S)=(^{im}MS^+/^{om}MS^+)$. Thus, the equilibrium potential of MS^+ is non-zero and leads to the non-zero value of V_m^o. The maximum value of the current in model II is determined by the rate at which MS^+ can diffuse to the membrane through the unstirred layer, not by the rate at which free carrier can diffuse from one to the other membrane surface.

In model III (Fig. 3d) the rate coefficients for the chemical reaction on the membrane surface, ^mk_R and ^mk_D, were made low compared to ^mk_S. These values of the parameters also yield I-V curves which simulate the experimental data. At zero current, the rate limiting step in the movement of carrier through the membrane is dissociation of $^{im}MS^+$ and $^{ia}MS^+$. Since the chemical reactions on the membrane surface are not in equilibrium, $^{im}MS^+ > {}^{om}MS^+$ despite the fact that $^{im}S \approx {}^{om}S$ due to the relatively high value of ^mk_S. As with model II, the saturation current at high values of V_m is limited by the rate at which MS^+ can diffuse to and from the membrane through the unstirred aqueous layers.

For several reasons, neither model II nor model III provide a completely adequate explanation for the action of PV on bilayer membranes. First, both models yield saturation currents which are limited by the unstirred layers and are significantly higher than the observed saturation current with PV. Second, the observed dependence of the membrane conductance on salt concentration, both at low and high values of V_m, is not simulated by the models. The observed conductances are proportioned to the salt concentration raised to the 0.4 power (Ting-Beall et al., 1974) over the range from 0.01 to 1.0.M. Both models yield conductances which are decreasingly dependent on salt concentration as the magnitude of the salt concentration increases. In order to obtain a model which shows the observed relation between conductance and salt concentration, it will probably be necessary to introduce a polymerization reaction of the type which free PV undergoes in chloroform solutions.

Fig. 4 shows the dependence of the zero-current, zero-voltage conductance (G_o) of sheep red cell lipid bilayers on the concentration of cyclic peptide carriers present in both bathing solutions. The slope of the line of these log-log plots is unity for PV but significantly less than unity for PVPA and PVAV. PVPA is almost as potent as PV but PVAV is much less active. Thus, the introduction of asymmetry into PV by replacing the 3 L-valine residues with L-alanine, as in PVPA, fails to improve substantially the effectiveness of the compound as ion carrier across bilayers. This substitution also fails to alter the ionic selectivity which is the same sequence for PVPA as for PV. By contrast, substitution of 3 L-alanine residues for 3 L-proline residues, as in PVAV, almost completely destroys the effectiveness of the compound as an ion carrier. Possibly the capacity of the amide bonds involving alanine nitrogen atoms to participate in intramolecular hydrogen bonds reduces the probability of forming the alternating sequence of such bonds that is required to stabilize the "bracelet" conformation of the complexes between cyclic peptide and ions.

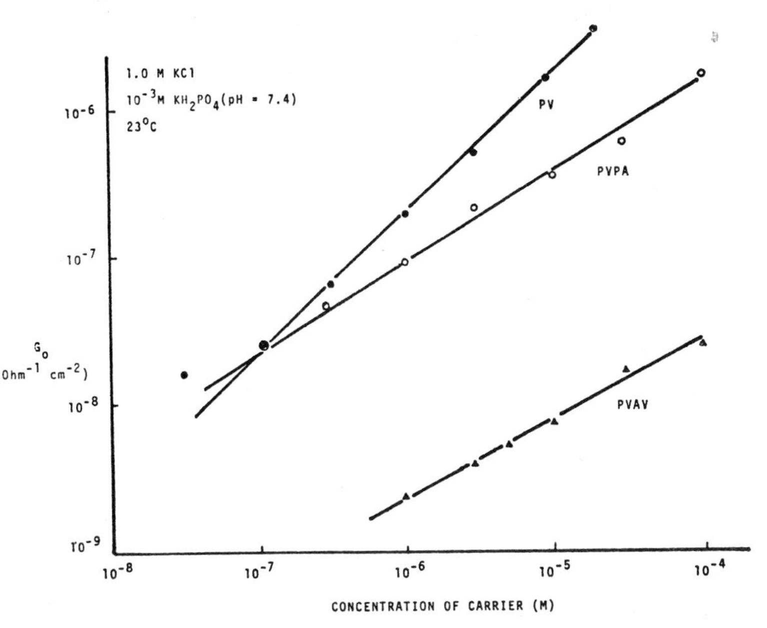

Figure 4. The dependence of the zero-current, zero-voltage conductance (G_0) of sheep red cell lipid bilayers on the concentration of cyclic peptide carriers present in both chambers containing 1.0 M KCl and 10^{-3} M KH_2PO_4(pH 7.4).

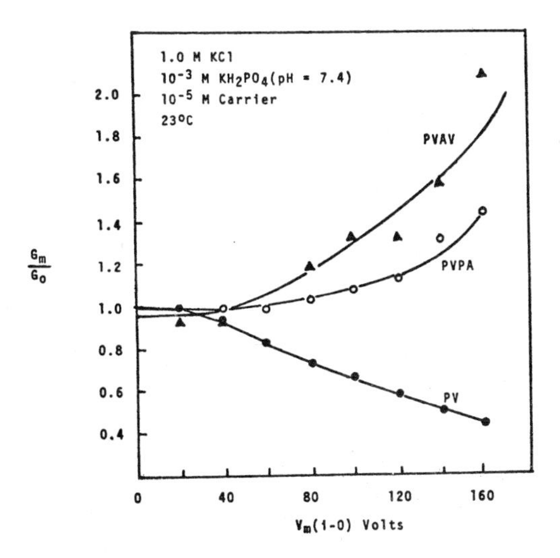

Figure 5. Conductance ratio (G_m/G_0) as a function of voltage (V_m) for cyclic peptide carriers. $G_m = I_m/V_m$ and $G_o = (dI_m/dV_m); V_m, I_m = 0$

Fig. 5 shows an interesting difference between the actions of PV, on the one hand, and PVPA and PVAV, on the other hand, on the conductance of bilayers. The ratio of the chord conductance ($G_m = I_m/V_m$) to the zero-current, zero-voltage conductance ($G_o = (dI_m/dV_m);V_m, I_m=0$) is plotted as a function of V_m. For PV, the I-V curve is of the saturating type and, therefore, G_m/G_o decreases monotonically with increasing V_m. However, both PVPA and PVAV produce I-V curves with slopes that increase with increasing V_m. This result suggest that the rate of translocation of MS^+ becomes rate limiting for charge transport mediated by these carriers. This behavior could be due to a reduction in the partition coefficients, β_s and β_{ms}, when alanine residues are substituted for proline or valine residues.

In summary, the differences between valinomycin and the recently synthesized cyclic dodeca-peptide analogue, PV, as ion carriers across bilayer membranes seem to depend on several factors. First, PV has a higher affinity for monovalent cations in both polar and non-polar solvents. The association constant of PV for K^+ in water is high enough so that most of the carrier in the aqueous phases bathing the membrane is in the form of ion complexes when the salt concentration is greater than 0.1 M. Second, the rates of dissociation of PV-ion complexes are at least 3 orders of magnitude slower than the corresponding rates for val-ion complexes. Third, free PV undergoes an aggregation reaction in low polarity solvents and, perhaps, also on membranes.

REFERENCES

Benos, D. J. and Tosteson, D. C. (1974). The Physiologist, 17, 180.

Davis, D. G., Gisin, B. F. and Tosteson, D. C. (1974). In preparation.

Gisin, B. F., Davis, D. G., Kimura, J., Tosteson, M.T. and Tosteson, D.C. (1972). Proc. IV Inter. Biophys. Cong. Symposium on Membrane Structure and Function, Moscow. Aug 7-14 (Abstr.)

Gisin, B. F. and Merrifield, R. B. (1972). J. Am. Chem. Soc. 94, 6165.

Grell, E., Funck, T. and Eggers, F. (1972). In: Progress in Surface and Membrane Science, eds. J. F. Danielli, M. D. Rosenberg and D. A. Cadenhead (Academic Press. Inc. New York) p. 139.

Ivanov, V. T., Laine, I. A., Abdulaev, N. D., Senyavina, L. B., Popov, E.M., Ovchinnikov, Yu. A. and Shemyakin, M. M. (1969). Biochem. Biophys. Res. Comm. 34, 803.

Pinkerton, M., Steinrauf, L. K. and Dawkins, P. (1969). Biochem. Biophys. Res. Comm. 35, 512.

Ting-Beall, H. P., Tosteson, M. T., Gisin, B. F. and Tosteson, D. C. (1974). J. Gen. Physiol. 63, 492.

Tosteson, D. C. (1973). In: Transport Mechanisms in Epithelia, Proceedings of the Alfred Benzon Symposium V, eds. H. H. Ussing and N. A. Thorn (Academic Press, N. Y.) p. 28.

MECHANISM OF Ca-DEPENDENT K-TRANSPORT IN HUMAN RED CELLS

G. Gárdos, Ilma Szász and B. Sarkadi

Department of Cell Metabolism, National Institute of Haematology and Blood Transfusion, Budapest, Hungary

INTRODUCTION

Under physiological circumstances K and Na ions exchange between human erythrocytes and plasma at a rate of 1.5-2.0 mEq/l RBC/h /Raker et al., 1950; Sheppard and Martin, 1950/. In certain experimental conditions, however, the permeability of the cell membrane greatly and selectively increases to some ions: NaF, iodoacetate + purine nucleosides and propranolol respectively, induce rapid K-efflux from the cells unaccompanied by equimolar Na-influx /Wilbrandt, 1940; Gárdos, 1956; Ekman et al., 1969/. This selective K-efflux is a Ca-dependent process: in the presence of Ca-chelators K-efflux stops, while it starts again on addition of Ca /Gárdos, 1958, 1959/.

According to the classification of Lew /1974/ slightly modified by us /Szász and Gárdos, 1974/, there are four phases in the development of the Ca-dependent rapid K-transport:

1. ATP depletion and/or alteration of the ATP compartmentation.
2. Ca-influx and its appropriate compartmentation.
3. Reaction of Ca ions with the receptors specific for K-transport.
4. K-movement through the membrane.

1. ATP depletion in fact is mainly a precondition for eliminating the ATP molecules protecting the specific membrane Ca-receptors. Propranolol brings about only a slight

decrease in the ATP-level, but most probably by altering ATP-compartmentation, it makes the receptors responsible for the K-transport available to Ca ions /Szász and Gárdos, 1974/.

2. Normal erythrocytes contain Ca ions only at a very low concentration. For the rapid K-transport, however, intracellular Ca is needed /Romero and Whittam, 1971; Riordan and Passow, 1971; Blum and Hoffman, 1972/. Hence the Ca-permeability of the membrane plays a significant role in the development of the phenomenon. Both intensive ATP-depletion or ATP-release from the membrane increase Ca-permeability and promote this way the rapid K-transport. The effect of a great variety of drugs can be explained by their impact on Ca-permeability /Szász and Gárdos, 1974/.

3. Based on Ca-binding experiments several membrane Ca-receptors of different type can be distinguished, which are connected with various RBC functions. Based on drug-sensitivity and binding intensity the Ca-receptor responsible for the Ca-pump, for the shape changes, or the rapid K-transport can be differentiated. Latter was proved to represent a loose interaction i.e. Ca ions can be eliminated from the receptors simply by diluting the medium or by washing the cells with physiological saline /Szász and Gárdos, 1974/.

4. When studying the characteristics of the K-transport two main problems emerge:

a./ Is it a simple diffusion or is it carrier-mediated?
b./ Which membrane component is responsible for it?

Many investigations have been published in connection with both problems. Blum and Hoffman /1971/ and Riordan and Passow /1973/ reported that rapid K-transport can be saturated, while Manninen /1970/ observed a counter-transport like ^{42}K-movement. Hoffman based on his experiments with ouabain and other ATPase inhibitors suggested that the membrane component responsible for rapid K-transport is the altered Na-K-ATPase /Hoffman and Knauf, 1973/.

RESULTS

Based on the study of the first three phases described above we concluded that in order to examine the kinetic relations of the rapid K-transport itself, care should be taken of achieving the maximum interaction between the specific membrane receptors and Ca ions. One of the methods meeting this demand was the complete phosphate-ester depletion of the cells that was attained by a 5 hour incubation with Na-bisulfite for 2,3-DPG depletion and subsequently a 2 hour incubation with iodoacetate and inosine for ATP depletion. The other method was the treatment of fresh or phosphate-ester depleted cells with 0.5 mM propranolol.

Three basic experimental set-ups were used by us:

1. The net K-loss of the completely phosphate-ester depleted and propranolol treated cells was measured. An immediate rapid K-efflux develops on addition of 0.5 mM $CaCl_2$ if these cells are incubated in the standard medium containing 155 mM NaCl and 5 mM KCl. At the same time the permeability of the erythrocytes to Na remains low. The rapid K-efflux is followed by Cl and water migration, cells undergo a marked shrinking. The change of the K-concentration of the medium and the haematocrit are in direct proportion with the rate of the net K-efflux. The rate of net K-loss was found to be identical with completely phosphate-ester depleted and propranolol treated cells, respectively.

2. Blum and Hoffman /1971/ with cells depleted from ATP by iodoacetate, and Manninen /1970/ with propranolol treated cells observed that if ^{42}K was added to an incubation medium of low K-concentration, in the presence of Ca ^{42}K was rapidly taken up by erythrocytes and ^{42}K-concentration in the cells reached a higher level than that of the medium. The ^{42}K-concentration in phosphate-ester depleted cells was found by us to increase up to twice of that of the medium, followed by a subsequent equilibration of ^{42}K. On propranolol treatment K-accumulation accelerated and the maximum cell/medium ^{42}K ratio became as high as five /Fig. 1/.

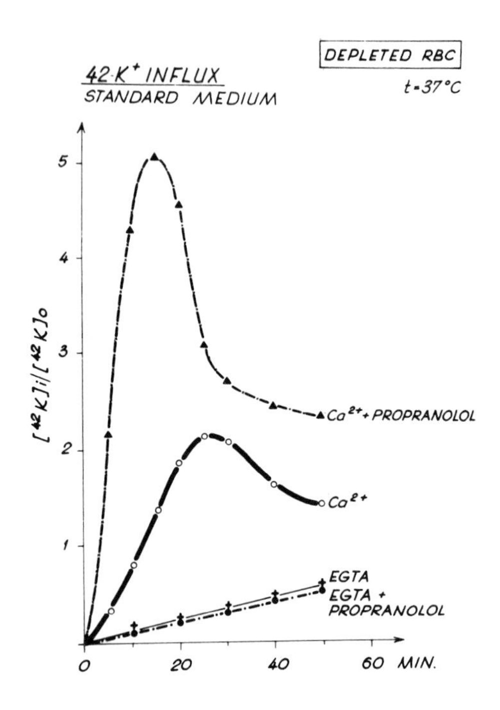

Fig. 1. Effect of 2.0 mM $CaCl_2$, 1.0 mM EGTA and 0.5 mM propranolol on the ^{42}K-influx into phosphate-ester depleted erythrocytes. Standard medium: 155 mM NaCl, 5 mM KCl.

This type of ^{42}K-movement reminds of the counter-transport processes often observed in nonelectrolyte transport investigations. In those cases according to widely expounded views, the phenomenon is carrier mediated. However, in the case of the Ca-dependent rapid K-transport Glynn and Warner /1972/ computed that the process may develop without carrier mediation if the permeability coefficient of K ions becomes higher than the value of net Cl permeability. The valinomycin experiments of Hunter /1971/ and Tosteson et al. /1973/ and the calculations of Lassen /1972, 1973/ based on direct electrical measurements have shown that the net Cl permeability is lower by several order of magnitude than the extremely high exchange Cl permeability. If in the course of the Ca-dependent rapid K-transport the value of K permeability exceeds that of the net Cl permeability the movement of K ions is significantly influenced by the mobility of the Cl ions accompanying them. The rate of net K-efflux will be proportional with the net Cl permeability and at the same time a significant potential difference generates. The potential is negative in the inside and positive at the extracellular side. Its extent is determined by the ratio of K and Cl permeability. Under these circumstances ^{42}K added to the medium migrates rapidly towards the inside of the cell due to the potential difference and is concentrated there.

3. Our third experimental set-up corresponds to the so called K-exchange condition. In this case the K-concentration of the incubation medium is kept identical with the intracellular K-concentration and ^{42}K-efflux rate is measured. In this system ^{42}K-efflux depends only on its own gradient and the K-permeability of the membrane. Hence K-permeability can be directly calculated from the half time and the rate constants, respectively.

Blum and Hoffman /1971/, as well as Riordan and Passow /1973/ reported that in case of the "exchange condition" i.e. with identical intracellular and external K-concentrations the rate constant of the ^{42}K-flux - after an initial rapid increase - decreased significantly with increasing K-concentration, i.e. a saturation effect was observed.

In our own experiments we measured the K-efflux in propranolol treated ghosts at identical intracellular and medium K-concentrations. The medium was made isosmotic with choline chloride. The value of the rate constant increased significantly between 1 and 5 mM, but increasing the K-concentration to 150 mM no saturation effect appeared, K-permeability remained constant /Fig. 2/.

It should be mentioned that in the presence of propranolol the Ca-membrane receptor interaction eliciting the rapid K-transport develops rapidly and to a full extent

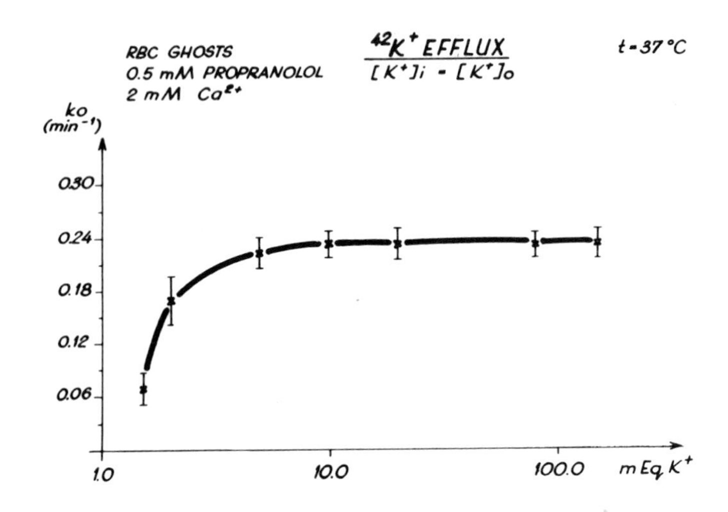

Fig. 2. Effect of external K-concentration on the rate of ^{42}K-efflux from propranolol treated erythrocyte ghosts.

while this process is less effective in case of depleted ghosts. Lew /1974/ demonstrated that the high K-concentration of the medium inhibits the Ca permeability of depleted cells. This effect might result in an apparent saturation effect which disappears in the presence of propranolol. Under the exchange condition drug effects and kinetical investigations were carried out at 150 mM intracellular and external K-concentrations. The half time of ^{42}K-efflux is 6.5 minutes in depleted cells, while it is reduced to half of this value by propranolol /Fig. 3/.

Summarizing the results obtained with phosphate-ester depleted erythrocytes in the three experimental set-ups it can be concluded that propranolol increases K permeability in depleted cells, it does not influence the net K-efflux, while it enhances the rapid ^{42}K-influx /Fig. 4/.

All this agree with the assumption that net K-efflux is regulated by the slower net Cl transport, while the rapid isotope K-influx by the potential difference proportional with the ratio of K and Cl permeabilities /Glynn and Warner, 1972; Kregenow and Hoffman, 1972; Hoffman and Knauf, 1973/.

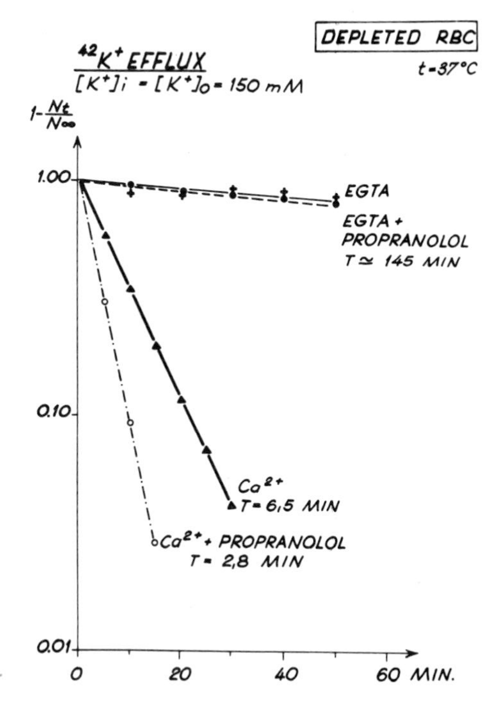

Fig. 3.
Effect of propranolol on the ^{42}K-efflux from phosphate-ester depleted erythrocytes under exchange conditions.
For concentrations see Fig. 1.

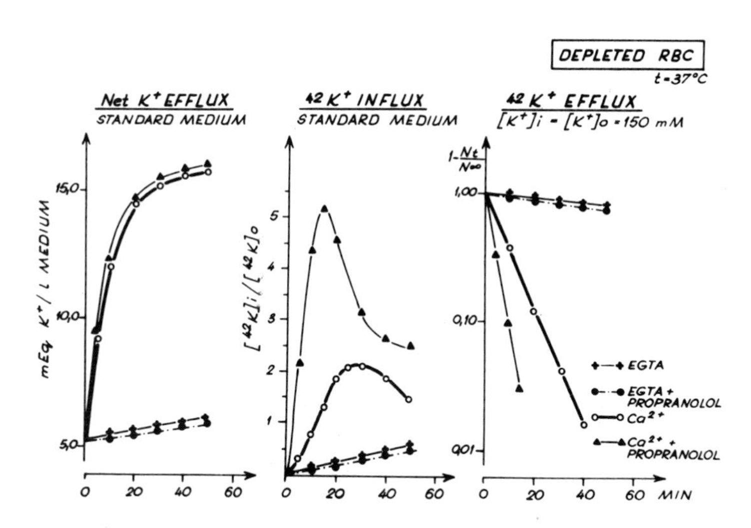

Fig. 4. Effect of propranolol on the Ca-dependent K-transport in phosphate-ester depleted cells. For concentrations see Fig. 1.

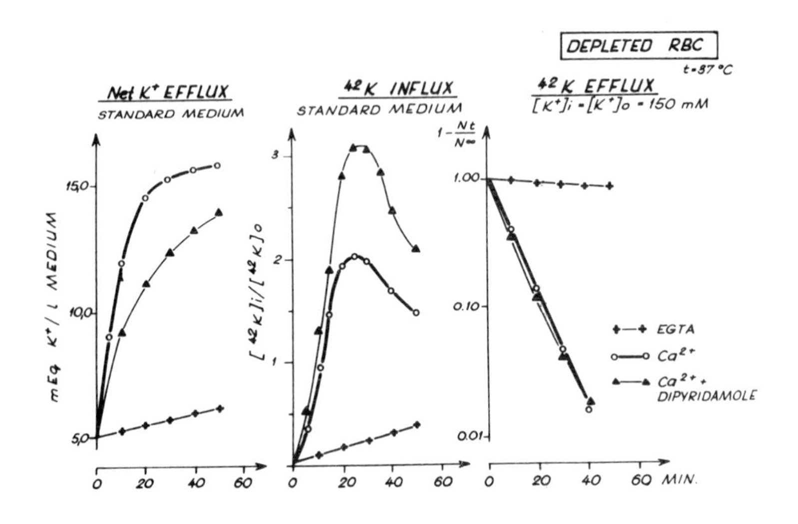

Fig. 5. Effect of 0.1 mM dipyridamole on the Ca-dependent K-transport in phosphate-ester depleted erythrocytes. For concentrations see Fig. 1.

The effect of certain drugs on the Ca-dependent rapid K-transport in the three experimental set-ups were studied. Those drugs were selected which had been proved not to influence significantly the Ca permeability or Ca binding of the membrane.

For studying the connection between the net Cl permeability and the K-transport the effect on the K-movement of the well-known Cl-transport inhibitor dipyridamole was checked. 0.1 mM Dipyridamole inhibited significantly the Ca-dependent net K-efflux, while enhancing greatly the ^{42}K-influx from the standard incubation medium. Under the exchange condition dipyridamole proved to be ineffective /Fig. 5/.

Hence dipyridamole was shown not to influence K-permeability itself but to inhibit net K-efflux by inhibiting the Cl-transport and promoting ^{42}K-uptake by increasing the potential difference depending on the mobility ratios of Cl and K ions. In order to find further support for this assumption Cl ions were substituted by organic anions of lower

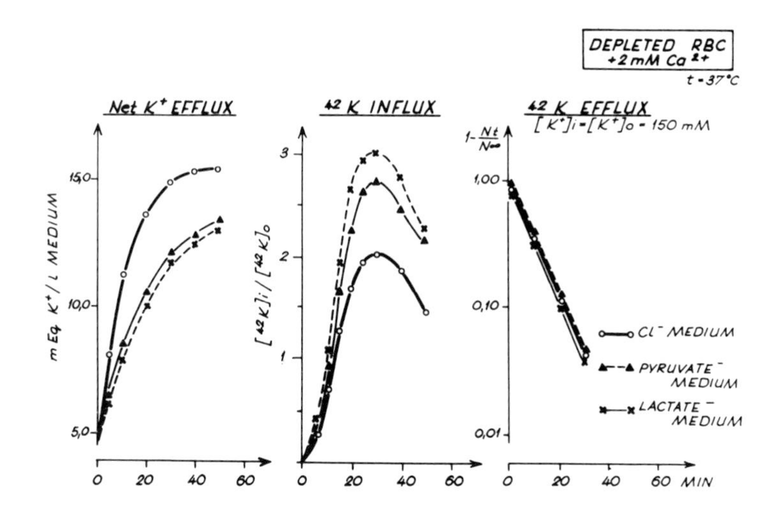

Fig. 6. Effect of various anions /155 mM/ on the Ca-dependent K-transport in phosphate-ester depleted erythrocytes.

permeability in the media. Fig. 6 shows that media containing organic anions had an effect on the Ca-dependent rapid K-transport of completely depleted cells identical with that of dipyridamole. While they did not affect the ^{42}K-transport in case of the exchange condition, they inhibited the net K-efflux and increased the ^{42}K-influx from the standard medium. Any specific effect of the anions used in the experiments can be excluded by the fact that pyruvate, lactate and malonate all had the same effect.

From among the further drugs tested: Ouabain had no effect in any of the three experimental set-ups with completely depleted or propranolol treated cells. On the other hand oligomycin had a strong inhibitory effect in all the three systems. Its inhibitory effect on the ^{42}K-transport under the exchange condition - considering that it did not influence Ca-permeability and Ca-binding - refers to a direct inhibition of the K-transport. As a consequence the net K-efflux and the rate of ^{42}K-influx decreased as well. Certain SH-reagents were found to exhibit an oligomycin-like effect on the Ca-dependent

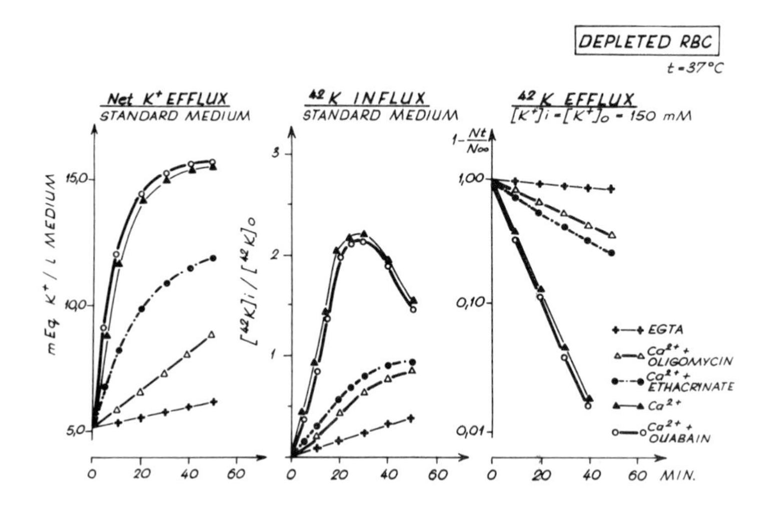

Fig. 7. Effect of various ATPase inhibitors - 0.1 mM ouabain, 0.1 mg/ml oligomycin and 1 mM ethacrynate - on the Ca-dependent K-transport in phosphate-ester depleted erythrocytes. For further conditions see Fig. 1.

K-transport /Fig. 7/. Mersalyl and PCMB behaved like ethacrynate. It is important to remark when speaking about the effect of the SH-reagents that the examined compounds inhibit the rapid K-transport at a low concentration which does not affect yet the passive Na-K-transport. The impact of the SH-reagents on the rapid K-transport could be suspended in every system with SH-protecting agents like cystein and mercaptoethanol.

The effect on the rapid K-transport of red blood cells of the large tetraethylammonium /TEA/ cation blocking the K current of the nerve membrane was also tested. TEA-Cl added to the medium inhibited the rapid ^{42}K-influx from standard medium. The rate of isotope flux measured at identical internal and external K-concentrations was also definitely reduced by TEA. The observed impacts agree well with the characteristics of TEA-effects on nerve membrane where TEA also inhibits

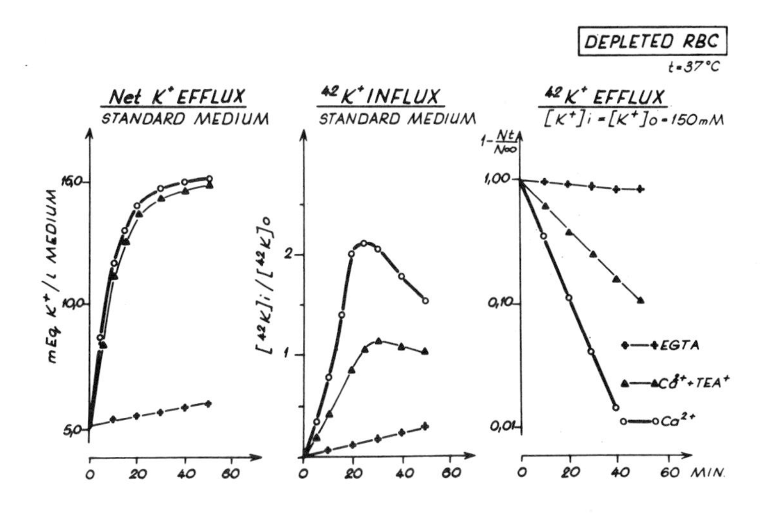

Fig. 8. Effect of 8 mM TEA-Cl on the Ca-dependent K-transport in phosphate-ester depleted erythrocytes. For concentrations see Fig. 1.

K-transport from the side of its application, as if "blocking" the K-channels /Fig. 8/.

DISCUSSION

The presented experimental results can be interpreted as follows:

1. In erythrocytes by intensive phosphate-ester depletion or by applying propranolol a selective rapid K-transport of maximum effectivity can be induced. The high K-permeability developing on the effect of Ca exceeds the net Cl-permeability of the red blood cell membrane significantly. From the measured net K-transport the value of the net Cl-permeability can be calculated /Sarkadi et al., 1974/. The calculated values agree well with literary data obtained as a result of other approaches.

2. With kinetic analysis of the rapid K-transport no evidence was found by us for carrier mediation. Up to the isotonic K-concentration, no saturation effect was found.

The isotope flux reminding of counter-transport in the course of the Ca-dependent rapid K-transport most probably is the consequence of the different mobilities of K and Cl ions. Carrier-mediation need not be suggested.

3. In the field of the membrane constituents responsible for the rapid K-transport no unambiguous evidence have been found. According to Hoffman's opinion /Blum and Hoffman, 1971/ this K-movement is a function of the altered Na-K-activated ATPase. In our experiments rapid K-transport was not inhibited by ouabain in completely depleted and propranolol treated cells. Our investigations agree better with Lew's /1971, 1974/ suggestion according to which in Hoffman's experiments the effect of ouabain is the result of the ATP sparing effect in not completely ATP-depleted cells. At the same time it still might draw the attention to the role of the Na-K-activated ATPase that oligomycin and low concentrations of SH-reagents - the inhibitors of this ATPase - inhibit effectively the Ca-dependent rapid K-transport.

4. According to our investigations a close parallelism can be revealed between the rapid K-transport of erythrocytes and the selective K-permeability increase in the nerve membrane. Both processes demand Ca ions and both can be blocked selectively by TEA-Cl. The relationship of the two processes is further supported by our experiments according to which Na-permeability can also be enhanced in erythrocytes by acetylcholine analogues. This effect is transitory, soon it turns into an inhibition of the Na-uptake and it proved to be tetrodotoxin sensitive /Szász and Gárdos, 1974/.

As a conclusion: Ca-dependent specific rapid K-transport of the erythrocyte membrane represents a useful model-system for the biochemical and biophysical study of living membranes. Some characteristics common with transport phenomena of excitable membranes might also be suspected.

SUMMARY

The effects of various drugs on the Ca-dependent K-movement in phosphate-ester depleted erythrocytes were tested. Propranolol was found to increase ^{42}K-efflux under exchange conditions and ^{42}K-influx from low K medium, while it did not influence the rate of the net K-efflux. Dipyridamole reduced net K-efflux, enhanced ^{42}K-influx, while it did not affect ^{42}K-efflux under exchange condition. The same changes were observed if Cl ion in the medium was substituted with anions of lower permeability: pyruvate, lactate or malonate. Ouabain was ineffective, while oligomycin and ethacrynate inhibited net K-outflow, ^{42}K-influx and ^{42}K-efflux under exchange condition. TEA-Cl inhibited K-transport from the side of its application. The mechanisms of these drug effects - lowered anion permeability, altered membrane potential, reaction with membrane components, blocking of K-channels - are discussed.

ACKNOWLEDGEMENTS

The excellent technical assistance of Mrs Eva Irmai and Mrs Susan Andrási is gratefully acknowledged.

REFERENCES

Blum, R.M. and Hoffman, J.F. /1971/. J. Membrane Biol. 6, 315.

Blum, R.M. and Hoffman, J.F. /1972/. Biochem. Biophys. Res. Comm. 46, 1146.

Ekman, A., Manninen, V. and Salminen, S. /1969/. Acta Physiol. Scand. 75, 333.

Gárdos, G. /1956/. Acta Physiol. Acad. Sci. Hung. 10, 185.

Gárdos, G. /1958/. Biochim. Biophys. Acta 30, 653.

Gárdos, G. /1959/. Acta Physiol. Acad. Sci. Hung. 15, 121.

Glynn, I.M. and Warner, A.E. /1972/. Brit. J. Pharmacol. 44, 271.

Hoffman, J.F. and Knauf, P.A. /1973/. In "Erythrocytes, Thrombocytes, Leukocytes", Gerlach, E., Moser, K., Deutsch, E. and Wilmanns, W. eds., Georg Thieme, Stuttgart, p. 66.
Hunter, M.J. /1971/. J. Physiol. /London/ 218, 49 P.
Kregenow, F.M. and Hoffman, J.F. /1972/. J. Gen. Physiol. 60, 406.
Lassen, U.V. /1972/. In "Oxygen Affinity and Red Cell Acid Base Status", Rorth, M. and Astrup, P. eds., Munksgaard, Copenhagen, p. 291.
Lassen, U.V., Pape, L. and Vestergaard-Bogind, B. /1973/. In "Erythrocytes, Thrombocytes, Leukocytes", Gerlach, E., Moser, K., Deutsch, E. and Wilmanns, W. eds., Georg Thieme, Stuttgart, p. 33.
Lew, V.L. /1971/. Biochim. Biophys. Acta 249, 236.
Lew, V.L. /1974/. In "Comparative Biochemistry and Physiology of Transport", Bloch, K., Bolis, L. and Luria, S.E. eds., North-Holland, Amsterdam, in press.
Manninen, V. /1970/. Acta Physiol. Scand. Suppl. 355, 1.
Raker, J.W., Taylor, I.M., Weller, J.M. and Hastings, A.B. /1950/. J. Gen. Physiol. 33, 691.
Riordan, J.R. and Passow, H. /1971/. Biochim. Biophys. Acta 249, 601.
Riordan, J.R. and Passow, H. /1973/. In "Comparative Physiology", Bolis, L., Schmidt-Nielsen, K. and Maddrell, S.H.P. eds., North-Holland, Amsterdam, p. 543.
Romero, P.J. and Whittam, R. /1971/. J. Physiol. /London/ 214, 481.
Sarkadi, B., Schubert, A., Szász, I. and Gárdos, G. /1974/. Abstr. Commun. 9th Meet. Fed. Europ. Biochem. Soc., Budapest, p. 241.
Sheppard, C.W. and Martin, W.R. /1950/. J. Gen. Physiol. 33, 703.
Szász, I. and Gárdos, G. /1974/. FEBS Letters 44, 213.
Tosteson, D.C., Gunn, R.B. and Wieth, J.O. /1973/. In "Erythrocytes, Thrombocytes, Leukocytes", Gerlach, E., Moser, K., Deutsch, E. and Wilmanns, W. eds., Georg Thieme, Stuttgart, p. 62.
Wilbrandt, W. /1940/. Pflügers Arch. 243, 519.

ELECTRICAL AND PERMEABILITY CHARACTERISTICS OF THE *AMPHIUMA* RED CELL MEMBRANE

U.V. Lassen, L. Pape and B. Vestergaard-Bogind

Zoophysiological Laboratory B, University of Copenhagen, Copenhagen, Denmark

The red cell membrane is, as are other cell membranes, mainly composed of various lipids arranged as a bimolecular leaflet (see Bangham, 1972). Also proteins are, however, shown to be integral constituents of the plasma membrane (see Kaplan and Criddle, 1971). Whereas the capacitance of the red cell membrane, as discussed below, is similar to that of an artificial black lipid membrane, the membrane resistance is considered highly different. This difference may be related to the presence of proteins in the membrane and to protein-lipid interaction. Detailed information about the electric resistance may thus give valuable information for understanding the membrane structure. This presentation will deal exclusively with electric properties and ion transport across the red cell membrane. Furthermore, although the red cell is non-excitable, the present account will present evidence that perturbation of the membrane can result in time-dependent phenomena.

Plasma membranes from dog and human erythrocytes have been shown to have an electric capacitance of about 0.8 $\mu F/cm^2$ (Fricke, 1926). The giant nucleated erythrocytes of a salamander (*Amphiuma means*) have a membrane capacitance in the same order of magnitude (Chang, 1964). Thus the capacitance of the erythrocyte membrane is similar to that found in membranes of other types (see Katz, 1966). This may be taken to indicate that the arrangement of the lipids in the red cell membrane is essentially similar to other cell membranes and also to artificial black lipid membranes including those made of red cell lipid (Andreoli et al., 1967, Bangham, 1972). In other respects, however, the red cell membrane has properties that are both quantitatively and qualitatively different from most other cells.

The electrical resistance of a cell membrane is a measure of the degree to which movement of ions is restricted by the presence of the membrane. The resistance of lipid bilayers is high, usually in the order 10^8 to 10^9 Ωcm^2, indicating that ions only with difficulty enter and diffuse across the lipid phase. As discussed e.g. by Bangham (1972), this may be related to the high energy of activation for most of the dominant physiological ion species. A knowledge of the resistance of cell membranes is thus important for understanding the way in which the membrane regulates physiological functions.

Johnson and Woodbury (1964) attempted to measure the cross resistance of single human red cells, which were sucked into a capillary. The obtained values of about 10 Ωcm^2 were similar to the apparent membrane resistances measured by Lassen and Sten-Knudsen (1968) and thus many orders of magnitude lower than the resistance of the unmodified lipid bilayer. The small membrane resistance reported in these studies were in the same range as the equivalent chloride resistance in beef and human red cells calculated by Tosteson (1959) on the basis of the velocity of tracer exchange. Dalmark and Wieth (1970) confirmed and extended data on the presence of an extremely rapid exchange of labelled chloride across the human red cell membrane. It was therefore natural to ascribe the reported low resistance of the erythrocyte membrane to the high permeability of chloride ions. Under normal conditions chloride ions are in equilibrium across the red cell membrane. In the case of the _Amphiuma_ red cell, this was demonstrated by the good agreement between the measured membrane potentials and the equilibrium potentials for chloride, as determined at different pH values (see Lassen, 1972). This would agree with the contention that the membrane potential under all conditions reflected the Nernst potential for chloride, particularly if the membrane as suggested had a very high permeability to chloride relative to other ions. Using antibiotic ionophores it was reported, however, that the _net transport_ of chloride (as opposed to tracer exchange) occurred at a relatively slow rate (Chappell and Crofts, 1966, Harris and Pressman, 1967). This type of study was performed in a quantitative way by Hunter (1971) and by Tosteson et al. (1972). With due correction for the influence of changes in membrane potential (as determined indirectly) by the antibiotic valinomycin the net transport of chloride was found to be 10^4 - 10^5 times slower than the one-to-one exchange measured with labelled chloride. This suggests that rapid diffusion of chloride across the red cell membrane does not occur as originally indicated, but that the membrane possesses a transport system allowing a rapid and strict one-to-one exchange for chloride. As shown by Gunn et al. (1973) bicarbonate utilizes the same transport system.

But once having accepted the fact that only a minute fraction of chloride passes the red cell membrane by simple diffusion it is unlikely that the membrane resistance can have such low values as mentioned above. The membrane permeabilities for Na and K are low and would account for specific resistances in the order 10^7 Ωcm^2 (Wieth, 1971). Reinvestigation of the membrane resistance and other parameters by physical methods, e.g. intracellular microelectrodes was therefore desirable at this point. The use of intracellular electrodes is problematic because red cells and other cells in suspension do not seal well around the microelectrode tip. Therefore it is preferable to use large cells, where the relative leak imposed on the cell becomes smaller. For this reason Hoffman and Lassen (1971) undertook microelectrode studies of giant red cells from the salamander Amphiuma means which have a surface area of some 30 times that of the human red cell. In these cells the "chloride resistance" calculated from exchange of labelled chloride is about 1 Ωcm^2 and the maximal resistance measured by the microelectrode technique about 2000 Ωcm^2 (see Lassen, 1972). This relatively high value for the membrane resistance confirmed the contention that only a small fraction of chloride passing the membrane could be carrying a net charge. But even these measurements most likely yielded values for the resistance that were too low due to the leak in the membrane around the microelectrode tip. Sucking the ends of single Amphiuma red cells into two thin pipettes and insulating the free part of the cell by oil, Rathlev in our laboratories (unpubl.) was able to make a better estimate of the desired parameter. Depending on the assumptions used for calculation of the data, the minimal specific resistances are 6×10^4 or 2×10^5 Ωcm^2. These figures are in accordance with those obtained by the indirect methods using valinomycin to increase the K-permeability of the human red cell membrane. The immediate consequences of these findings are that less than 10^{-5} of the chloride passes the membrane without the obligatory passage of a monovalent anion (e.g. chloride) in the opposite direction.

The above considerations have emphasized the difficulties in measuring the transmembrane resistance in red cells. Membrane potential measurements are, with some limitations, possible when dealing with the relatively large Amphiuma red cells (see Lassen et al. 1971, 1974). If it is true that chloride does not dominate the conductance of the membrane, even a moderate change of the permeability of other ions like K is expected to change the membrane potential from the normal -14 mV (pH 7.2, 17° C). Table 1 (Rønne, unpubl.) shows the influence of valinomycin on the membrane potential of Amphiuma red cells.

Table 1.

Influence of valinomycin on the membrane potential of Amphiuma red cells. Ringer's solution containing NaCl: 115 mM, KCl: 2.5 mM, $CaCl_2$: 1.8 mM, bovine albumin: 1 g/l, MOPS-buffer: 10 mM.

Valinomycin	Membrane potential
0	-14 mV
1.7×10^{-7} M	-15 -
3.1×10^{-7} -	-18 -
6.3×10^{-7} -	-25 -
1.3×10^{-6} -	-30 -
2.5×10^{-6} -	-39 -
5.0×10^{-6} -	-40 -

As seen from table 1 increasing concentrations of valinomycin result in more negative membrane potentials. Under similar circumstances the K-flux is increased up to 5 times (Hoffman and Lassen, 1971). These observations confirm that the permeability to net-flux of chloride is low and in the same range as the Na and K permeabilities. It is possible to obtain a quantitative measure of the transference number for chloride by varying the extracellular Cl concentration in exchange for a non-permeating anion. In the experiment shown in fig. 1 (Rønne, unpubl.) para-amino-hippurate was used as the impermeant anion (Hoffman and Lassen, 1971, Hoffman and Laris, 1974). The measured membrane potentials are plotted against the chloride concentrations of the medium (log scale). The experiments were performed at 1 mM or 50 mM K in the medium. In the curves marked "0 valinomycin" the potential change for a 10-fold change in chloride concentration is about 20 mV, corresponding to a transference number for chloride of 0.3. It should be noted that the 50-fold change in K-concentration between the two parallel lines (without valinomycin) only has a small influence on the potential at a given chloride concentration. This can readily be explained by the similar permeabilities of Na and K in the unmodified red cell membrane. In the presence of valinomycin, the membrane potential is no longer changed by alterations of chloride contents of the media. In this situation the K-permeability dominates the membrane as demonstrated by a marked depolarization when going from 1 mM to 50 mM K in the medium.

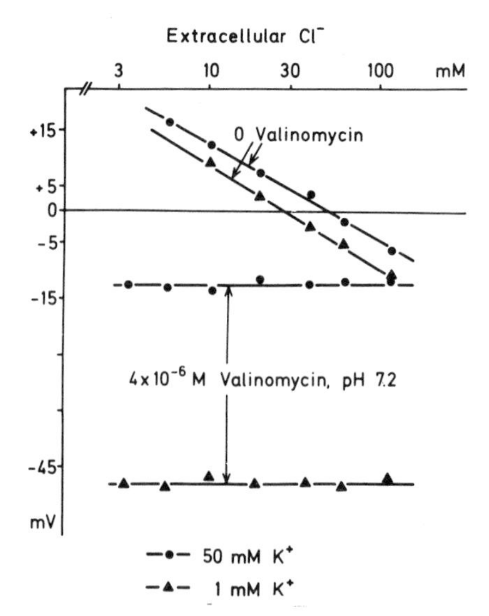

Fig. 1. Influence of external chloride concentration on membrane potentials of Amphiuma erythrocytes. Abscissa: Cl-concentration in mM (log scale), ordinate: membrane potential in mV. The experiments were performed at 1 or 50 mM K (K replacing Na in the medium) and ± 4 x 10^{-6} M valinomycin. pH 7.2, 17° C.

The above experimental evidence leaves us with a picture in which the membrane conductances of the red cell (as exemplified by the Amphiuma erythrocyte) to K, Na and Cl are about equal. This is experimentally a fortunate situation because absolute or relative changes in the permeability of one of these ions can then be detected as changes in membrane potential. The remainder of this contribution will be concerned with induced changes in the membrane potential.

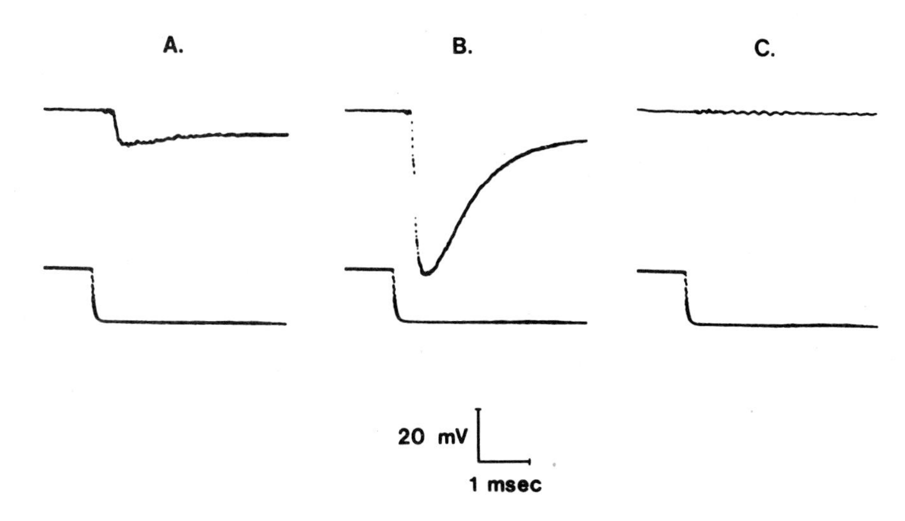

Fig. 2. Oscilloscope tracings from micropuncture of an Amphiuma erythrocyte. Upper traces show the potentials recorded by the microelectrode, lower traces monitor the high voltage pulse that via a piezoelectric "gun" causes advancement of the electrode. Frame A: first puncture of the cell membrane, frame B: second puncture of the same cell (30 sec after first puncture), frame C: advancement of electrode without contact to a cell. Abscissa: bar indicates 1 msec, ordinate: bar indicates 20 mV (upper traces). pH 7.2, 17° C. (Lassen et al., 1974).

During investigations of the membrane potential of Amphiuma erythrocytes it was observed that the normal potential of about -15 mV had changed to much more negative values if a cell was micropunctured, the electrode withdrawn after about one sec, and the cell repunctured a second time after 30 sec. This sequence of events is demonstrated in fig. 2. The three sets of oscilloscope tracings (A - C) show the recorded potentials (upper traces) and the high voltage pulses for the electromechanical transducer used to advance the microelectrode (lower traces). In frame A is shown the typical time course of the potential changes obtained upon penetration of a cell membrane. Following an abrupt drop, the potential reaches a negative peak value which with some precautions can be taken as the membrane potential of the intact cell. From this value the potential decays to a stable level that is determined by a complex diffusion zone around the tip of the electrode. Between frame A and frame B, the electrode had been withdrawn from the cell for nearly 30 sec and on the second puncture (same cell) the membrane potential, as indicated by the value of the negative peak in upper trace in frame B, is now close to -60 mV. Frame C shows that potential changes are negligible if the electrode tip is advanced without contact to a cell. In the course of an additional 2 minutes the potentials were back to the original value, about -15 mV.

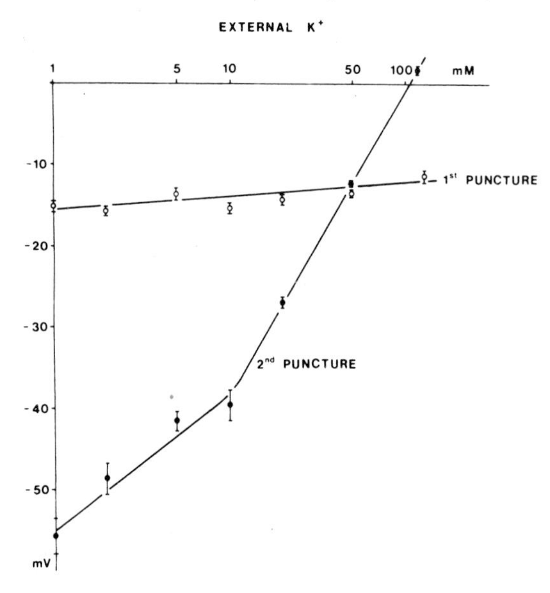

Fig. 3. Membrane potential as a function of the logarithm of external K concentration (K plus Na in medium kept constant at 120 meq/l). The initial membrane potential of the cells are shown on the curve marked "1st puncture". Repeated puncture of the cells 30 sec after the first puncture revealed hyperpolarization at the lower K concentrations (2nd puncture) that decreased as higher K concentrations were employed. Abscissa: log of K concentration of medium (mM), ordinate: membrane potential in mV. pH 7.2, 17° C (Lassen et al., 1974).

A closer study of the time course of the hyperpolarization following micropuncture revealed that the most negative potentials were obtained 30 - 40 sec after withdrawal of the electrode at the first micropuncture (Lassen et al., 1974). This highly reproducible phenomenon was independent of pump-poisons like ouabain and metabolic inhibitors. Thus a change in passive permeability of the membrane was a likely explanation for a change in the diffusion potential across the membrane. Of the major ions involved, only K has an equilibrium potential that is more negative than the membrane potentials observed during the transient hyperpolarization. For this reason the K-dependence of the normal membrane potential and the potential on repuncture after 30 sec was investigated. In fig. 3 the curve marked "Ist puncture" shows that the membrane potential of the unperturbed cell is nearly uninfluenced by the external K over a concentration range of more than two decades. In contrast to this, the potential on the second puncture (30 sec after the first on the individual cells) is highly K dependent: with increasing K concentration the membrane is progressively depolarized. The maximal slope corresponds to a transference number for K of 0.6. This means that 60 per cent of the current across the membrane is carried by K under these circumstances.

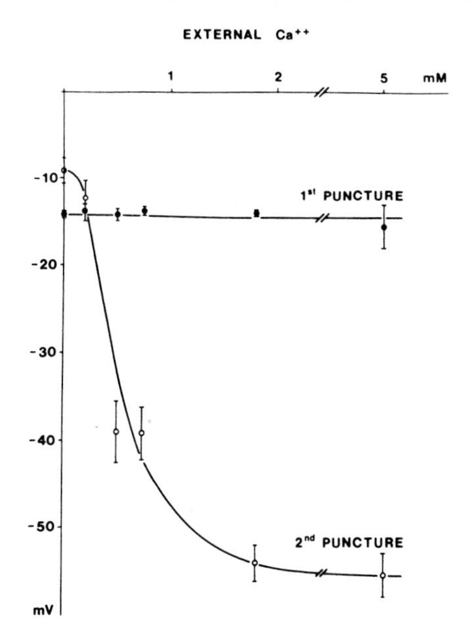

Fig. 4. Effect of external Ca concentrations on initial membrane potentials (Ist puncture) and potentials recorded 30 sec after the initial micropuncture (2nd puncture). Abscissa: Ca concentration of medium in mM, ordinate: membrane potential in mV. pH 7.2, 17° C (Lassen et al., 1974).

At this point we ran into the difficulty of explaining the surprising change in K permeability induced by a preceding micropuncture. As it is not possible to measure membrane conductance under these conditions, it is theoretically possible that the first micropuncture by some unknown mechanism reduced the Na and Cl permeabilities, leaving K as the main permeant ion. Another mechanism is more likely, however. In 1940 Wilbrandt demonstrated that fluoride poisoning led to a massive K loss from human red cell. Gárdos (1958) made the key observation that the K loss from poisoned red cells only took place if there was Ca ions present in the medium.

Blum and Hoffman (1972) and Riordan and Passow (1971) demonstrated that induction of a high K permeability required Ca on the inside of the membrane. The existence of this so-called Gárdos effect led us to investigate the influence of Ca in the medium on the hyperpolarization. In fig. 4 it is demonstrated that hyperpolarization is only observed when the Ca concentration of the medium is above 0.5 mM. This effect of Ca becomes maximal in a concentration range of 2 - 5 mM. The micropuncture induced hyperpolarization of the Amphiuma erythrocyte may be related to entry of Ca ions through the hole in the membrane caused by penetration of the electrode. This hole remains electrically open for about 5 sec (Lassen et al., 1974). During this interval of time the large inward gradient of the Ca concentration will undoubtedly lead to influx of Ca. Experiments with direct injection of CaCl into the cells support this explanation. Thus the presence of a Gárdos effect (Ca-induced K-loss) is not limited to strongly poisoned red cells but can be evoked in normal cells if their internal Ca concentration is raised.

In the above mentioned studies mechanical damage to the cell membrane was a prerequisite for eliciting hyperpolarization. It might be speculated that the newly formed membrane covering the hole caused by the initial puncture had a high K permeability which for a short period dominated the total conductance of the membrane. In the experiments to be mentioned below, the goal was to cause Ca entry into the cells by a sudden rise in extracellular Ca concentration. Fig. 5 shows results from an experiment where the cells were transferred from Ringers with the normal 1.8 mM Ca to one containing 15 mM Ca. The diagram shows the distribution of the measured membrane potentials in intervals of 100 sec from the moment the cells were subjected to the high Ca concentration. In the first of the time intervals there is a clear tendency for more than half of the population of cells to have membrane potentials in the range -40 to -60 mV. Another frequently occurring potential was in the normal potential range of -10 to -20 mV. In the next intervals of 100 sec, there was still a marked tendency for the potentials to be distributed in the same two areas, but with increasing incubation time in the 15 mM Ca medium more and more cells resumed the normal potential. From this and a large number of similar experiments we conclude that it is possible to evoke a hyperpolarization of the Amphiuma red cell membrane by suspending the cells at a higher Ca concentration.

Despite continued presence of the high Ca concentration the hyperpolarization is transient. The appearance of the frequency distribution showing two clear maxima, one in the normal range and one in the range -40 to -60 mV with relatively few in between gives independent support to the notion (see Riordan and Passow, 1973) that the effect of Ca on the permeability of red cells is of an all or non nature. The disappearance of the hyperpolarization despite the maintained presence of Ca resembles the effect of lead on poisoned human red cells. The increased permeability to K disappears during prolonged incubation in lead containing media (Passow and

Tillman, 1955, Riordan and Passow, 1973). In contrast to these lead experiments, however, the Ca induced hyperpolarization of Amphiuma red cells can be repeated after washing the cells in Ringer with a normal concentration.

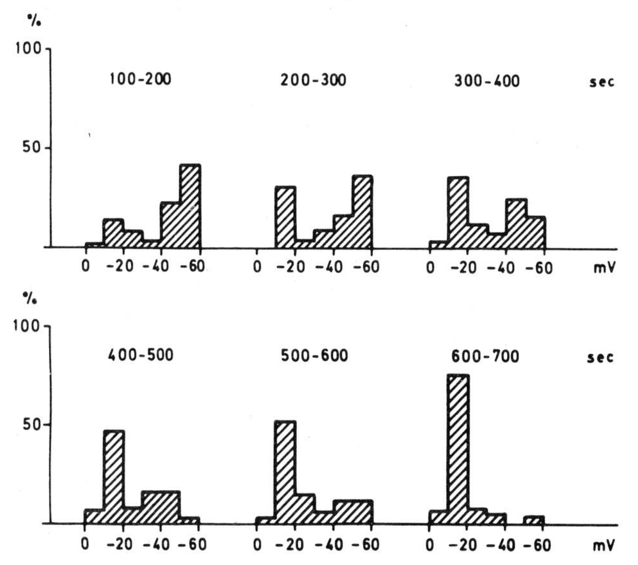

Fig. 5. Frequency of membrane potentials obtained in Amphiuma red cells after suspension in Ringers with 15 mM Ca. The individual cells were only punctured once, and the values thus are representative of membrane potentials of mechanically unperturbed cells. Abscissae: membrane potentials in mV, ordinates: frequency of individual potential values in intervals of 10 mV. The numbers on top of each histogram indicate the time (in sec) after transfer of the cells from a medium containing the normal 1.8 mM Ca to one containing 15 mM Ca.

Following the arguments presented in the case of hyperpolarization after micropuncture, it is likely that the 15 mM Ca induced hyperpolarization is related to a rise in K permeability. An experimental verification of this assumption is shown in fig. 6, where the mean of potentials measured within the first 4 minutes after suspension of Amphiuma erythrocytes in 15 mM Ca medium is plotted against the log of the K concentration of the extracellular phase. From the slope of the line it can be concluded that the hyperpolarization as expected is caused by a relative increase in K permeability. The slope of about 40 mV per 10-fold change in K concentration corresponds to a transference number for K of 0.6. This is a minimal figure, because the mean of the potentials includes values from a fraction of the cells that had normal potentials (-15 mV) (compare fig. 5, first histogram).

This time-dependent induced hyperpolarization has some

resemblance to time-dependent phenomena in excitable tissues. In such systems there is usually a threshold value for the perturbation needed to provoke a response. It was therefore of interest to determine the Ca concentration dependence of the induced hyperpolarization of the _Amphiuma_ red cell membrane. Fig. 7 shows the membrane potentials obtained within 3 min after transfer of the cells from the normal medium containing 1.8 mM Ca to media with the Ca concentrations indicated on the abscissa. From the figure it can be seen that hyperpolarizations occur if the Ca concentration exceeds a "threshold value" of 6 mM. At even higher Ca concentrations, the relative number of hyperpolarized cells

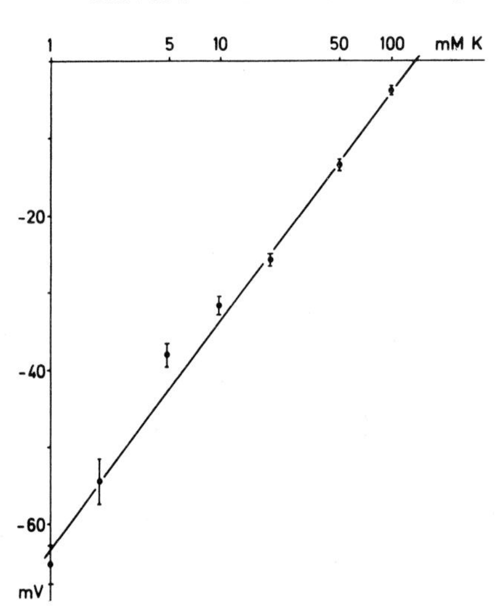

Fig. 6. Hyperpolarization of the _Amphiuma_ red cell membrane following suspension in 15 mM Ca medium. The graph shows the mean of the potentials measured less than 4 minutes after suspension in 15 mM Ca medium with different K concentrations. Abscissa: K concentration in mM (log scale), ordinate: membrane potential in mV.

increases, but there is a clear tendency for the measured potentials to be distributed either at the normal potential of less than -20 or at potentials of -40 to -70 mV. Thus in this set of experiments there is also an indication of an an all or none phenomenon.

It was obvious to ask whether _Amphiuma_ erythrocytes would show a Gárdos phenomenon if depleted of ATP after metabolic poisoning. In collaboration with Professor Gárdos we attempted to poison these red cells with a number of known poisons of glycolysis and respiration. It was not possible to reduce the level of intracellular ATP to the very low levels that are needed in human red cells for attainment of a Gárdos effect. As an alternative we attempted to potentiate the effect of raised Ca concentration by adding histamine to the medium. This substance has been shown by Passow (1963) and by Gárdos and Szász (1968) to enhance the Ca induced K loss. This would, in the case of membrane potentials in the Amphiuma red cell, correspond to a larger hyperpolarization and/or a larger fraction of the cell population being hyperpolarized. Fig. 8 shows that 5 mM histamine has little or no effect on the hyperpolarization induced by 15 mM Ca. If anything there is a small (hardly significant) decrease in the fraction of hyperpolarized cells.

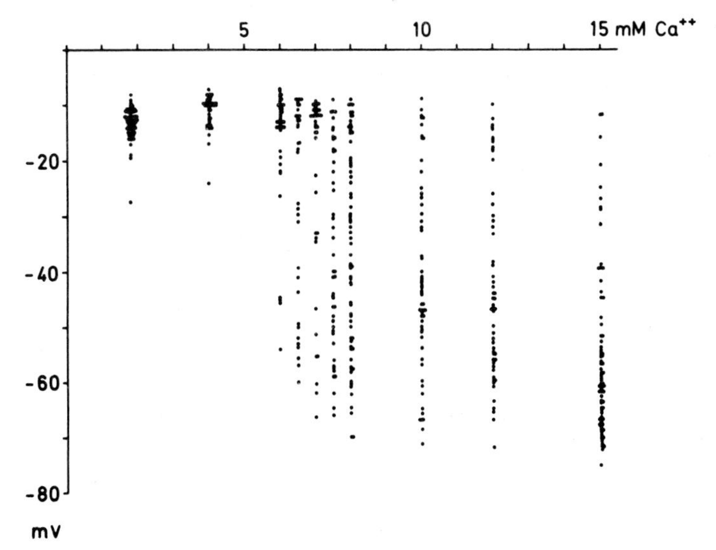

Fig. 7. Membrane potentials of Amphiuma red cells as a function of Ca concentration of the medium. Abscissa: Ca concentration of final suspension medium (mM), ordinate: individual membrane potentials recorded (mV). Data from initial 4 min after transfer from 1.8 mM Ca medium to media with the Ca concentrations indicated on the abscissa.

Even though the effect of histamine in the employed concentration was dubious, we found it of value to test the effect of antihistamines on the Ca induced hyperpolarization of the Amphiuma red cell. Gárdos and Szász (1968) showed that a typical antihistamine, promethazine, could prevent the Ca induced K loss in poisoned human red cells. As demonstrated in fig. 9, a very low concentration of promethazine (10 μM) almost completely prevented the Ca induced hyperpolarization. In concentrations of promethazine close to 1 mM, the cells appeared deformed, acquiring a bizarre saddle shape. No morphological changes were observed with low concentrations of promethazine or with 5 mM histamine. The typical antihistamines are substituted ethylamines. Another of this type of drugs is mepyramine, the structure of which is shown in fig. 10. This substance has an effect on the Amphiuma red cell which is similar to that reported for promethazine.

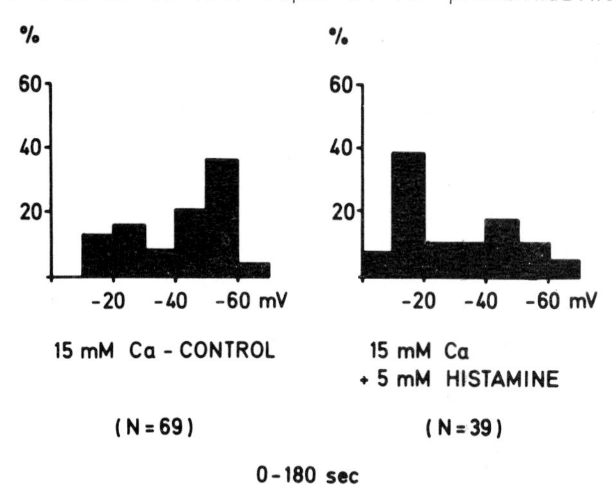

Fig. 8. Effect of histamine on the hyperpolarization of Amphiuma erythrocytes caused by suspension in 15 mM Ca medium. Abscissae: membrane potential in mV, ordinates: per cent of observations in each group of 10 mV. Data from the first three min after suspension in 15 mM Ca.

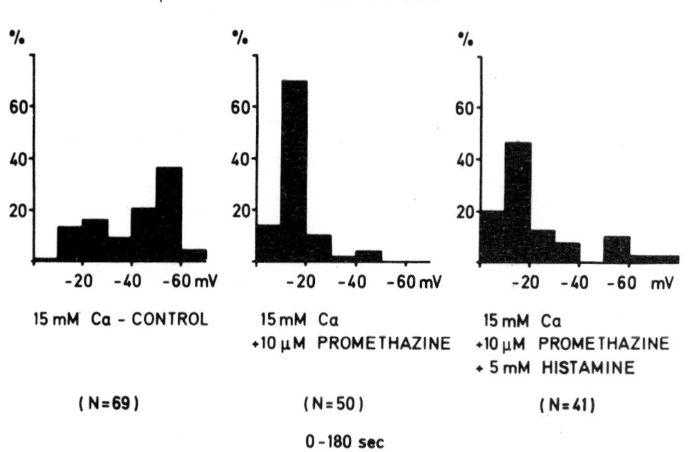

Fig. 9. Effect of promethazine and of promethazine + histamine on the Ca induced hyperpolarization of the Amphiuma red cell. As in the previous figure, the cells were suspended in medium containing 15 mM Ca and the potentials recorded within the first 3 min of incubation. Abscissae: membrane potentials in mV, ordinates: per cent of observations in each group of 10 mV.

It is noteworthy that both of the tested antihistamines had a marked inhibitory effect on the Ca induced hyperpolarization whereas histamine by itself did not enhance the effect of Ca. It was therefore important to see if histamine could prevent the inhibitory action of promethazine. The results of such an experiment is shown in the right hand frame of fig. 9. Again the effect of adding 5 mM histamine is hardly visible. This led us to consider the possibility that the effect of promethazine was different from its action as an antihistamine. Promethazine shares its phenothiazine structure with a number of other pharmacologic agents, the most well known being chlorpromazine (an antipsychotic drug, for chemical structure see fig. 11) (see Goodman and Gilman, 1970).

HISTAMINE

PROMETHAZINE

MEPYRAMINE (=PYRILAMINE)

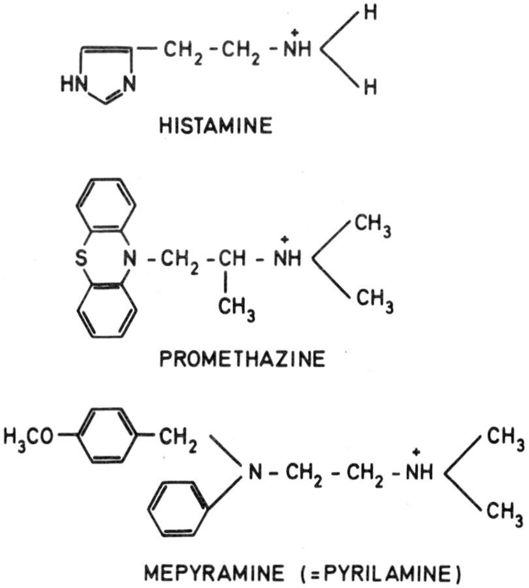

Fig. 10. Structure of histamine and two potent antihistamines, promethazine and mepyramine.

Although chlorpromazine has at best a weak antihistamine action, it nevertheless shares a number of pharmacologic effects with promethazine and the commonly used local anesthetic amines (see Goodman and Gilman, 1970, Seeman, 1972). When employed in the present system, chlorpromazine in a low concentration (10 μM) completely blocked the hyperpolarization after suspension in 15 mM Ca medium (fig. 12). In fig. 12 it should be noted that the hyperpolarization is very marked in the control experiment. Even under apparently similar experimental conditions the hyperpolarization after suspension in 15 mM Ca is of different duration. Thus the time needed for reversal of the potentials to normal values vary from 3 to 15 min.

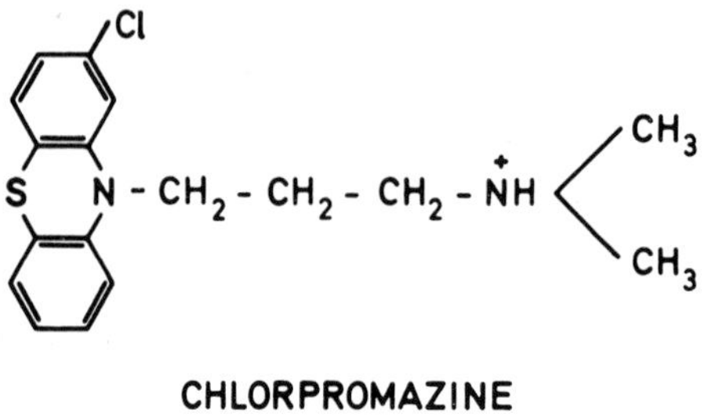

CHLORPROMAZINE

Fig. 11. Structure of chlorpromazine. Note similarity to the structure of promethazine (fig. 10).

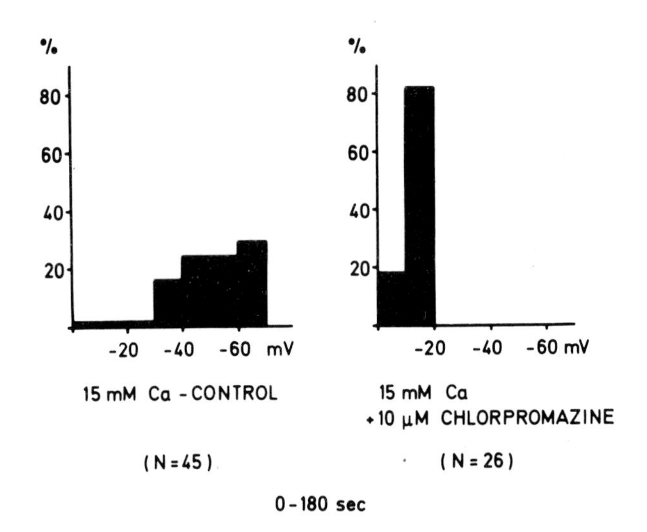

Fig. 12. Effect of chlorpromazine on the hyperpolarization following suspension of Amphiuma red cells in 15 mM Ca medium. Abscissae: membrane potentials in mV, ordinates: per cent of observations in each group of 10 mV. Data from the first three min after suspension in 15 mM Ca.

The effect of Ca on K permeability in red cells seems to be related to the level of intracellular Ca (Blum and Hoffman, 1972, Romero and Whittam, 1971). This is confirmed in the present experiments, where a mechanical damage of short duration of the cell membrane allows entry of Ca with a resulting transient K conductance increase. The normal permeability of the membrane to K is thus regulated indirectly via the efficiency of the ATP driven transport of Ca out of the cell (Schatzmann and Vincenzi, 1969). Most of the Ca bound to red cell membranes is bound to protein (Forstner and Manery, 1971). It is possible that this binding changes the protein conformation in the membrane, thus directly modifying the K permeability. It is also of interest that of the 16 per cent of

Ca bound to the lipid fraction of ghosts, the majority is associated with the phospholipid fraction (Forstner and Manery, 1971) of which especially phosphatidylserine is located mainly on the inside of the membrane. If the lipid bound Ca in the inward facing side of the membrane is involved in regulation of the K permeability, it is of interest that the hyperpolarization of the membrane will cause a driving force for the negatively charged phosphatidylserine to "jump" from the inner leaflet of the lipid bilayer to the outer. Such a model has been proposed by Kornberg and McConnell (1971) who coined the phrase "phospholipid flip-flop". The importance of the mechanism was questioned by the same authors in the case of lipid vesicle membranes because of the extremely slow turnover (0.04 hr^{-1}). However, McNamee and McConnell (1973) report half-times for the flip-flop of 3.8 to 7 min at 15° C when studying vesicles from the *Electrophorus* electric organ. McLaughlin and Harary (1974) emphasize the importance of distribution of surface charges for the field inside the membrane and thus for the eventual operation of a phospholipid flip-flop. Despite the seductive similarity of the half-times in the study of McNamee and McConnell to the time intervals needed for disappearance of the hyperpolarization of the *Amphiuma* red cell in 15 mM Ca, it is obviously too early to draw parallels to the Ca "inactivation". But studies along these lines may prove useful, especially as the necessary experimental techniques are available.

The effect of chlorpromazine and the related promethazine probably has a connection to the binding of anesthetic amines to biomembranes where two amine molecules seem to compete with and displace one Ca^{++} from fixed negative sites of the membrane resulting in an increasing disordering of the membrane lipids (see review by Seeman, 1972). These amines may thus block entry of Ca into the cells and prevent reaching of the K permeability regulating sites on the inside of the membrane. However, this may not be the only explanation since promethazine in concentrations of 50 µM can partially prevent micropuncture induced hyperpolarization (Gárdos, Lassen and Pape, unpubl.). In this situation Ca has access to the cell interior despite an eventual block of the transmembrane pathway. The general interest in these substances as well as the low concentrations needed to evoke strong effects makes it desirable to pursue this line of experimentation.

ACKNOWLEDGEMENTS

The research was supported by the Danish Science Research Council (511-1704-72, 511-1305-72, 511-2650-73). Skillful technical assistance by Kirsten Abel and Annelise Honig and expert assistance in preparation of the manuscript by Elisabet Krenchel and Villy Rasmussen is gratefully acknowledged.

REFERENCES

Andreoli, T.E., Bangham, J.A. and Tosteson, D.C. (1967). J. Gen. Physiol. 50, 1729.

Bangham, A.D. (1972). Ann. Rev. Biochem. 41, 753.

Blum, R.M. and Hoffman, J.F. (1972). Biochem. Biophys. Res. Comm. 46, 1146.
Chang, Y.C. (1964). Rev. de Biol. (Lisboa) 5, 119.
Chappell, J.B. and Crofts, A.R. (1966). In Regulation of Metabolic Processes in Mitochondria. Tager, J.M., Papa, S. Quagliariello, E. and Slater, E.C. eds. B.B.A Library, 7, 293.
Dalmark, M. and Wieth, J.O. (1970). Biochim. Biophys. Acta, 219, 525.
Forstner, J. and Manery, J.F. (1971). Biochem. J. (Lond.). 124, 563.
Fricke, H. (1926). J. Gen. Physiol. 9, 137.
Gárdos, G. (1958). Biochim. Biophys. Acta. 30, 653.
Gárdos, G. and Szász, I. (1968). Acta Biochim. Biophys. Acad. Sci. Hung. 3, 13.
Goodman, L.S. and Gilman, A. (1970). The Pharmacologic Basis of Therapeutics (4' ed.). McMillan Co., N.Y. 156 and 635.
Gunn, R.B., Dalmark, M., Tosteson, D.C. and Wieth, J.O. (1973). J. Gen. Physiol. 61, 185.
Harris, E.J. and Pressman, B.C. (1967). Nature, 216, 918.
Hoffman, J.F. and Lassen, U.V. (1971). Abstract, XXV Int. Congr. Physiol. Sci. (München).
Hunter, M.J. (1971). J. Physiol. (Lond.) 218, 49P.
Johnson, S.L. and Woodbury, J.W. (1964). J. Gen. Physiol. 47, 827.
Kaplan, D.M. and Criddle, R.S. (1971). Physiol. Rev. 51, 249.
Katz, B. (1966). Nerve, Muscle and Synapse. McGraw-Hill, N.Y., 47.
Kornberg, R.D. and McConnell, H.M. (1971). Biochemistry 10, 1111.
Lassen, U.V. (1972). In Oxygen Affinity of Hemoglobin and Red Cell Acid Base Status. Rørth, M. and Astrup, P. eds. Munksgaard, Copenhagen. 291.
Lassen, U.V., Nielsen, A.-M.T., Pape, L. and Simonsen, L.O. (1972). J. Membrane Biol. 6, 269.
Lassen, U.V., Pape, L. and Vestergaard-Bogind, B. (1974) J. Membrane Biol. in press.
Lassen, U.V. and Sten-Knudsen, O. (1968). J. Physiol. (Lond.) 195, 681.
McLaughlin, S. and Harary, H. (1974). Biophys. J. 14, 200.
McNamee, M.G. and McConnell, H.M. (1973). Biochemistry 12, 2951.
Passow, H. (1963). In Cell Interface Reactions. Brown, H.D., ed. Scholar's Library, N.Y. 57.
Passow, H. and Tillmann, K. (1955). Pflügers Archiv. 262, 23.
Riordan, J.R. and Passow, H. (1971). Biochim. Biophys. Acta 249, 601.
Riordan, J.R. and Passow, H. (1973). Comparative Physiology Bolis, L., Schmidt-Nielsen, K. and Maddrell, S.H.P. eds. North-Holland Publishing Co. 543.
Romero, P.J. and Whittam, R.R. (1971). J. Physiol. (Lond.) 214, 481.
Schatzmann, H.J. and Vincenzi, F.F. (1969). J. Physiol. (Lond.) 201, 369.
Seeman, P. (1972). Pharmacol. Rev. 24, 583.
Tosteson, D.C. (1959). Acta Physiol. Scand. 46, 19.
Tosteson, D.C., Gunn, R.B. and Wieth, J.O. (1972). In Erythrocytes, Thrombocytes, Leukocytes, G. Thieme Publ., Stuttg., 62.
Wieth, J.O. (1971) Erythrocyters Selektive Ionpermeabilitet. Dissertation, FADL's Forlag, Copenhagen. 124.

MEMBRANE PROTEINS AND ANION EXCHANGE IN HUMAN ERYTHROCYTES

H.PASSOW, H.FASOLD, L.ZAKI, B.SCHUHMANN and S.LEPKE

Max-Planck-Institut für Biophysik and Biochemisches Institut der Universität, Frankfurt/Main

The present talk is concerned with attempts to identify membrane constituents which participate in the control of anion movements across the red cell membrane and to explore the role of such constituents in the molecular mechanism of anion translocation.

The obvious approach for the identification of anion control sites would consist of labeling with a specific modifier, isolating the labeled membrane constituent, and establishing the relationship between binding of the modifier and effect on anion permeability. An attempt with this technique has recently been made by Cabantchik and Rothstein (1974).

Our own attempt (Zaki and Passow, 1973) was made with a different technique. We labeled the erythrocyte membrane with a rather non-specific modifier, dinitrofluorobenzene (DNFB),and tried to identify a membrane constituent whose reactions with this non-specific modifier could be prevented by pretreatment with a more specific modifier, 4-isothiocyano-4'-acetamido stilbene-2,2'-disulfonic acid (SITS). This technique is rather indirect and its quantitative application involves inaccuracies due to difference formation. However, as will be shown below, it has advantages in the study of the function of the anion control sites.

Fig. 1 a depicts the modifiers which we used as tools for the identification of anion control sites. DNFB is a non-

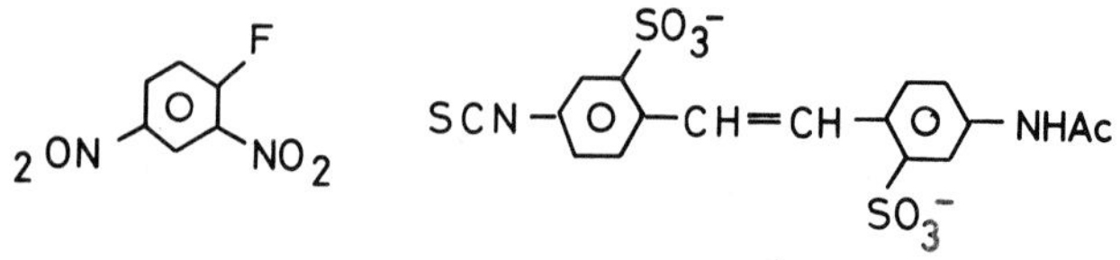

DNFB SITS

Fig. 1 a

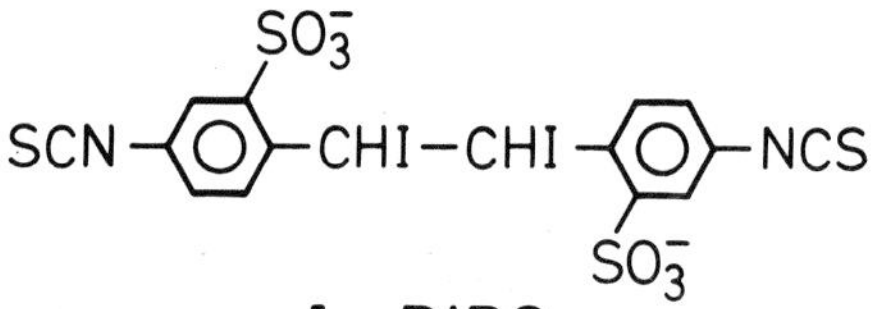

I_2 DIDS

Fig. 1 b

specific irreversibly acting amino reagent which penetrates rapidly across the red blood cell membrane and inhibits anion exchange (Passow, 1969, Poensgen and Passow, 1971). SITS is a disulfonic acid whose anionic sites combine reversibly with positively charged groups of proper spacing. Its SCN group reacts irreversibly with amino groups. SITS does not pene-

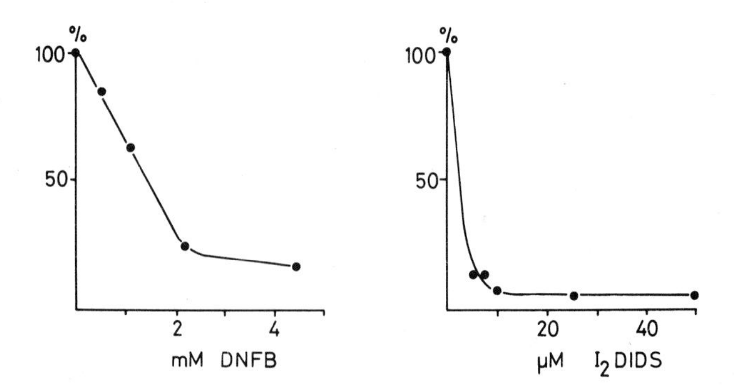

Figure 2 Effects of varying concentrations of DNFB and I_2DIDS on sulfate equilibrium exchange in intact erythrocytes. Ordinate: rate constant in percent of control value without inhibitor.

trate (Maddy, 1964). SITS itself as well as many of its derivatives are highly specific inhibitors of anion exchange (Knauf and Rothstein (1971). Fig. 2 shows the effects on sulfate equilibrium exchange of dinitrofluorbenzene and a derivative of a stilbene disulfonic acid (I_2DIDS, see fig. 1 b) which carries two isothiocyanate groups.

The SDS polyacrylamide gel electropherogram of the red blood cell membrane in fig. 3 demonstrates possible candidates for anion control sites. The peak at the lower end of the gel represents the lipids. Lipids do not seem to play a major role in facilitating anion exchange since anion movements can be inhibited by maleylation without significant modification of membrane lipids (Obaid, Rega and Garrahan, 1972) and since the proteolytic enzyme pronase inhibits anion exchange without producing detectable hydrolysis of amino lipids in the membrane (Passow, 1971). It is most likely therefore that a

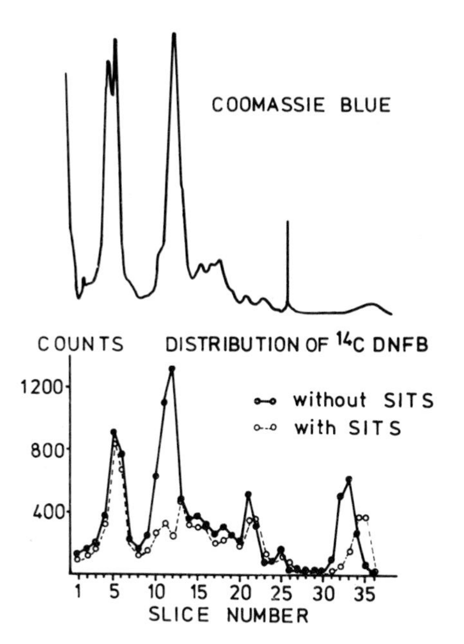

Fig. 3

Distribution of ^{14}C DNP residues (lower tracings) on polyacrylamide gel electropherograms of red cell membranes isolated from cells which had been dinitrophenylated in the presence or absence of SITS. The distribution of the Coomassie blue stain (upper tracing) is independent of the presence or absence of SITS.

protein is involved in anion control.

Fig. 3 also shows the labeling pattern after exposure to ^{14}C DNFB (6o' at 37°C, pH 7.4) as observed in membranes derived from untreated red blood cells and from red blood cells which had been exposed to SITS (25' at 37°C, pH 7.4) prior to dinitrophenylation. Among the major membrane proteins only those in band 3 (nomenclature of Steck 1974) are protected against dinitrophenylation by SITS. Since the proteins in this band are the only ones at which an interaction between two different inhibitors of anion exchange can be observed we conclude that they are possibly involved in anion control. Under the conditions employed for the separation of the proteins on the gel in fig. 3, band 3 contains proteins of a molecular weight of about 95.ooo Daltons (as determined at the front of the band) and a glycoprotein of smaller molecular weight. Separation of the glycoprotein from the other proteins by the technique of Hubard and Cohn (1972) revealed that the 95.ooo Dalton proteins rather than the glycoprotein are protected against dinitrophenylation by SITS.

A comparison of DNFB binding to SITS protectable DNFB binding sites with the inhibition of sulfate equilibrium exchange (fig. 4) shows that binding has the tendency to continue after anion exchange is maximally inhibited. This finding is supported by additional experiments in which the saturation of the SITS protectable binding sites by non-radioactive DNFB was tested by the addition of small amounts of ^{14}C DNFB. This method does not allow the calculation of numbers of binding sites; but it is a sensitive indicator of saturation. The uppermost curve in fig. 4 shows that pretreatment with increasing concentrations of non-radioactive DNFB reduces the binding of a subsequently added small amount of ^{14}C DNFB. However, even after pretreatment with DNFB concentrations which far exceed those necessary to produce maximal inhibition of anion exchange there is still a fixation of ^{14}C DNFB. This finding confirms that there are more SITS protectable DNFB binding sites than are involved in the control of anion exchange.

Anion exchange is maximally inhibited when o.8 - 1.2 · 10^6 sites are dinitrophenylated at the proteins in band 3 of each single cell. Since each SITS molecule can protect up to three DNFB binding sites this corresponds to 26o.ooo - 4oo.ooo SITS binding sites. There is no evidence for a cross linking of band 3 protein molecules by SITS. Hence these figures represent the upper limit of the number of protein molecules in band 3 which could possibly participate in the control of anion permeability. It is significantly smaller than the total number of molecules/cell in this band. According to Steck (1974) this number is o.9 - 1.o · 10^6. Hence if SITS protectable DNFB binding sites should actually comprise the anion control sites, the separation of the labeled protein species from the other proteins in band 3 has still to be accomplished.

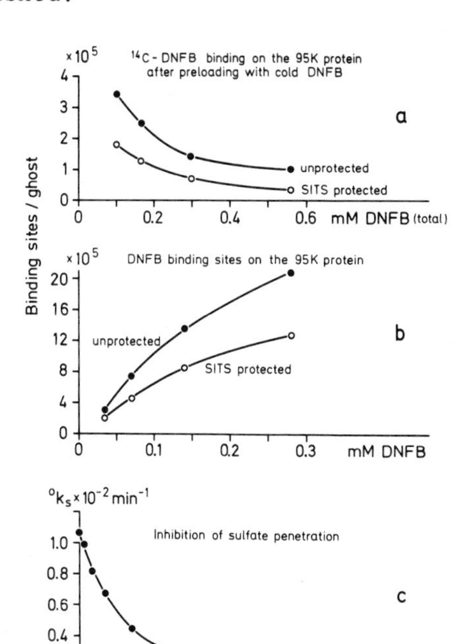

Fig. 4

Inhibition of sulfate equilibrium exchange and binding of ^{14}C DNFB to resealed red cell ghosts after dinitrophenylation at 37°C for one hour at the concentration indicated on the abscissa. The ordinate of fig.(c) represents the rate constant of sulfate exchange as measured at 3o°C

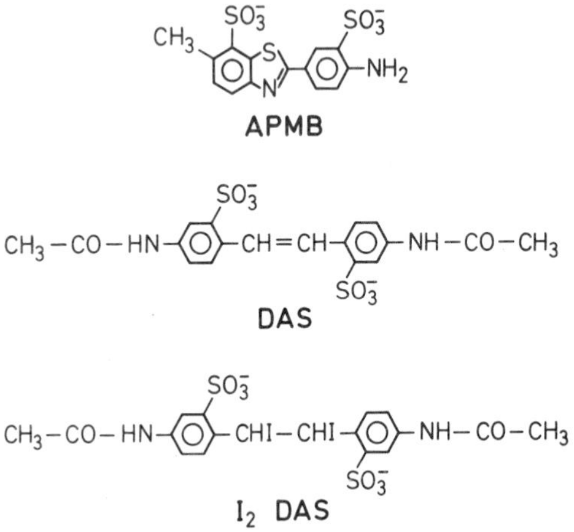

APMB

DAS

I_2 DAS

Fig. 5

In the previous estimate of the maximal number of anion control sites it was assumed that in addition to the SCN group the sulfonyl groups of SITS are also capable of protecting the anion control sites against dinitrophenylation. A direct confirmation of this inference can be obtained by studying the capacity of reversibly acting disulfonic acids on DNFB binding. Fig. 5 shows the structure of some disulfonic acids which have been used as tools in this type of work. Fig. 6 demonstrates that the inhibition produced by these agents disappears completely after washing the cells. Dinitrophenylation in the presence of such agents leads in fact to a protection of the DNFB binding sites in band 3 of the SDS polyacrylamide gel electropherograms. This protection is nearly as complete as with irreversible inhibitors like I_2DIDS (table 1). At the same time the inhibition of anion exchange by DNFB is at least partially prevented (fig. 7). Obviously the interactions between the reversibly inhibiting disulfonic acids and the anion control sites are strong enough to decrease the rate of dinitrophenylation to such an extent that a considerable protection of the anion exchange mechanism is afforded. This result adds further substance to the claim that some band 3 proteins are involved in anion transfer. They also show that the charged groups which interact with the disulfonic acids cannot be guanidino groups but must be groups which are susceptible to alkylation by DNFB. They support our previous claim (e.g. Schnell and Passow, 1969) that amino groups control anion exchange.

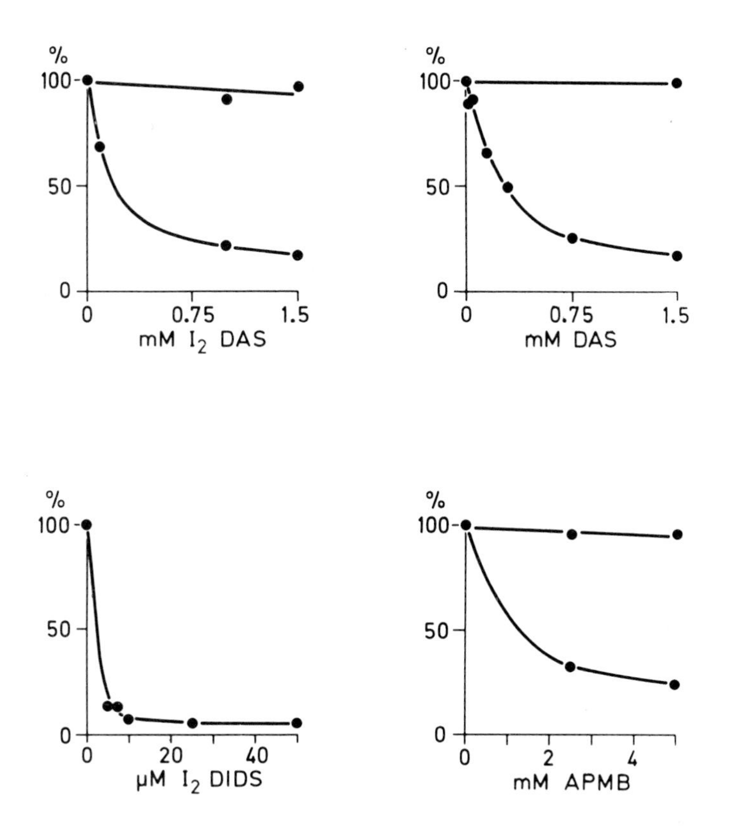

Fig. 6 Effects of varying concentrations of various disulfonic acids on sulfate equilibrium exchange in intact erythrocytes. Ordinate: rate constant in percent of control value without inhibitor. pH: 7.1, temperature: 30°C. The upper curves represent rate constants after removal of the inhibitor by washing.

Table 1

Effects of 4-4'-diacetamido-2-2'-stilbene disulfonic acid (DAS) and 4-4'-diisothiocyano-2-2'-diiodo stilbene disulfonic acid (I_2DIDS) on ^{14}C DNFB binding to resealed type II ghosts from human red cells

	number of DNFB binding sites on 95 K protein	number of protected DNFB binding sites on 95 K protein
control	$23.4 \cdot 10^5$	-
1.5 mM DAS	$13.4 \cdot 10^5$	$10.0 \cdot 10^5$
25 μM I_2DIDS	$12.7 \cdot 10^5$	$11.7 \cdot 10^5$
50 μM I_2DIDS	$10.2 \cdot 10^5$	$13.2 \cdot 10^5$

DNFB concentration in the medium 0.28 mM

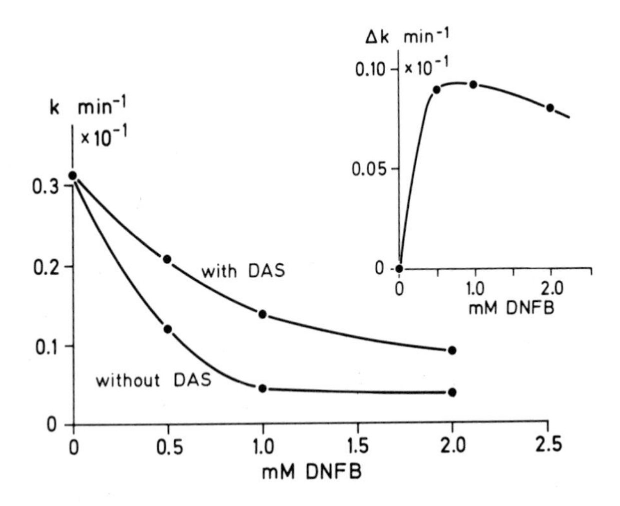

Fig. 7 Protection of anion equilibrium exchange against the effect of dinitrophenylation at the DNFB concentrations indicated on the abscissa by the presence of DAS. "With DAS" and "without DAS": dinitrophenylation in the presence or absence of 1.5 mM DAS, respectively. Inset: Difference of rate constants as measured in the presence and absence of DAS.

A further indication for a participation of the SITS protectable sites in anion control comes from experiments on the action of trypsin.

External trypsin has no detectable effect on the location of the band 3 proteins in SDS polyacrylamide gel electropherograms. Internal trypsin digests spectrin and most other membrane proteins. After lysis of the trypsin-loaded ghosts in trypsin inhibitor containing media and subsequent washings, the degradation products of most proteins are released and no longer detectable on polyacrylamide gels (see fig. 8).However, the band 3 proteins are retained and found to be split into two fragments of molecular weights of 58.ooo ± 1.6oo and 48.ooo ± 2.2oo (standard errors of the means, 5 determinations each). The sum is close to the mean molecular weight of the band 3 proteins which we find to be 1o6.ooo ± 1.5oo Daltons. This value for the peak of the stained band is higher than that reported by Cabantchik and Rothstein (1974), but still within the range to which the tail of the band 3 proteins extends in the experiments of Steck (1974). Besides these two fragments and some undigested band 3 protein, we find one additional major band at 34.7oo ± 8oo Daltons, and a minor band close to the bromophenol blue front. The origin of these bands is unclear. The described observations with resealed ghosts support similar observations obtained by Steck (1972) with inside out vesicles. They indicate that after digestion of the proteins at the inner membrane surface by trypsin one is left with vesicles which resemble liposomes made from red cell lipids containing in the bilayer a few membrane proteins, including fragments of band 3 proteins.

A study of the effect of internal trypsin on the SITS protectable DNFB binding sites shows that their location on SDS polyacrylamide electropherograms is shifted from their original position in the single band at 1o6.ooo Daltons to the two bands at 58.ooo and 48.ooo Daltons. Each of the new bands carries about half of the SITS protectable DNFB binding sites (fig. 8). These findings indicate that the anion control

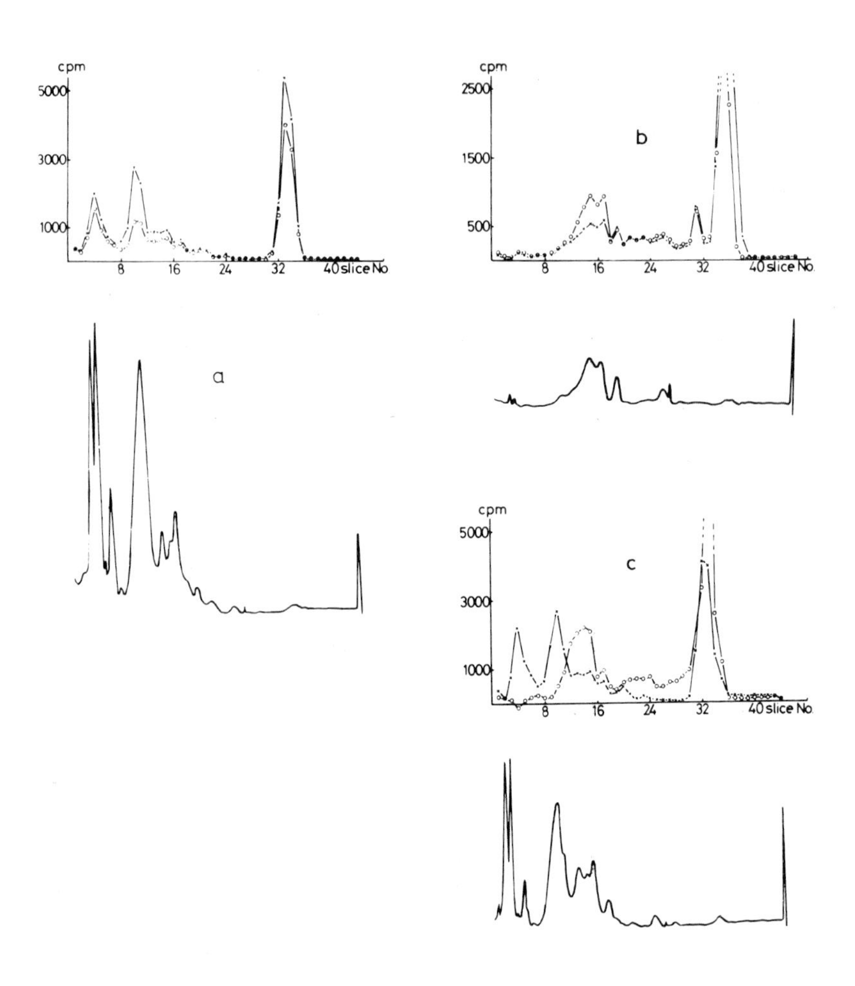

Fig. 8 Protection by pretreatment with SITS (o.25 mM)against dinitrophenylation (o.28 mM, pH 7.2, 3o', 37°C) in untreated resealed ghosts and in resealed ghosts exposed to internal trypsin (25 µg/ml) for 6o min. Double labeling with ^{3}H DNFB and ^{14}C DNFB. Upper curves: radioactivity, lower curves: Coomassie blue staining. (a) untreated ghosts dinitrophenylated in the presence (o) or absence (+) of SITS. Note protection of band 3 proteins. (b) internally trypsinized ghosts dinitrophenylated in the presence (+) or absence of SITS (o) (c) mixture of membranes from intact ghosts (+) and internally trypsinized ghosts (o), both dinitrophenylated in the absence of SITS.Note shift of band 3 proteins to lower molecular weight.

sites are either amino groups of one or several of the band 3 proteins which span the membrane or else that they are constituents of molecules which do not themselves span the membrane but cannot be dissociated in SDS from a supporting protein whose peptide chain extends from the anion control peptides at the outer surface to the inner surface of the membrane.

External trypsin has no effect on anion transfer (Passow, 1971). Trypsin which has been incorporated into red blood cell ghosts at a concentration at which the described effects on the band 3 proteins are observed exerts some inhibition. The degree of inhibition could not yet be determined accurately. This is partly due to the fact that extensive digestion of the membrane proteins is associated with a fragmentation of the ghosts. The resulting vesicles which reseal for sulfate ions are smaller than the ghosts from which they are derived. This leads to an increase of the ratio surface/volume which in turn affects the numerical value of the permeability constant. Since the size of the individual vesicles varies considerably, this ratio is difficult to determine and its contribution to the measured rate is difficult to assess. Another uncertainty is associated with possible variations of the permeabilities of the individual vesicles. In view of these difficulties we cannot directly compare the measured half times for sulfate equilibrium exchange in untreated and internally trypsinized ghosts (90 ± 12 min and 114 ± 18 min., respectively, 6 determinations at 30°C, pH 7.1). If the described increase of the surface/volume ratio upon trypsination is taken into account, it is qualitatively apparent that trypsin inhibits anion permeability to a greater extent than it would appear on the basis of the indicated half times. In the present context, we shall not discuss the implications of this finding. Instead we should like to direct attention to the observation that disulfonic acids and phlorizin (similar observations were made with DNFB) are capable of inhibiting anion exchange in untreated and internally trypsinized ghosts to nearly the same extent. (fig. 9). This suggests that the SITS protectable DNFB binding sites are indeed involved in the control of anion exchange and indicates that the anion control sites continue to function after cleavage of the band 3 proteins by internal trypsin.

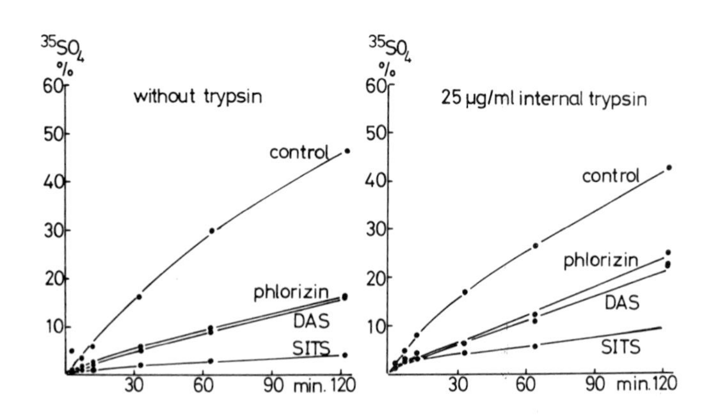

Fig. 9 Effects of phlorizin (2.5 mM), DAS (1.5 mM), or SITS (o.25 mM) in the external medium on SO_4 equilibrium exchange in untreated and internally trypsinized (6o', 37°C) ghosts. The release of radioactivity from $^{35}SO_4$ loaded ghosts into the external medium is represented by plotting the concentration of $^{35}SO_4$ in the medium (in percent of the equilibrium value) against time. The $^{35}SO_4$ movements had been measured at 3o°C, pH 7.1.

A considerable fraction of the sulfate movements as measured by means of $^{35}SO_4$ at Donnan equilibrium does not contribute to the conductance of the red blood cell membrane and hence may not be associated with a transfer of sulfate ions as such. This can be inferred from the experiments represented in table 2. In this table sulfate permeabilities as calculated from sulfate net movements which accompany valinomycin-induced potassium movements across the red blood cell membrane are compared with the rate of SO_4^{2-}/SO_4^{2-} equilibrium exchange

Table 2

Fraction of sulfate equilibrium exchange which could maximally contribute to the conductance of the membrane

Human red cells washed free of Cl^- and equilibrated in media containing 115 mM SO_4^{2-} 3o°C. Averages of 7 (pH 6.6) and 5 (pH 7.8) determinations.

Method	$P_S/{}^ok_s$	
	pH 6.6	pH 7.8
I	small	o.14
II	o.19	o.38
III	o.o4	o.27

across the intact cation impermeable membrane. In this context, the term "permeability" (P_S) refers to the diffusion of sulfate ions as such and the term "exchange" (${}^ok_s$) to the sum of the movements of sulfate ions and of electrically neutral complexes between sulfate and a carrier. The permeabilities had been calculated on the assumptions:
(1) SO_4^{2-}/SO_4^{2-} exchange which does not contribute to the membrane conductance is independent of sulfate movements which do contribute.
(2) sulfate/sulfate exchange is independent of membrane potential.
(3) the net efflux of K^+ as well as of SO_4^{2-} in the presence of valinomycin obeys the Goldman equation.
Method I is based on assumptions (1), (2), (3), method II on (1) and (2), and method III on (1) and (3).

The assumptions which lead to the derivation of the figures in table 2 are not the only ones which could explain the discrepancy between the rate of anion exchange and the permeabilities derived from anion net movements across the cation permeable membrane. For example it would seem feasible that

Sulfate penetrates exclusively in the form of an electrically neutral complex after combination with a charged carrier molecule and that the steady net flow of sulfate across a cation permeable membrane is maintained by a back flow in the opposite direction of the unloaded, charged carrier molecule. Under these conditions, a net transfer of SO_4^{2-} takes place which does not require the transfer of SO_4 ions as such. Obviously, the assumptions which lead to the figures in table 2 tend to give maximum numbers for the penetration of sulfate ions as such. Whatever the merits of these figures, they show that, at Donnan equilibrium that fraction of the sulfate movements which contributes to the conductance of the membrane is much larger for sulfate than for halides where this fraction is about $1 : 10^4$ (Hunter, 1971). Nevertheless, an exchange which does not contribute to the conductance always prevails.

This conclusion raises the question whether the disulfonic acid protectable anion control sites participate in an electrically neutral sulfate exchange as mobile carrier sites which shuttle back and forth between the two surfaces of the cell membrane (similar to valinomycin in cation transfer across a lipid bi-layer), or if these sites act as stationary control sites whose function does not require transitions from one membrane surface to the other.

In order to elucidate this question we studied the sidedness of the effect of reversibly binding disulfonic acids on sulfate equilibrium exchange and DNFB binding to the proteins in band 3. If a diffusible carrier should be involved one would expect the disulfonic acids to inhibit anion exchange and DNFB binding from either surface. With DAS this is not observed. When DAS is used, dinitrophenylation of the band 3 proteins as well as inhibition of anion exchange is only reduced if the agent is present at the outer surface, but not if present at the inner surface (fig. 1o). However, when APMB is used, inhibition of anion exchange is produced at either surface (fig. 11). The sidedness of the effect of APMB on DNFB binding to band 3 proteins is not yet clear.

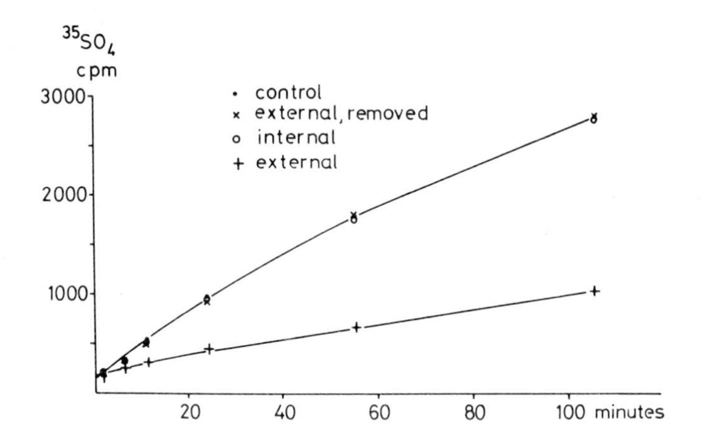

Fig. 1o Sidedness of action of DAS on sulfate equilibrium exchange in resealed ghosts. "Control": no DAS present. "Internal" and "external": 1.5 mM DAS present on inside or outside, respectively. "External removed": 1.5 mM DAS added after resealing and subsequently removed by washing to show reversibility of effect. Ordinate: $^{35}SO_4$ in the medium in percent of equilibrium value. Abscissa: time in min. 3o°C.

The inhibition of anion exchange by APMB at the outer as well as at the inner surface is compatible with the assumption of a carrier mediated anion transfer while the result with DAS would seem to contradict the involvement of carrier transport. Obviously, the situation is complicated and requires further clarification.

The involvement of a mobile carrier in anion exchange would not be easily reconciliable with the assumption that such carrier molecule has a molecular weight of about 1oo.ooo Daltons. It is even difficult to visualize pairs of lysine residues of a large molecule being able to oscillate between

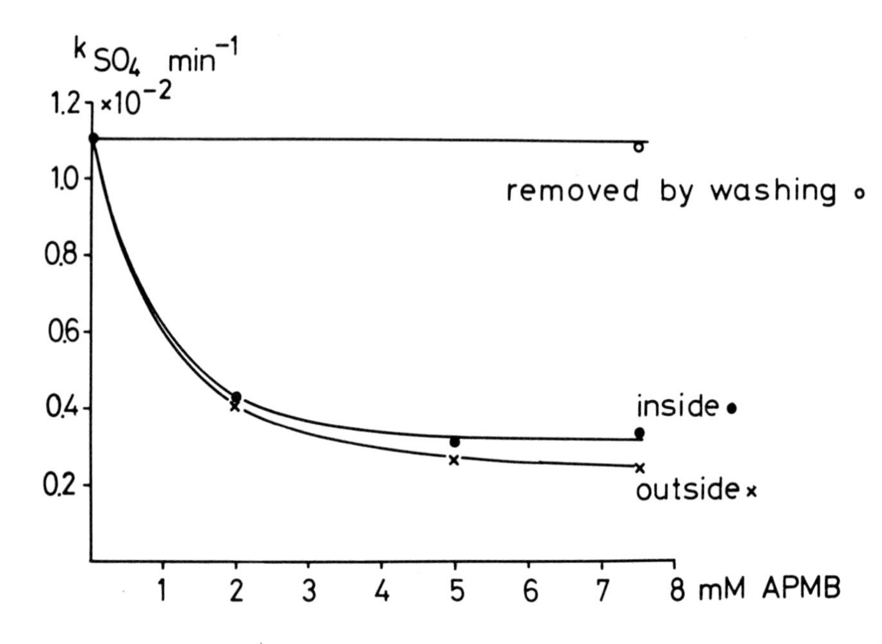

Fig. 11 Sidedness of action of APMB on sulfate equilibrium exchange in resealed ghosts. Ordinate: rate constant for sulfate exchange; abscissa: concentration of APMB in mM. "Inside" and "outside" refer to APMB inside or outside the ghosts. "Removed by washing" refers to ghosts to which the agent had been added at the end of the resealing period but subsequently been removed by washing to demonstrate the reversibility of the effect of the externally applied agent.

the two surfaces of the lipid bilayer. If such residues should play a role as carriers, then they ought to be located in some special region of the large protein molecule where they are accessible to APMB from either surface. Alternatively one may doubt that the carrier molecule is identical with one of the proteins of 100.000 Daltons; instead it could be a very much smaller molecule which migrates for some unknown reasons together with the band 3 protein on SDS polyacrylamide gels. Finally, the possibility cannot be completely ruled out that in addition to band 3 some other protein has been labeled by SITS but escaped detection and that this protein rather than a component of band 3 constitutes the anion carrier.

Summary

(1) Two different inhibitors of anion exchange, 4-acetamido-4'-isothiocyano-2,2'-stilbene disulfonic acid (SITS) and dinitrofluorobenzene(DNFB) were observed to interact at proteins found in band 3 of SDS polyacrylamide electropherograms of human red cell membranes but not at other major membrane proteins. This suggests that band 3 proteins are involved in the control of anion exchange.

(2) Reversibly binding disulfonic acids such as 4,4'-diacetamido stilbene disulfonic acid or 2-(4'-aminophenyl)-6-methyl benzenethiazol-3',7-disulfonic acid (APMB) partially protect band 3 proteins against dinitrophenylation and anion exchange against inhibition by DNFB.

(3) Trypsin which had been incorporated into resealed red blood cell ghosts digests most of the major membrane proteins at the inner surface of the cell membrane. It splits the band 3 proteins into 2 moieties of 58.000 and 48.000 Daltons without releasing these fragments from the membrane. Sulfate equilibrium exchange is inhibited but the susceptibility to further inhibition by SITS, DAS, phlorizin, and DNFB persists. Both fragments of the band 3 proteins carry SITS protectable DNFB binding sites.

(4) In resealed ghosts, APMB inhibits sulfate equilibrium exchange at either surface of the membrane while DAS only inhibits at the outer surface.

(5) Determinations by several different methods of that fraction of sulfate equilibrium exchange which may proceed by means of a carrier transport mechanism and does not contribute to the conductance of the membrane yielded figures of between 6o % and 86 % at pH 7.8 and between 8o % and more than 99 % at pH 6.5.

References

Cabantchik, Z.I. and Rothstein, A., 1974, J.Membrane Biol. 15, 2o7, 227.
Hubard, A.N. and Cohn, Z.A., 1972, J.Cell.Biol. 55, 39o.
Hunter, M.J., 1971, J.Physiol. 218, 49P.
Knauf, P.A. and Rothstein, A., 1971, J.Gen.Physiol., 58, 19o.
Lepke, S. and Passow, H., 1973, Biochim.Biophys.Acta, 298,529.
Maddy, H., 1964, Biochim.Biophys.Acta, 88, 39o.
Obaid, A.L., Rega, A.F. and Garrahan, P., 1972, J.Membrane Biol., 9, 385.
Passow, H., 1969, Progr.Biophys.Mol.Biol.,19, 424.
Passow, H., 1971, J.Membrane Biol., 6, 233.
Poensgen, H. and Passow, H., 1971, J.Membrane Biol., 6, 21o.
Schnell, K.F. and Passow, H., 1969, Experientia, 25, 46o.
Steck, T.L., 1972, in: Membrane Research, C.F.Fox, ed. Academic Press Inc. New York, p.71.
Steck, T.L., 1974, J.Cell Biol., 62, 1.
Zaki, L. and Passow, H., 1973, Abstr. 9th Intern.Congr. Biochem., Stockholm, p.287.

REGULATION OF SODIUM TRANSPORT IN EPITHELIAL CELLS

K. Janáček

Laboratory for Cell Membrane Transport, Institute of Microbiology, Prague, Czechoslovakia

INTRODUCTION

The survival value of active sodium transport by cell membranes manifests itself in a number of ways which may be divided into two categories. The first of them includes those faculties of the cell which make use of the free energy stored in ionic gradients, created by the sodium pump, and of the regulated ion composition of the "milieu interieur" of the cell itself; the second involves transcellular sodium transport. Phenomena belonging to the first category may be observed in non-polar and polar cells alike. We can note the following:

Energy stored in the form of the difference of electrochemical potential of sodium ions across the cell membrane permits the occurrence of excitability phenomena and in some cases enables the cell to accumulate such nutrients as carbohydrates and amino acids by co-transport with sodium ions returning passively into the cell. It was also possible to demonstrate (Glynn and Lew, 1970) that the energy of the sodium gradient may be used for ATP synthesis. Finally, the high intracellular potassium concentration produced by the operation of the sodium or sodium-potassium pump is an activating factor for a number of intracellular enzymes and one can imagine that a lowering of the potassium concentration in poikilotherm cells during winter might act as an additional regulating factor reducing the basal metabolism even beyond the direct temperature effects.

Apart from these and similar functions contributing in a

more direct way to the well-being of the cells themselves, in more developed organisms we find specialized epithelial cells of a polar character, in which the active sodium transport not only endows them with the above-mentioned faculties, but also performs a function belonging to the second category, i.e., performs the transcellular sodium transport which serves the homeostasis of the "milieu interieur" of the organism as a whole.

The two functions of the sodium pump, viz., the building up of the sodium gradient across the cell membrane, and the transcellular sodium transport, are subject to a number of regulating mechanisms, the survival value of which, in turn, rests in optimalization of the gradients and of the transcellular flows. The aim of the present paper is to describe some of the features of these regulating mechanisms, referring mostly to one of the favourite model objects in studies of ion transport, the epithelial cell of an ion-transporting layer, like frog or toad bladder or skin.

The first Figure summarizes schematically various mechanisms regulating sodium transport in the epithelial cell. The

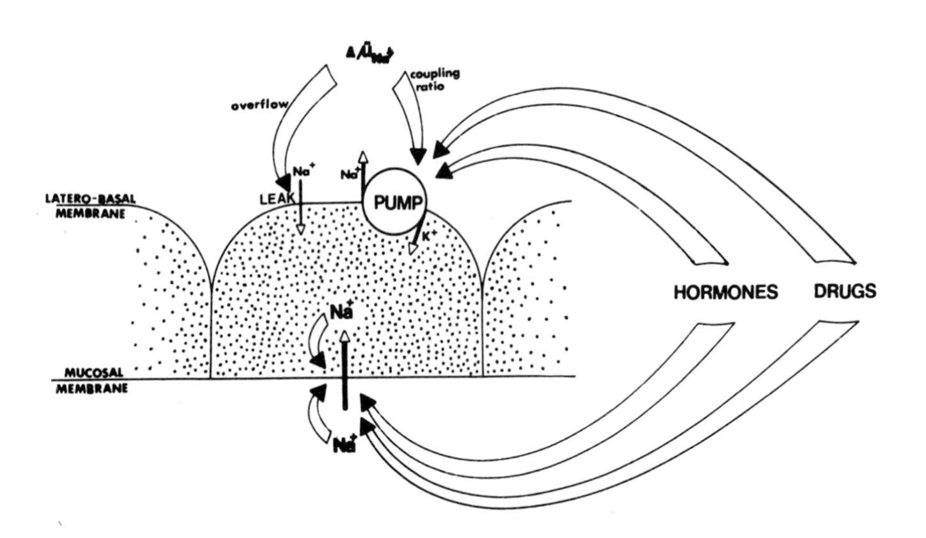

Fig. 1. Modes of regulation of sodium transport in epithelial cell.

modes of regulation are symbolized by eight arrows and will be presently discussed one by one. On the top left we recognize the most primitive of the feedback regulating mechanisms, that of an overflow; the greater the difference of the sodium electrochemical potential, the greater the passive inflow or leak of sodium counteracting the operation of the sodium pump. Further to the right the arrow expresses the belief that the value of the electrochemical potential difference governs the coupling ratio, in other words, the electrogenicity of the pump. Still further to the right the arrow symbolizes the disputed stimulation of the pumping mechanism by hormones, independent of the hormonal effects on the mucosal membrane at the opposite pole of the cell. Still more to the right another arrow depicts the regulation by agents foreign to the organism and administered by the experimentator or physician. A great number of inhibitory substances reducing the pumping rate are known and the time may soon come when we shall discover drugs capable of stimulating more or less directly the sodium pump itself. Doubtless a better insight into the operation of active transport mechanisms will be achieved when the chemical nature of such substances is known and their mode of action established. The bottom part of the figure symbolizes the regulating factors characteristic for polar cells performing transcellular transport and possessing mucosal membranes. Apart from the well-established effects of hormones and drugs on sodium permeability of the mucosal membrane we see the effects of sodium ions themselves on the same permeability. It will be seen that a reduction in sodium permeability results from an increase of the sodium concentration at either of its sides. Let us now turn our attention to a closer inspection of the individual regulating modes.

REGULATING MECHANISMS OPERATIVE AT LATERO-BASAL MEMBRANES

We already mentioned the simplest of the feedbacks, the leak. I would like to point out that unless the leak is coupled with accumulation of nutrients, it represents a waste of metabolic energy for useless recycling of sodium ions through the pump and leak. It may well be, however, that the leak is very small;

its quantitative estimation is not a straightforward matter. It may be insufficient, e.g., to measure the unidirectional influx of sodium ions into the epithelial cell and assume it will proceed exclusively through the passive leak, the reason for this being that the pump is a reversible mechanism and hence there must be a back flow through the pumping mechanism as well. The reversibility of the sodium pump is demonstrated not only by Glynn's ATP synthesis by the reversal of the pump which has already been mentioned; it is already implicit in the notion of the critical energy barrier introduced by Conway (Conway and Mullaney, 1961). As shown by Conway a net extrusion of sodium from muscle previously leached in potassium-free saline can proceed only when the difference of sodium electrochemical potential across the membrane, which can be varied by changing sodium and potassium concentrations in the medium, does not exceed a certain level. As a reversible process, the active sodium extrusion is necessarily subject to the principle of Le Chatelier and is opposed and finally balanced by the gradient which it creates.

The capability of the sodium electrochemical potential gradient to regulate the rate of pumping, however, does not seem to be exhausted by this simple mechanism and I would like to present some arguments in favour of the hypothesis that the coupling ratio of the pump on latero-basal membranes is dependent on the same gradient.

The dependence of the coupling ratio or electrogenicity of the sodium-potassium pump on the opposing gradient was discovered by Kostyuk and co-workers (1972) in the membrane of snail neurones. An iontophoretic microinjection of sodium into the giant neurones of Helix pomatia results in a measurable pump current. The pump current can be inhibited by ouabain or by cooling and is decreased with increasing hyperpolarization of the neurone. The authors concluded that when the membrane is hyperpolarized the rate of sodium active transport remains constant, but the rate of potassium active transport in the opposite direction is increased. Hence the pump in a hyperpolarized membrane is more coupled and less electrogenic.

Available data suggest that an analogous regulatory mechanism is operative in latero-basal membranes. In Fig. 2, A, we see on the latero-basal membrane a tightly coupled sodium-potassium exchange pump, as postulated in the well-known model of the frog skin cell by Koefoed-Johnsen and Ussing (1958). The inherent

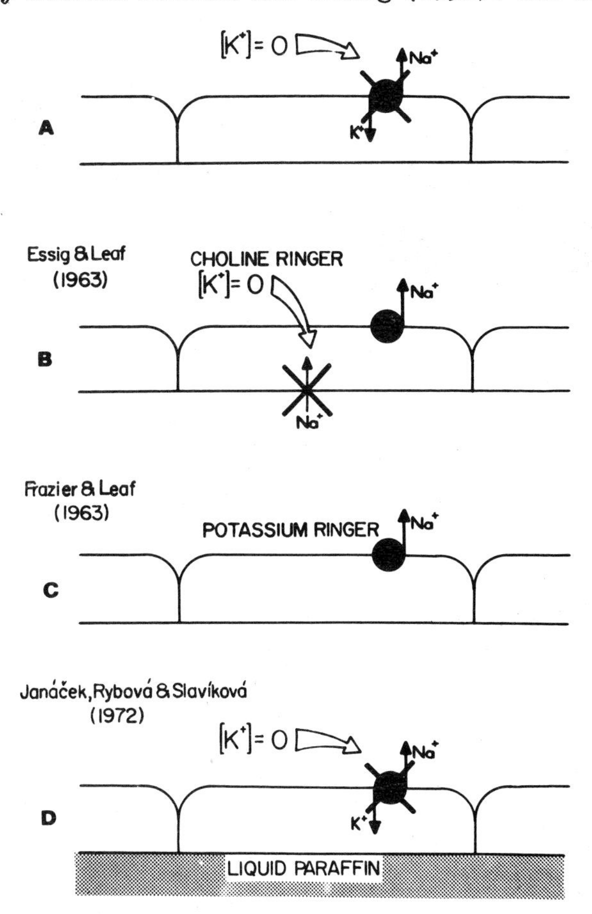

Fig. 2. Electroneutral and electrogenic pump in epithelial cell

property of such a tightly coupled pump is that the transport in the absence of potassium ions in the medium adjacent to the membrane comes to a stop and the phenomenon was indeed observed by a number of authors. This, however, is not the whole story. Essig and Leaf (1963) studied the phenomenon in toad bladder and found that with potassium-free sodium Ringer at the latero-basal membranes the sodium content of the cells increases and that of potassium decreases, and that the observed depression of the sodium transport could be just as well explained by the resulting marked decrease in the sodium permeability of the mucosal membrane. The arrow in Fig. 2, B shows this inhibitory effect of potassium absence, mediated by an increase in the intracellular sodium to potassium ratio. We shall see later that a similar external concentration change also brings about a decrease in mucosal sodium permeability. When, however, the authors used a potassium-free choline solution at the latero-basal membranes, the electrolyte content of the cells was much less influenced and the sodium transport much less depressed. The obvious conclusion was that the sodium pump is not a tightly coupled sodium-potassium pump, but rather an electrogenic pump. The conclusion was further supported by the finding of Frazier and Leaf (1963) that with potassium solutions at the latero-basal membranes the membrane potential across these membranes cannot be explained by an unequal distribution of any ion and thus, under these conditions again, the pump works in an electrogenic manner (Fig. 2, C).

Let us now, however, consider the experiment shown in Fig. 2, D. The nonpolar preparation of frog bladder was used, in which the mucosal membranes are blocked by liquid paraffin and changes of the mucosal permeability play no role. The electrolyte composition of the cells in this preparation is governed solely by the latero-basal membranes and inhibition of the pump manifests itself by an increase in the sodium content of the tissue. In this preparation the absence of potassium at the latero-basal membranes is an equally potent inhibitory factor as are high concentrations of ouabain (Janáček et al., 1972). Thus we see a tightly coupled sodium-potassium pump at work again. Moreover, the results of a careful flux analysis carried out on toad bladder in normal Ringer solution by Finn and Nellans (1972) are consistent with the operation of one-to-one sodium-potassium pump.

What is the reason that the pump is electrogenic when it transports sodium into a choline or potassium solution and tightly coupled when it transports the same ion into a sodium solution? In the former case the active transport of sodium is actually downhill, in the latter it has a considerable barrier of sodium electrochemical potential to overcome. Coupled electroneutral transport is not opposed by electrical forces and hence represents a device which, to use the terms of network thermodynamics, lessens at the same energy rate the effort, or force variable. Change in the coupling ratio represents a positive feedback, acting against Le Chatelier´s depression of the active sodium transport.

Next in the hierarchy of sodium transport regulations come regulations by hormones. In ion-transporting epithelial cells the rapid and transient stimulation of transport by neurohypophysial hormones, like vasopressin, oxytocin and vasotocin, and the slow and protracted stimulation by corticosteroids, of which aldosterone is the most potent and which is mediated by protein synthesis, are probably the most important.

Considerable evidence was accumulated and reviewed, e.g., by Orloff and Handler (1964, 1967) in favour of the theory that the action of the neurohypophysial hormones fits the well-known general scheme of hormonal regulation with cyclic adenosine monophosphate as a second messenger. The scheme is the discovery of Sutherland and his school, and is outlined in Fig. 3.

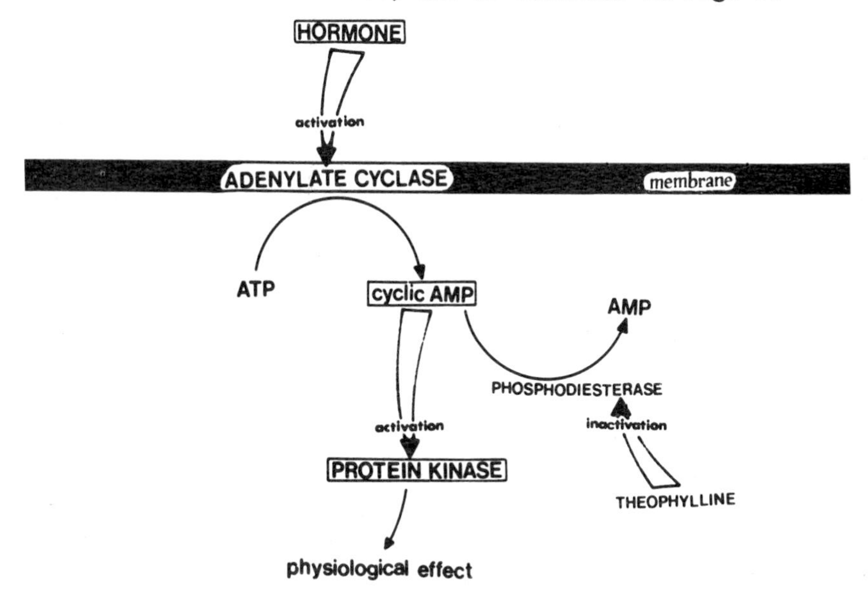

Fig. 3. Mediation of hormone action by cyclic AMP

The binding of hormone to a specific receptor site activates adenylate cyclase and, as a result of this, the intracellular level of cyclic 3',5'-adenosine monophosphate is increased. The cyclic AMP, in turn, activates a protein kinase and phosphorylation of specific proteins results in a physiological effect. Cyclic AMP is continually converted by phosphodiesterase into inactive 5'-adenosine monophosphate. Phosphodiesterase itself is inactivated by methylxanthines, theophylline being most effective. As a result of this, theophylline also increases the intracellular level of cyclic AMP and, hence, mimics various hormonal effects.

Not all authors accept this mechanism of action of neurohypophysial hormones; Cuthbert and Painter (1968) showed that antidiuretic hormone, cyclic AMP and theophylline added to frog skin influence its various electrical characteristics differently and concluded that it is not necessary to involve cyclic AMP in order to explain the effects of ADH or theophylline. Some kind of intracellular compartmentalization of cyclic AMP, however, appears to be required to explain differences between the physiological effects of various hormones and even various hormonal analogues and in such a complex situation it can be hardly assumed that specific hormonal effects can be perfectly simulated by exogenous cyclic AMP or theophylline.

The binding of neurohypophysial hormones to receptors in frog skin was studied by Bockaert and co-workers (1972) using oxytocin highly tritiated at the tyrosine residue. There are four different modes of incorporation of the radioactivity from the hormone in frog skin, of which three were shown by the authors to be unrelated to the physiological response. (1) The first probably corresponds to tyrosine incorporation into proteins, for it can be prevented by preincubation with unlabelled tyrosine or with inhibitors of protein synthesis, puromycin or cycloheximide. (2) Binding by covalent disulfide bonds which could be released by dithiothreitol and which shows no competition with O-methyltyrosine-carba-1-oxytocin possessing no disulfide bonds. (3) Binding to sites with an apparent K_D of $5 \cdot 10^{-8}$M and high capacity of some $20 \cdot 10^{-12}$ mol per g tissue. (4) Finally, there is a reversible binding with a high affinity (apparent K_D 2.5 · 10^{-9}M) and a low capacity (1 to 2 · 10^{-12} mol per g tissue) which is related to the physiological response; the same affinity was obtained from the dependence of the response of the potential difference across the same preparation on the dose of the hormone. The time course of the binding is identical with that of the response but precedes it slightly and competition for these receptor sites with oxytocin analogues corresponds to their relative physiological potencies.

Although the sequence of events following hormone binding is, to my knowledge, not yet known with such quantitative accuracy, I should like to mention at least the most interesting negative feedback mechanism discovered by Jard and Bernard (see the discussion remark of Morel, 1973) by which, apparently, the physiological effect of the hormone, the enhanced sodium transport, reduces the intracellular concentration of cyclic AMP in isolated frog skin epithelium. When no sodium is present at the outer surface of the epithelium, so that no physiological effect on sodium transport is obtained, the intracellular level of cyclic AMP increases continually after the addition of the hormone and reaches a very high steady-state level; in the presence of sodium it reaches a peak at about 1 min and then drops to a relatively low steady-state level. The regulatory effect of the hormone on sodium transport is thus regulated itself and kept within proper limits.

Whatever the chemical nature of the reactions influenced by the cyclic AMP-activated protein kinase, the question of importance is whether the enhancement of sodium transport by neurohypophysial hormones is due to an increase in the sodium inflow across the mucosal membrane of the epithelial cells, or whether there is, at the same time, a stimulation of the sodium pump at the latero-basal membrane, independent of the hormonal effects at the mucosal membrane. My belief that the latter alternative is correct is based on the following experimental evidence.
1st: In the above-mentioned nonpolar preparation of frog bladder shown in Fig. 4 (Janáček and Rybová, 1967, 1970) in which sodium transport across the mucosal membrane is nonexistent, oxytocin brings about a highly significant decrease in the tissue sodium content.

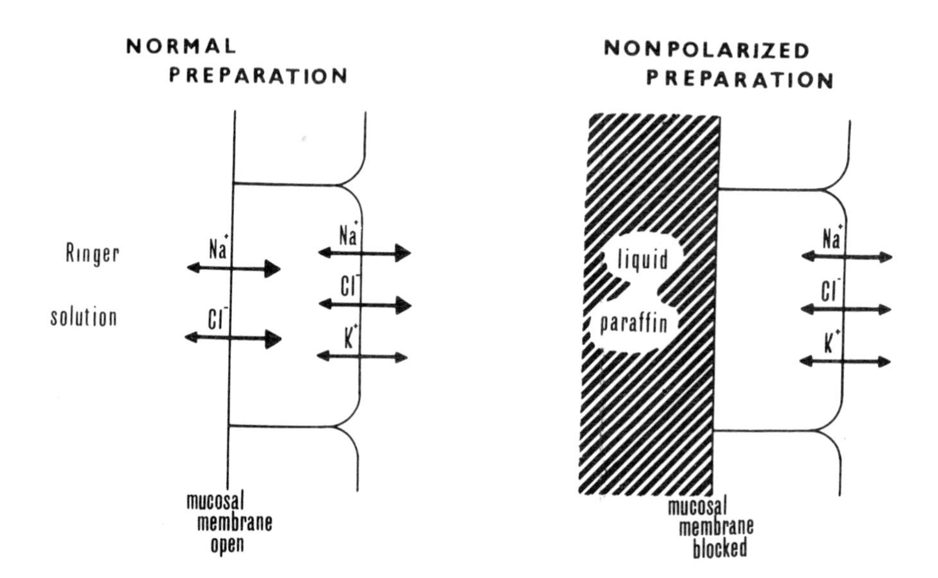

Fig. 4. Nonpolar preparation of epithelial layer.

2nd: In toad bladder, amphotericin added from the mucosal side increases the cellular sodium content without affecting the transcellular sodium transport - the pump is obviously already saturated. Still, an addition of vasopressin to the latero-basal membranes of the amphotericin-treated bladder brings about a significant decrease in the cellular sodium content (Finn, 1968).

3rd: The kinetic analysis of sodium transport in toad bladder, developed by Finn and Rockoff (1971), shows a dual effect of vasopressin on sodium transport (Finn, 1971); there is both an effect at the mucosal membrane resulting in an increase of the sodium transport pool and an increase in the rate constant of the flux of sodium from cells across the latero-basal membranes, not related to the transport pool increase.

Contrary to the idea of the dual action of neurohypophysial hormones, an argument is sometimes advanced, based on the experimental finding of Civan and Frazier (1968). As much as 98% of the resistance decrease brought about by vasopressin in toad bladder occurs at the mucosal membrane. However, in view of what was discussed earlier - that under the physiological condition of an uphill sodium transport the sodium-potassium pump at the latero-basal membrane operates in an electroneutral manner - one would not assume that an enhanced performance of such a pump can bring about a decrease in electrical resistance.

Let us now turn our attention to the regulation of the transcellular sodium transport by the representative of corticosteroids, aldosterone. The time course of aldosterone stimulation is different from the rapid effect of neurohypophysial hormones, there being time lag between 1 and 2 hours and the maximum response being reached only after some 6 hours. According to considerable accumulated experimental evidence, as reviewed by Edelman (1973), the action of aldosterone is probably not mediated by cyclic AMP but rather involves the following steps: (1) Migration of d-aldosterone from the capillary into the cytoplasm of the target cell; (2) binding of the native steroid to an aporeceptor; (3) temperature-dependent transformation of the steroid-aporeceptor complex into an active steroid-receptor complex; (4) migration of the active complex into the nucleus; (5) second-stage modification of the active complex and binding to specific genomic sites; (6) activation or depression of transcription which results in an induction of the synthesis of specific proteins; (7) augmentation of sodium transport by the action of the aldosterone-induced proteins.

Like neurohypophysial hormones, aldosterone enhances both the sodium entry across the mucosal membrane and, independently, the operation of the pump at the latero-basal membrane, as suggested by its sex-dependent effect on the sodium content of the frog bladder tissue (Janáček et al., 1971). As shown schematically in Fig. 5, aldosterone in the male urinary bladder brings about a highly significant decrease in the sodium content, whereas in the female bladder the sodium content increases with the same treatment. Hence it appears that in the former case the independent effect on the pump prevails over that on the mucosal membrane, whereas in the latter the situation is reversed. Still, even with female bladders an independent stimulation of the pump by aldosterone is present, since aldosterone decreases the sodium content in the nonpolar preparation of the female bladder.

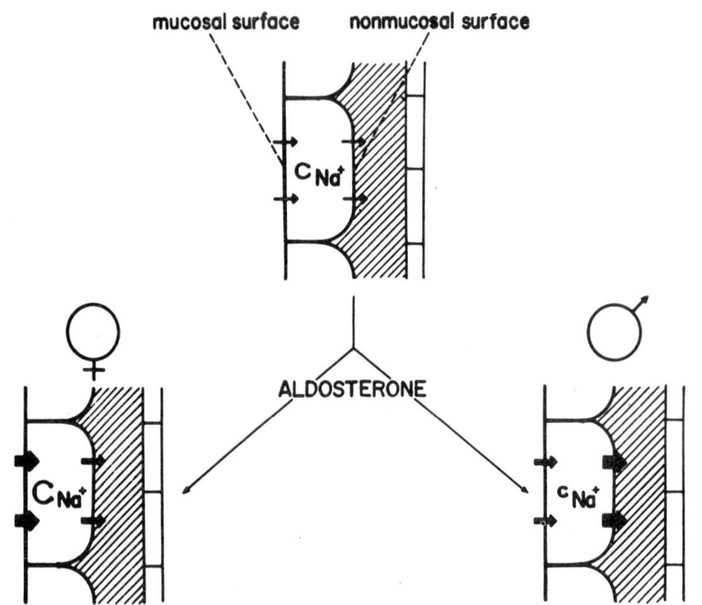

Fig. 5. Sex-dependent effect of aldosterone on sodium content.

Of other hormones influencing the sodium pump I would like to mention insulin, which, as found by Herrera (1965, 1967), enhances the transcellular sodium transport with a simultaneous tendency of the sodium pool within the tissue to decrease.

Substances foreign to the organism and inhibiting the active sodium extrusion from epithelial cells by the latero-basal membranes can be divided into two classes: inhibitors which, like cyanide or 2,4-dinitrophenol, interfere with the supply of ATP for the pump, and substances inhibiting directly the sodium pump, like ouabain and related compounds. To my knowledge there is as yet no drug that unequivocally stimulates the sodium pump directly. It is interesting to note that the natural activation of the pump by potassium cannot be simulated by thallium ions; although these ions activate the sodium-potassium ATPase from frog skin (Maslova et al., 1971), they inhibit the sodium transport across the skin in a way resembling the inhibition by ouabain (Natochin, 1971).

REGULATING MECHANISMS OPERATIVE AT MUCOSAL MEMBRANES

In cells performing transcellular transport, further possibility of transport regulation arises; the rate of entry of the transported substance into the cells may be regulated. As already mentioned, a decrease in the sodium permeability of the mucosal membrane with increasing sodium-potassium ratio was observed in toad bladder, when sodium transport was inhibited by potassium-free media. An analogous phenomenon is observed in toad skin, when the extrusion of sodium from cells by the pump is prevented by cyanide or ouabain (Larsen, 1973). It cannot yet be said with certainty whether it is the increasing sodium concentration or the decreasing potassium concentration which results in the reduced sodium permeability of the mucosal membrane, but the physiological significance of the regulatory mechanism is obvious; it acts as a negative feedback on sodium entry into epithelial cells. The same mechanism may be responsible for the decrease of sodium permeability of analogous membranes in toads adapted to high salinities, as found by Katz (see discussion remarks by Lindeman and Maetz, 1973).

The same phenomenon of a decrease in the sodium permeability of the outward-facing membrane of frog skin and mucosal membrane of frog bladder after replacement of potassium in the external medium with sodium was discovered and thoroughly studied by Lindemann and Gebhardt (1973), using a "fast-flow" chamber. The sodium permeability change has a relatively low time constant of several seconds in the skin and some 50 milliseconds in the bladder.

Turning now to the hormonal regulation of sodium entry into epithelial cells it may be stated that the entry is enhanced by neurohypophysial hormones; the sodium content of the cells invariably increases when sodium is present at the mucosal side of the bladder or outer surface of the skin and hormones are added to the opposite side of the preparation. The enhanced entry of sodium across the mucosal membrane of the frog bladder in the presence of aldosterone was already mentioned and the same occurs in frog skin where the process is assisted by the moulting process, the detachment of stratum corneum, induced by aldosterone (Nielsen, 1973).

Herrera's finding (1965, 1967) that insulin at the latero-basal membranes of toad bladder brings about a decrease in the tissue sodium pool and hence stimulates the pump was already mentioned; recently we confirmed the phenomenon with frog bladder and, moreover, observed an opposite phenomenon (an increase in the tissue sodium content) when insulin was present at the two sides of the preparation. Hence it seems that insulin, too, enhances the mucosal sodium permeability, but only when it is applied at the mucosal membrane.

The effects of noradrenaline are interesting. In low doses it stimulates sodium transport across the frog skin without affecting the transport across frog bladder (Bastide and Jard, 1968). In frog bladder, noradrenaline increases the sodium inflow across the mucosal membrane but apparently does not stimulate the pump at latero-basal membranes, having little or no effect on the sodium content of the nonpolar preparation (Janáček et al., 1973). Thus, once more, as in Finn's experiments (1968), the pump appears to be already saturated and an enhanced sodium entry unable to promote a higher rate of pumping. It may be that in frog skin the sodium ions entering after noradrenaline treatment reach unsaturated pumps in deeper cell layers.

Finally, coming to drugs which influence the entry of sodium into epithelial cells, the effects of guanidine compounds are perhaps of greatest interest. Amiloride (amino-pyrazine-guanidine) is known to be a powerful competitive inhibitor of sodium entry (Bentley, 1968; Nagel and Dörge, 1970). On the other hand, according to Zeiske and Lindemann, there are guanidine compounds (benzimidazole-guanidine being most effective) that stimulate the sodium entry by preventing the decrease of sodium permeability, which is normally caused by sodium at the mucosal membrane and which was discussed previously (Lindemann, 1974).

REFERENCES

Bastide, F. and Jard, S. (1968). Biochim. Biophys. Acta 150, 113.

Bentley, P.J. (1968). J. Physiol. 195, 317.

Bockaert, J., Imbert, M., Jard, S. and Morel, F. (1972). Exp. and Molecular Pathology, 8, 230.

Civan, M.M. and Frazier, H.S. (1968). J. Gen. Physiol. 51, 589.

Conway, E.J. and Mullaney, M. (1961). In: Membrane transport and metabolism, eds. A. Kleinzeller and A. Kotyk (Academic Press, New York or Publishing House of the Czechoslovak Academy of Sciences) p. 117.

Cuthbert, A.W. and Painter, E. (1968). J. Physiol. 199, 593.

Edelman, I.S. (1973). In: Transport mechanisms in epithelia, Alfred Benzon symposium V, eds. H.H. Ussing and N.A. Thorn, Munksgaard, Copenhagen and Academic Press, New York, p. 185.

Essig, A. and Leaf, A. (1963). J. Gen. Physiol. 46, 505.

Finn, A.L. (1968). Amer. J. Physiol. 215, 849.

Finn, A.L. (1971). J. Gen. Physiol. 57, 349.

Finn, A.L. and Nellans, H. (1972). J. Membrane Biol. 8, 189.

Finn, A.L. and Rockoff, M.L. (1971). J. Gen. Physiol. 57, 326.

Frazier, H.S. and Leaf, A. (1963). J. Gen. Physiol. 46, 491.

Glynn, I.M. and Lew, V.L. (1970). J. Physiol. 207, 393.

Herrera, F.C. (1965). Amer. J. Physiol. 209, 819.

Herrera, F.C. (1967). Acta Cient. Venezolana, Supl. 3, 28.

Janáček, K. and Rybová, R. (1967). Nature, 215, 992.

Janáček, K. and Rybová, R. (1970). Pflügers Arch. 318, 294.

Janáček, K., Rybová, R. and Slavíková, M. (1971). Pflügers Arch. 326, 316.

Janáček, K., Rybová, R. and Slavíková, M. (1972). Biochim. Biophys. Acta 288, 221.

Janáček, K., Rybová, R. and Slavíková, M. (1973). Physiol. Bohemoslov. 22, 237.

Koefoed-Johnsen, V. and Ussing, H.H. (1958). Acta Physiol. Scand. 41, 657.

Kostyuk, P.G., Krishtal, O.A. and Pidoplichko, V.I. (1972). J. Physiol. 226, 373.

Larsen, E.H. (1973). In: Transport mechanisms in epithelia, Alfred Benzon symposium V, eds. H.H. Ussing and N.A. Thorn, Munksgaard, Copenhagen and Academic Press, New York, p. 131.

Lindemann, B. (1974). Personal communication.

Lindemann, B. and Gebhardt, U. (1973). In: Transport mechanisms in epithelia, Alfred Benzon symposium V, eds. H.H. Ussing and N.A. Thorn, Munksgaard, Copenhagen and Academic Press, New York, p. 115.

Lindemann, B. and Maetz, J. (1973). Ibid.,p. 159.

Maslova, M.N., Natochin, J.V. and Skul'skii, I.A. (1971). Biokhimiya, 36, 871 (in Russian).

Morel, F. (1973). In: Transport mechanisms in epithelia, Alfred Benzon symposium V, eds. H.H. Ussing and N.A. Thorn, Munksgaard, Copenhagen and Academic Press, New York, p. 229.

Nagel, W. and Dörge, A. (1970). Pflügers Arch. 317, 84.

Natochin, J. (1971). In: Biofizika membran, materials of symposium held in Kaunas, p. 636 (in Russian).

Nielsen, R. (1973). In: Transport mechanisms in epithelia, Alfred Benzon symposium V, eds. H.H. Ussing and N.A. Thorn, Munksgaard, Copenhagen and Academic Press, New York, p. 214.

Orloff, J. and Handler, J.S. (1964). Amer. J. Med. 36, 686.

Orloff, J. and Handler, J.S. (1967). Amer. J. Med. 42, 757.

SYNTHESIS OF ADENOSINE TRIPHOSPHATE BY SODIUM, POTASSIUM ADENOSINE TRIPHOSPHATASE

R. L. POST, K. TANIGUCHI and G. TODA

Vanderbilt University, Nashville, TN USA

INTRODUCTION

In order to characterize better the mechanism of the sodium and potassium ion pump of the plasma membranes of animal cells we have been investigating the sodium and potassium ion-transport adenosine triphosphatase (Na, K-ATPase) which expresses the activity of the pump in preparations of broken or leaky membranes. Specifically we have investigated the sequence of addition and release of ligands and have become interested in the interaction free energies of their binding (Weber, 1974). In particular there appears to be an interaction between the binding of Na^+ and the free energy of hydrolysis of a phosphate group at an active site of phosphorylation. In order to distinguish better between effects due to binding and those due to translocation of ions we have preferred to work with leaky membranes, across which concentration gradients do not persist. It has been an advantage to conduct the hydrolytic reaction of this enzyme in the backward direction. In a first step the enzyme was phosphorylated from inorganic phosphate (P_i). In a second step addition of ADP with a high concentration of Na^+ produced ATP. Low concentrations of K^+ inhibited synthesis.

The Pump - The pump is composed of phosphatidyl serine and at least one protein (molecular weight about 100, 0C0) and possibly also a glycoprotein (molecular weight about 50, 000) which are intrinsic to the plasma membrane. The pump has access to the solutions in contact with the intracellular and extracellular faces of the membrane. Per cycle it transports approximately 3 Na^+ outward and 2 K^+ inward, thus generating an electric current (Kerkut and York, 1971). For net outward transport Na^+ is a unique substrate but for net inward transport K^+ has as congeners Li^+, NH_4^+, Rb^+, Cs^+ and Tl^+. Per cycle the pump hydrolyses the terminal phosphate bond of one molecule of intracellular ATP with intracellular release of products. Intracellular Mg^{2+} is required. The pump is thus an endo-ATPase with intracellular Na^+ and extracellular K^+ as additional substrates and with extracellular Na^+ and intracellular K^+ as additional products (Fig. 1).

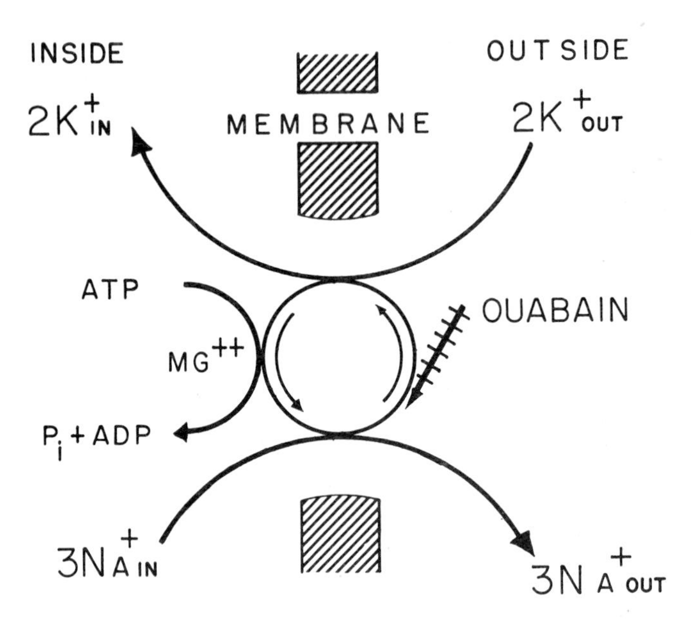

Fig. 1. Stoichiometry and sidedness of the sodium and potassium ion pump of plasma membranes. Ouabain is a specific cardioactive steroid inhibitor. (From Post, 1968).

In the presence of abnormally large concentration gradients of Na^+ and K^+, opposing transport in the forward direction, the pump runs backward and synthesizes ATP. In preparations of broken or leaky membranes, where gradients cannot persist, the activity of the pump is an ATPase (EC 3.6.1.3) which requires Mg^{2+}, Na^+ and K^+ simultaneously and which is inhibited by cardioactive steroids. This is the Na, K-ATPase. For a review see Hokin and Dahl (1972) or Schwartz et al. (1972).

The Phosphoenzyme - The larger protein in the pump is phosphorylated from ATP in the presence of Na^+ uniquely. In the presence of K^+ or its congeners it is phosphorylated from P_i and Na^+ is uniquely inhibitory (Table I). It is also phosphorylated from P_i when complexed with ouabain. In all cases Mg^{2+} is required for phosphorylation. In each case the active site of phosphorylation is the β-carboxyl group of a specific aspartyl residue (Post and Kume, 1973; Nishigaki et al. 1974; Post et al. 1975).

TABLE I

Effect of monovalent cations on phosphorylation of Na, K-ATPase from ATP or from P_i in the presence of Mg^{2+}. (From Post et al. 1965; Post et al. 1973).

Monovalent cation	Phosphoenzyme, %	
	From ATP	From P_i
Li^+	7	18
Na^+	100	0.4
NH_4^+	0	14
K^+	2	10
Rb^+	5	13
Cs^+	0	12
Tl^+	-	8
None	14	44*

*This reactive state of the phosphoenzyme is insensitive to K^+ or ADP.

A working hypothesis of the reaction sequence for phosphorylation and dephosphorylation is shown in Fig. 2. It treats phosphorylation from ATP or from P_i as functions of alternative reactive forms of the dephosphoenzyme. These are designated E_1 and E_2 respectively. The corresponding phosphorylated products are designated $E_1 \sim P$ and E_2-P. At physiological concentrations of Na^+ (up to 160 mM) $E_1 \sim P$ appears only transiently but a related reactive state is stabilized by partial inhibition of the enzyme with N-ethyl maleimide or oligomycin. Thus phosphorylation from ATP ordinarily yields sensitive E_2-P which equilibrates with P_i (Post et al., 1975). It is extremely sensitive to attack by K^+, as will be shown, but ordinarily it is insensitive to ADP. K-complexed phosphoenzyme equilibrates with P_i much more rapidly than do the other reactive states of E_2-P (Post et al., 1975).

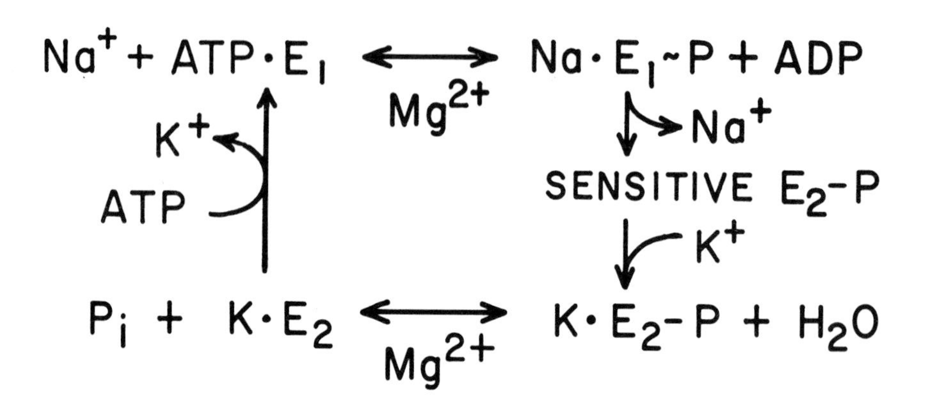

Fig. 2. A working hypothesis for a reaction sequence of the phosphoenzyme. For simplicity and lack of direct evidence the stoichiometry of Na^+ and K^+ has not been specified. Similarly the actions of Mg^{2+} are incomplete.

This hypothesis is based on evidence for isolated reaction steps. For example, the attack of ATP (or MgATP) on $K \cdot E_2$ is based largely on experiments of Post et al. (1972). But it is not established that this pathway is necessarily predominant under physiological conditions.

RESULTS

Sensitive Phosphoenzyme from ATP - The enzyme was in a crude membrane suspension obtained from a homogenate of guinea pig kidney by differential centrifugation. It was phosphorylated from [^{32}P]ATP in the presence of Mg^{2+} and Na^+ at 0° and neutral pH. In order to expose the rate of dephosphorylation, an excess of EDTA was added at zero time to chelate the free Mg^{2+}. The reaction was stopped with acid and the denatured membranes were washed and counted. Background labeling, appearing when K^+ replaced Na^+, was subtracted from the labeling in all of the samples. Dephosphorylation was exponential with a time constant of about 16 seconds. When K^+ was added at 4 seconds, dephosphorylation was immediate (Fig. 3). The sensitivity to K^+ is shown by the ratio of Na^+ to K^+, namely, 160 to 1.

Sensitive Phosphoenzyme from Inorganic Phosphate - The strategy for synthesis of ATP was first to prepare sensitive phosphoenzyme from P_i and second to treat this phosphoenzyme with ADP and Na^+. Because Na^+ inhibits phosphorylation from P_i in a steady state (Table I), the enzyme cannot be recycled and the recovery of ATP does not exceed the initial amount of phosphoenzyme.

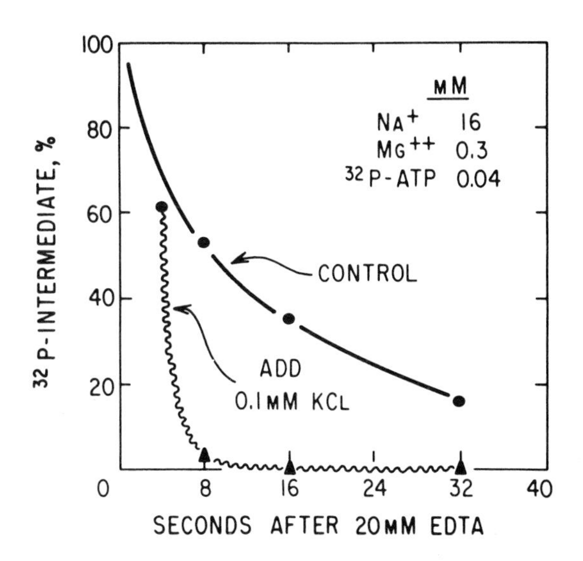

Fig. 3. Dephosphorylation of sensitive phosphoenzyme from ATP. (From Post, 1968).

The procedure for synthesis of sensitive E_2-P from P_i was discovered by Dr. Toda in this laboratory. The membranes are first washed in 1 mM $MgCl_2$ by centrifugation and resuspension at 4° and neutral pH and then the Mg^{2+} is removed. The effect of washing persists for at least 18 hours at 4°. It is not difficult to phosphorylate the native enzyme from P_i in the absence of K^+ (Table I). The difficulty is that Mg^{2+} rapidly attacks sensitive phosphoenzyme in the absence of monovalent cations. This attack makes the enzyme insensitive to K^+ (Post et al., 1975). Fortunately a low concentration of Na^+ protects sensitive phosphoenzyme from Mg^{2+} for about 30 seconds without itself immediately inhibiting phosphorylation from P_i.

To form sensitive phosphoenzyme, P_i and Na^+ were added to washed membranes followed by Mg^{2+} 5 seconds later to initiate phosphorylation at zero time. Under these conditions 30 per cent of the enzyme was phosphorylated and the level was stable for at least 21 seconds (Fig. 4). If the enzyme had not been washed with $MgCl_2$ previously, no significant amount of phosphoenzyme would have appeared. Addition of CDTA (cyclohexylenedinitrilotetraacetic acid) to the phosphoenzyme chelated the free Mg^{2+} and exposed a slow dephosphorylation. Addition of K^+ with the CDTA showed that half of the phosphoenzyme was sensitive and that half was not (Fig. 4). Other experiments showed that the amount of the insensitive component was negligible at zero time and increased progressively during the incubation. In this way it was possible to prepare sensitive phosphoenzyme from inorganic phosphate.

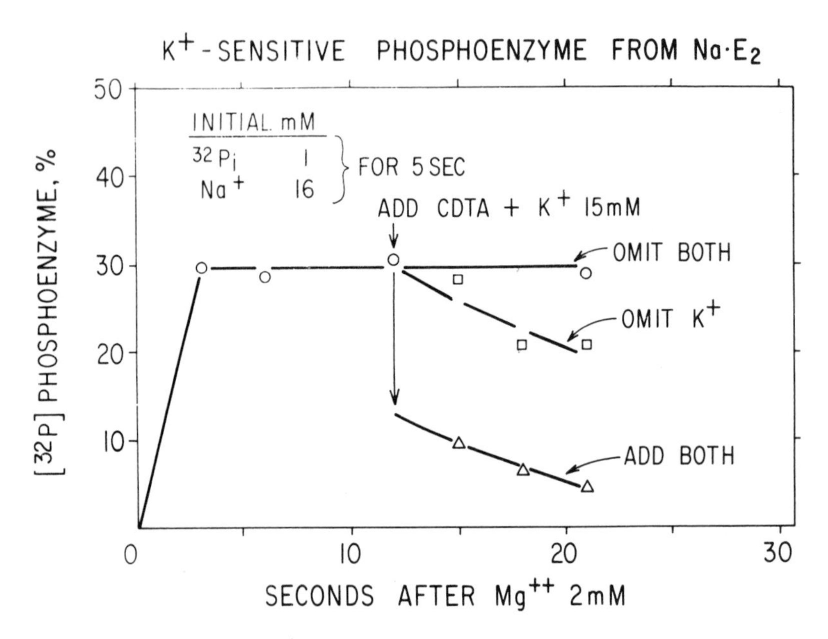

Fig. 4. Formation of K^+-sensitive phosphoenzyme from inorganic phosphate (unpublished). Phosphoenzyme was estimated by peptic digestion and paper electrophoresis according to Post et al. (1975).

Synthesis of ATP from Sensitive Phosphoenzyme - These experiments were performed by Dr. Taniguchi. The membranes were first treated with concentrated NaI to remove or inactivate a non-specific ATP synthetase. Then they were washed briefly with 1 mM $MgCl_2$. The membranes were incubated at neutral pH and 0° with 1 mM $^{32}P_i$ and 0.1 mM CDTA (to prevent premature phosphorylation). At zero time 0.5 mM $MgCl_2$ with 16 mM NaCl was added. At 4 seconds 1 mM ADP with 20 mM $(Tris)_3$CDTA and various concentrations of NaCl were added. At 6 seconds or later times the reaction was stopped with acid. Phosphoenzyme was estimated in the precipitate and $[^{32}P]$ATP was estimated in the supernatant (Post et al., 1974). The 20 mM CDTA was a convenience; it prevented destruction of the synthesized ATP and reduced the rate of synthesis by only 25%. In the absence of extra Na^+ the phosphoenzyme disappeared exponentially with a time constant of about 40 seconds. (It was stabilized partly by the ADP and partly probably by a greater freedom from traces of K^+ in the reaction mixture.) Very little ATP appeared. As the concentration of NaCl increased, so did the rate of dephosphorylation. As the concentration of NaCl increased, more ATP appeared. The increase in ATP corresponded fairly closely to the loss in phosphoenzyme. The concentration of NaCl for a half-maximal effect was remarkably high, about 0.5 M (Fig. 5).

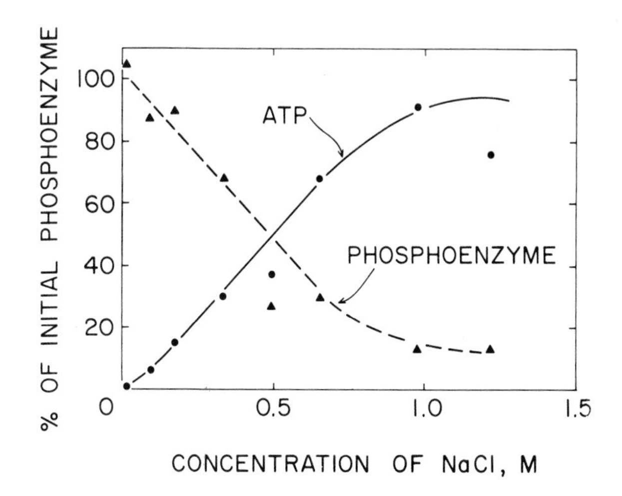

Fig. 5. Dephosphorylation of sensitive phosphoenzyme and synthesis of ATP in the presence of increasing concentrations of Na^+. The reaction was stopped 2 seconds after the addition of ADP and NaCl to the phosphoenzyme at 0° (unpublished).

Control Experiments - The functional significance of the synthesis is shown by experiments in which ATP did not appear. When the initial amount of NaCl (16 mM) was incubated with the enzyme for 10 minutes before addition of P_i, neither phosphoenzyme nor ATP appeared. When the initial amount of NaCl was replaced with KCl, K^+-complexed phosphoenzyme appeared (Table I) but ATP did not. When 0.25 mM ouabain was present in the initial reaction mixture and CDTA was omitted, a large amount of stable phosphoenzyme appeared but no ATP. It is well known that the complex of the enzyme with the inhibitor ouabain is easily phosphorylated from P_i (Hokin and Dahl, 1972; Schwartz et al., 1972). When 16 mM K^+ was added with 0.7 M Na^+ and CDTA to sensitive phosphoenzyme, the phosphoenzyme was immediately dephosphorylated and no ATP appeared. When ADP was added to the initial mixture, phosphorylation from P_i was inhibited and ATP did not appear.

DISCUSSION

Is Binding of Na^+ Sufficient or is Translocation of Na^+ Necessary? - Electron microscopic examination of the membrane preparation showed membrane fragments with free edges and apparently unbroken membrane vesicles. The vesicles could be resealed fragments of homogenized plasma membrane. In these vesicles addition of a high concentration of NaCl could possibly produce a transient concentration gradient of Na^+ across the membrane and induce a translocation of Na^+ which would provide energy for the synthesis. This mechanism requires not only binding of Na^+ to the pump on the exterior face of the membrane but also release of bound Na^+ from the pump on the interior face of the membrane. To test the necessity for a gradient we added agents intended to dissipate a gradient, namely, the ionophore gramicidin or the detergent Lubrol WX. In neither case was the recovery of ATP from the phosphoenzyme significantly affected. We also estimated the rate at which Na^+ reached the site at which it catalyzes phosphorylation from ATP. There was no significant delay within a period of 3 seconds. This is presumably the site from which it would have to be released if translocation were necessary for synthesis. We concluded that translocation was probably not necessary and that binding was sufficient.

Implications for the Mechanism - Since the active site of phosphorylation is the same for phosphorylation from ATP or from P_i, presumably a conformational change in the active center surrounding the active site phosphate group is responsible for the change in functional energy level involved in the distinction between $E_1 \sim P$ and E_2-P. Since Na^+ is unique as a substrate for net transport outward, as a catalyst for phosphorylation from ATP and as an inhibitor of phosphorylation from P_i (Table I), the 3 sites on the enzyme which recognize it (according to the stoichiometry of transport) must make an extremely precise fit to these ions and probably experience an induced conformational change in the process. Our experimental results suggest a strong interaction between conformational changes in the active centers for cation binding and those for phosphorylation (Fig. 6). Our preliminary estimate of the interaction free energy (Weber, 1974) indicates a value of about 10 Kcal both for the change in free energy of hydrolysis of the phosphate group which is induced by the binding of Na^+ and also for the change in binding free energy of Na^+ which is induced by phosphorylation. This is a remarkably large interaction free energy.

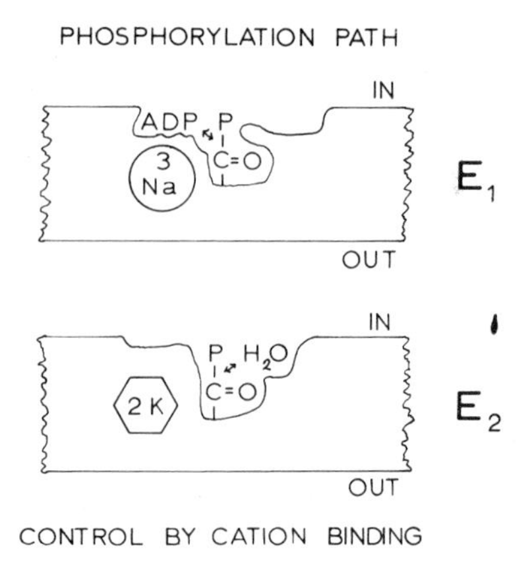

Fig. 6. Hypothetical interaction between conformation of active center for phosphorylation and that for cation binding in sodium and potassium adenosine triphosphatase (From Post et al., 1973).

SUMMARY

Adenosine triphosphate was synthesized from adenosine diphosphate and inorganic phosphate by partial reversal of the forward reaction of sodium and potassium ion-transport adenosine triphosphatase. In a transient reaction the recovery of adenosine triphosphate was less than or equal to the quantity of enzyme. In a first step, potassium-sensitive phosphoenzyme was synthesized from inorganic phosphate. In a second step, addition of adenosine diphosphate and up to one molar sodium ion produced adenosine triphosphate. This enzyme is intrinsic to plasma membranes. However, a concentration gradient of sodium ions across the membrane did not appear to be necessary for synthesis. It was suggested that free energy of interaction between phosphorylation and binding of sodium ions supplied the energy for the reaction.

ACKNOWLEDGEMENTS

This research was supported by grants from the National Institutes of Health, USPHS. They were No. 5R01-HL01974 from the National Heart and Lung Institute and No. 5P01-AM07462 from the National Institute of Arthritis and Metabolic Diseases.

REFERENCES

Hokin, L. E. and Dahl, J. L. (1972) in "Metabolic Pathways", 3rd Ed, Vol. 6, ed. L. E. Hokin (Academic Press, New York and London) p. 269.

Kerkut, G. A. and York, B. (1971) "The Electrogenic Sodium Pump", (Bristol:Scientechnica Ltd., Bristol).

Nishigaki, I., Chen, F. T. and Hokin, L. E. (1974) J. Biol. Chem. 249, 4911.

Post, R. L. (1968) in "Regulatory Functions of Biological Membranes", ed. J. Järnefelt (Elsevier Publ. Co., Amsterdam, London and New York) p. 163.

Post, R. L. and Kume, S. (1973) J. Biol. Chem. 248, 6993.

Post, R. L., Sen, A. K. and Rosenthal, A. S. (1965) J. Biol. Chem. 240, 1437.

Post, R. L., Hegyvary, C. and Kume, S. (1972) J. Biol. Chem. 247, 6530.

Post, R. L., Kume, S. and Rogers, F. N. (1973) in "Mechanisms in Bioenergetics", eds. G. F. Azzone, L. Ernster, S. Papa, E. Quagliariello and N. Siliprandi (Academic Press, New York and London) p. 203.

Post, R. L., Taniguchi, K. and Toda, G. (1974) Ann. N. Y. Acad. Sci. (in press).

Post, R. L., Toda, G. and Rogers, F. N. (1975) J. Biol. Chem. (in press).

Schwartz, A., Lindenmeyer, G. E. and Allen, J. C. (1972) in "Current Topics in Membranes and Transport", Vol. 3, eds. F. Bronner and A. Kleinzeller (Academic Press, New York and London), p. 1.

Weber, G. (1974) Ann. N. Y. Acad. Sci. 227, 486.

ENERGY RELATIONSHIPS IN FLIP-FLOP MODEL OF (Na,K)-ATPase TRANSPORT AND ATP SYNTHESIS FUNCTION

K. R. H. Repke, R. Schön and F. Dittrich

Biomembrane Section, Central Institute of Molecular Biology, Research Centre of Molecular Biology and Medicine, Academy of Sciences of GDR, Berlin-Buch, GDR

INTRODUCTION

(Na,K)-ATPase appears to function according to the mechanism of flip-flop enzymes (Repke and Schön, 1973; Repke et al., 1974b). As detailed by Repke and Schön (1973), the evidence for the flip-flop mechanism of the enzyme relies on the existence of two chemically and functionally identic subunits, the occurrence of half-of-the-sites reactivities and the presence of functional interdependencies of the half units. In the present paper, we wish to examine within the framework of the flip-flop conception at both the molecular and submolecular level two interdependent characteristic features of the (Na,K)-pump, its high thermodynamical efficiency and its easy reversibility which enable the pump to interconvert efficiently chemical and osmotic free energies (for reviews see Caldwell, 1969; Whittam and Wheeler, 1970).- Preliminary communications of the results were published recently (Repke et al., 1974b; Schön et al., 1974; Dittrich et al., 1974). A fuller account will be given elsewhere (Repke and Schön, 1974b).

RESULTS

1. Thermodynamical treatment of flip-flop model

The real understanding of the interconversion of the chemical free energy of ATP and the osmotic free energy of sodium and potassium ion gradients as produced by the (Na,K)-ATPase pump system of cell membrane is of course a matter of quantification that rests on the reasonably precise determination of the thermodynamical parameters of the important intermediary events of the process. Since the imbalances between the cation concentrations in the intra- and extracellular fluid appear to be maintained in a pump and leak system, its thermodynamical treatment would require the application of non-equilibrium thermodynamics. However, the available phenomenological equations which treat the coupled processes of energy conversion as occurring in a "black box" (for a review see Caplan, 1971), cannot describe the molecular mechanism of the pump as required.

Table 1.
Apparent dissociation constants and standard free-energy changes at 37°C of the interactions between various ligands and (Na,K)-ATPase pump system

Ligand		K'_{Diss} mole/l		ΔG^{o}_{obs} kcal/mole	Comments
ATP		$8.0 \cdot 10^{-7}$ [1]		8.7	high affinity site
		$3.0 \cdot 10^{-3}$ [1]		3.6	low affinity site
ADP	appr.	$4.0 \cdot 10^{-6}$ [1]	appr.	7.7	high affinity site
P_i	appr.	$1.0 \cdot 10^{-4}$ [2]	appr.	5.0	at 0°C
Na^+		$1.9 \cdot 10^{-4}$ [3]		5.3	inside membrane
Na^+		$3.1 \cdot 10^{-2}$ [3]		2.1	outside membrane
K^+		$1.4 \cdot 10^{-4}$ [4]		5.5	outside membrane
K^+		$9.0 \cdot 10^{-3}$ [3]		2.9	inside membrane
Mg^{++}		$8.4 \cdot 10^{-7}$ [1]		8.6 [5]	

[1] Concentrations showing half-maximal effect in supporting ouabain binding (Henke et al., 1974)

[2] From phosphoenzyme formation with P_i (Post et al., 1973)

[3] From $^{22}Na^+$ flux measurements (Garay and Garrahan, 1973)

[4] From $^{42}K^+$ flux measurements (Garrahan and Glynn, 1967)

[5] Because of the high free energy which would be required for the discharge of magnesium ions from the enzyme, the metal is assumed to be bound to the enzyme throughout all reaction stages

For the aim of this paper, it appears to be necessary to treat the pump as being in momentary microscopic equilibrium states at the diverse intermediary steps when effecting either the uphill coupled countertransports of sodium and potassium ions driven by ATP hydrolysis or the synthesis of ATP driven by downhill coupled countermovements of sodium and potassium ions. The task then is to transform the available kinetic data of the single events of the over-all process into thermodynamical descriptions of reactants, intermediates and products and to project the data on the reaction and transport steps of the (Na,K)-pump. So, the standard free energies (ΔG^{o}) were calculated from the equilibrium states of the diverse single processes as found under standard conditions at the enzyme or cell level (cf. table 1). Clearly, such equilibrium states are not encountered under

Table 2.

Relationship between the free-energy changes of single events within each of the 4 major reaction steps in (Na,K)-ATPase transport function as referred to the half units either separately (ΔG^o_{obs}) or thermodynamically coupled ($\Delta\Delta G^o_{obs}$)

Step[1]	One half unit: Event	ΔG^o_{obs} kcal/mole	Other half unit: Event	ΔG^o_{obs} kcal/mole	Both half units: $\Delta\Delta G^o_{obs}$ kcal/mole
1 → 2	Discharge of K^+	+2.9	Discharge of Na^+	+2.1	
	Binding of Na^+	−5.3	Binding of K^+	−5.5	
	Discharge of P_i	+7.2	Discharge of ADP	+7.7	
	Binding of ATP[1]	−8.7			
		−3.9		+4.3	+0.4
2 → 3	Translocation of Na^+ and reduction of Na^+-affinity	+3.2	Translocation of K^+ and reduction of K^+-affinity	+2.6	
	Formation of phosphoenzyme[2]	−5.0	Hydrolysis of phosphoenzyme[2]	−2.2	
		−1.8		+0.4	−1.4
3 → 4	Discharge of Na^+	+2.1	Discharge of K^+	+2.9	
	Binding of K^+	−5.5	Binding of Na^+	−5.3	
	Discharge of ADP	+7.7	Discharge of P_i	+7.2	
			Binding of ATP[1]	−8.7	
		+4.3		−3.9	+0.4
4 → 1	Translocation of K^+ and reduction of K^+-affinity	+2.6	Translocation of Na^+ and reduction of Na^+-affinity	+3.2	
	Hydrolysis of phosphoenzyme[2]	−2.2	Formation of phosphoenzyme[2]	−5.0	
		+0.4		−1.8	−1.4

[1] Step includes all events occurring between the consecutive reaction stages as visualized in fig. 1

[2] The total free-energy change for phosphoenzyme formation and hydrolysis was set to be equal to $\Delta G^o_{ATP} = -7.2$ kcal/mole (Rosing and Slater, 1972)

physiological conditions so that what results is an idealization for well-defined standard state equilibria of the various steps (cf. table 2). However, it is felt that the found relationship between the free-energy changes of the single events within each of the reaction steps on principle truly describes the thermodynamical behaviour of the chemiosmotic machinery.

The single events known to be involved in (Na,K)-ATPase reaction and pump function may be combined in a different manner to result in the essential transport steps. In table 2, for both half units sets of events are arranged in the conventional order for the diverse steps. Thus, it is suggested that the transport of sodium ions is combined with the formation of the phosphoenzyme and the transport of potassium ions with the hydrolysis of the phosphoenzyme. If (Na,K)-ATPase would not follow the flip-flop mechanism, but would function according to the widely accepted one unit sequential model (cf. Skou, 1972), the intermediary events should be accompanied by the free-energy changes calculated for one unit of the enzyme (cf. table 2). Thus, the net free-energy changes occurring in the consecutive reaction steps, would show maximum and minimum swings between -3.9 and +4.3 kcal/mole. Such high swings necessarily accompanied by large energy dissipations, are incompatible with the known high thermodynamical efficiency and easy reversibility of the (Na,K)-pump system. In fact, the occurrence of highly positive ΔG^{o}-values should even block the enzymic reaction at this point since the energy required to overcome the sink cannot be borrowed from the surroundings as thermal heat.

From thermodynamical reasons, too, we thus found us forced to challenge a flip-flop mechanism for the (Na,K)-ATPase pump system (Repke et al., 1974b). In dimeric flip-flop enzymes (for a review see Lazdunski, 1972), all relevant changes occurring in the one or the other half unit along the reaction pathway are chemically, mechanically and thermodynamically coupled and coordinated with the result of the efficient use of free energies and the control of activity. In (Na,K)-ATPase (cf. fig. 1), there are couplings between the half units as to ATP binding and ADP release (stages 1→2 and 3→4), phosphorylation of the enzyme and hydrolysis of the phosphoenzyme (2→3 and 4→1), binding of both sodium and potassium ions (1→2 and 3→4), translocation of both sodium and potassium ions across the membrane as well as reversal of the cation affinities of both binding sites (2→3 and 4→1). Of course, these diverse events should also thermodynamically be coupled.

As pointed out by Lumry (1974), local geometric changes of enzyme proteins resulting from chemical transformations and ligand interactions, may spread through the subunit across the interface into the contiguous subunit hence altering its catalytic and binding sites. Thus the free-energy changes produced by any modification at any site of the enzyme and any point of the reaction pathway can become thermodynamically effective via changes of protein conformation for the

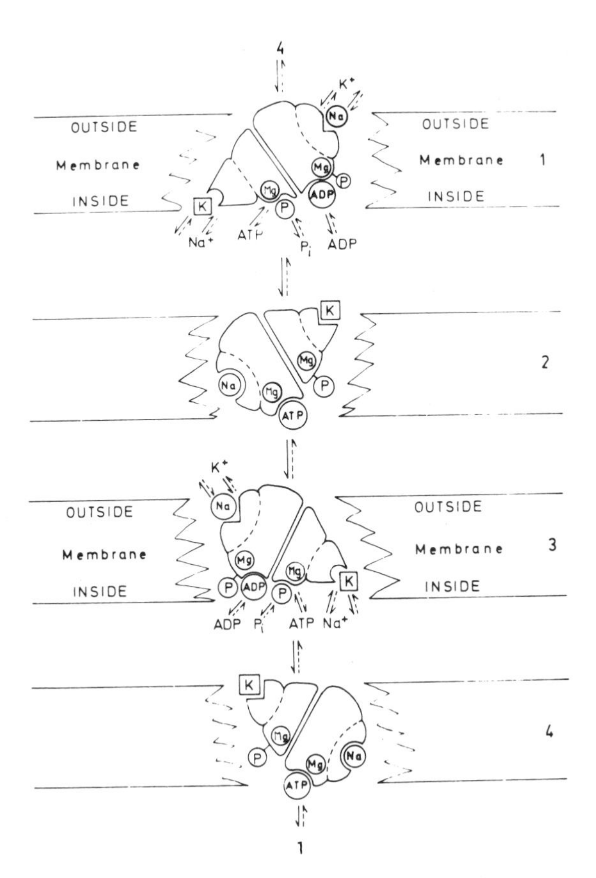

Fig. 1. Schematic visualization of the flip-flop mechanism of (Na,K)-ATPase function. From the multitude of possible intermediate stages, only four have been selected as the most relevant ones for the description of the enzyme function; they are at best average stages of unknown composition (cf. fig. 3). Stages 1 and 3 as well as stages 2 and 4 are mirror images. Of the three binding sites for sodium ions and of the two binding sites for potassium ions, only one of each is shown for the sake of clarity. The scheme is designed to describe the mechanistic implications of flip-flop behaviour only, but should not suggest any bearing on the real geometries of the system. Presumably, the changes of the geometries of the cation binding sites effecting the cation transports (stages 2 → 3 and 4 → 1) involve movements of peptide chains by a few Ångstrøms only so that the cation sites communicate freely with either the intra- or extracellular fluid.

whole enzyme. Free energies of diverse binding processes and chemical reactions can be pooled to effect transport steps with too large a free-energy requirement to be met by a single "loading" step.

Similarly, a large free energy of loading (e.g. ATP binding or formation of phosphoenzyme) can efficiently be utilized to drive a number of less costly processes of chemical, electrostatic and mechanical nature (e.g. translocation of cation binding sites across the membrane or change of their geometries altering their cation specificities, cf. fig. 1).

The guiding idea of our attempts to design a functional model of the (Na,K)-ATPase pump machinery was then the application of the above-outlined principle of free-energy complementarity to the earlier proposed flip-flop model (Repke and Schön, 1973). We emerged with the thermodynamical model shown in table 2 and the corresponding mechanistic model shown in fig. 1.

In the _conversion of ATP free energy into osmotic work_, the free-energy changes of the single events in both half units of (Na,K)-ATPase approximately compensate each other at the four major steps forming all together one complete reaction and transport cycle (cf. table 2 and fig. 1). The net free-energy profile for both half units together swings along the reaction coordinate between +0.4 and -1.4 kcal/mole (fig. 2). The flat net free-energy profile appears to be the basis for the reversibility of the (Na,K)-ATPase system, i. e. for its capacity to function either as transport ATPase or as ATP synthetase (for reviews cf. Caldwell, 1969; Whittam and Wheeler, 1970). The simultaneous presence of complementary free-energy maxima and minima in the different reaction steps reduces entropy production. This may be the basis for the high thermodynamical efficiency of the (Na,K)-pump computed to be 80% in squid axon (Caldwell, 1969). The entropic energy loss during the operation of the pump drives the chemical reactions and by this the cation translocations in the direction of ATP hydrolysis and uphill cation transports. This appears to account for the fact that in the cell the (Na,K)-ATPase transport and ATP synthesis system under steady conditions and at sodium and potassium ion gradients not too adverse to be overcome essentially works as an osmotic machine.

For the _conversion of osmotic free energy into chemical free energy of ATP_, the outlined model suggests the following implications (fig.1). Steep cation gradients translocate the sodium ion binding sites from membrane outside to inside and the potassium ion binding sites from membrane inside to outside and simultaneously convert these sites from low to high affinities for the proper cation. The resulting free-energy gains (cf. table 1) are utilized for the synthesis of ATP when a correspondingly low concentration ratio of ATP to ADP and orthophosphate also favours the backward running of the pump.

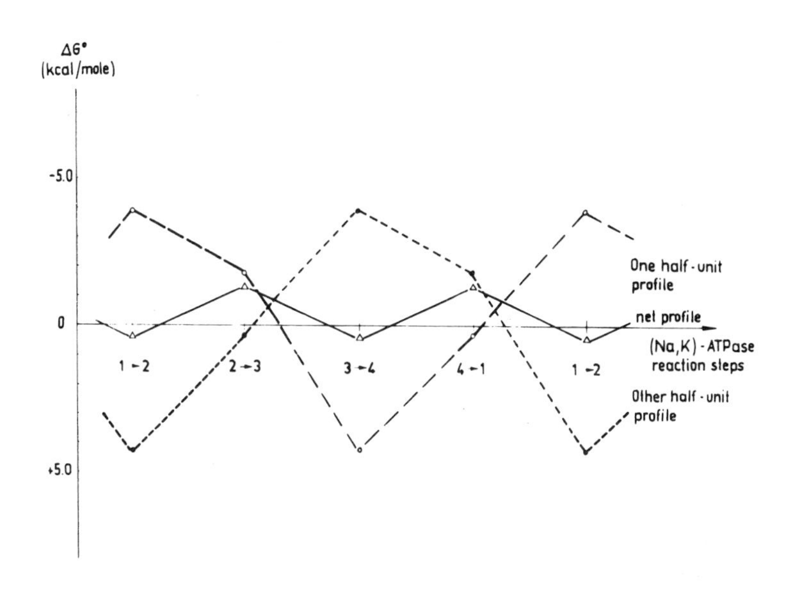

Fig. 2. Net free-energy profiles of the single events at the major reaction and transport steps of (Na,K)-ATPase pump system when the processes are occurring either separately and sequentially in each half unit or thermodynamically coupled and simultaneously. Data taken from table 2.

2. Mechanism of free-energy transfer and utilization

In effect, we wish to propose a mechanism for the conversion of the skalar chemical free energy of ATP into the vectorial mechanical free energy of the enzyme which is able to do osmotic work. The hypothetical mechanism is based on the fundamentals of phosphorus chemistry (Gillespie et al., 1971; Marquarding et al., 1973; Benkovic and Schray, 1973), the physical chemistry of protein behaviour in relation to physiological function (Lumry and Biltonen, 1969; Lumry, 1971, 1974) and, of course, the known facts of the intermediary reactions in ATP hydrolysis by (Na,K)-ATPase (cf. Post et al. 1973; Repke et al., 1974b).

According to molecular orbital calculations (Alving and Laki, 1972), considerable free energy is stored in ATP by the change from the tetrahedral arrangement of the orthophosphate ion to the unsymmetrical arrangements of the oxygen ligands found in the polyphosphate chain. The authors suggested that these "high energy rearrangements" are the result of perturbation of mutually repulsive electrostatic charges and a form of potential mechanical energy. Our thinking about the problem of energy conversion was guided firstly by the idea that a mechanism should exist for the most direct conversion of the high energy rearrangements of ATP into mechanical free energy of (Na,K)-ATPase and secondly by our experimental finding that already ATP binding to the enzyme in presence of sodium ions results in a conformational change of the enzyme traceable by ouabain binding which appeared to be the equivalent of mechanical free energy usable for osmotic work (Schönfeld et al., 1972). We came up with the proposal of a mechanism shown in figure 3 which will briefly be commented in the following as far as belonging to the major topic of this paper.

In exchange reactions of tetracovalent phosphorus compounds as they occur during the phosphorylation of the enzyme by ATP (stages 1→3) and the hydrolysis of the phosphoenzyme (3→1), pentacoordinated transition states and pentacovalent unstable intermediary stages (phosphoranes) are involved.- Energy transfer by ATP binding to active centre (1→1") results from the Berry pseudorotation of the ligands of ATP terminal phosphorus (1"→2). Stage 1" corresponds to a pentacovalent form of the high energy rearrangement of ATP. The bulky ADPO-residue lies here on the equatorial plane in a strained and thus thermodynamically unfavoured conformation. Its free energy becomes partially relieved by the ligand permutation which shifts the ADPO-residue to the thermodynamically favoured apical position (stage 2) thus making it a good leaving group. The free-energy release from such a process may be quite high since for the Berry pseudorotation of the ligands of the comparable model compound $P(OH)_3(O^-)_2$, ΔG^0 is calculated to be as high as -13 kcal/mole (primary data taken from Gillespie et al., 1971).

In the present special case, however part of ATP free energy is conserved by the formation of the enzyme carboxyl phosphate intermediate (2→3) which is dependent on Mg^{++} catalysis. The residual free energy of ATP becomes transferred to the enzyme during the hydrolysis of the phosphorylated intermediate (3→4"→1). In stage 4', the bond between the carboxyl residue and the P atom lies on the equatorial plane. Because of the voluminous nature of the ligand, the position is strained and thermodynamically unfavoured. This is the driving force for the permutation of the phosphorus ligands (4'→4") which results in the apical position of the carboxyl substituent making the phosphate group (stage 1) a good leaving group. The free-energy change of this ligand permutation (via Berry pseudorotation or turnstile rotation) appears to be lower than the former one (1"→2) (cf. table 2).

All steps of ATP hydrolysis, among them the permutation of the ligands of ATP terminal phosphorus involved in the phosphorylation of the enzyme as well as the permutation of the ligands of the carboxyl phosphate phosphorus involved in the dephosphorylation of the enzyme, occur in tight interactions with active centre groups and thus induce rearrangements of side chains and water molecules important for enzyme conformation. The linkage between such synchronous chemical and conformational processes may be effected over a considerable distance within a protein and the

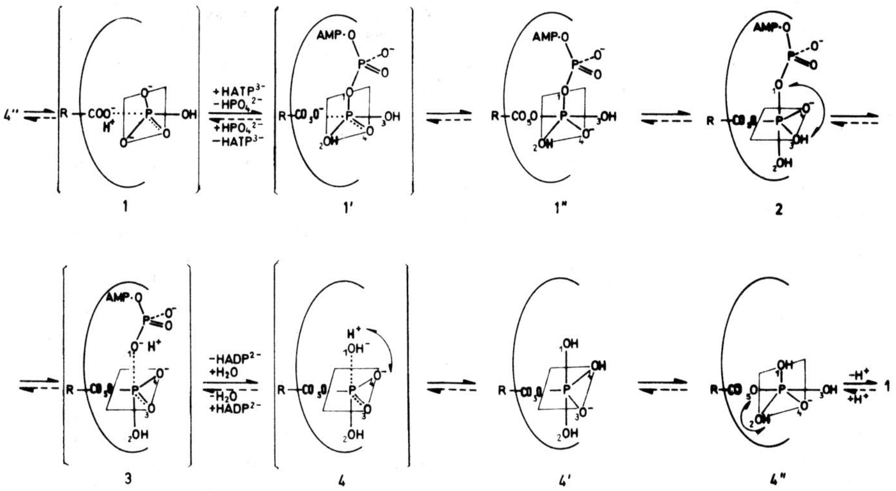

Fig. 3. Hypothetical reaction mechanism of ATP hydrolysis or synthesis by (Na,K)-ATPase. Only the active centre of one half unit of the enzyme is shown. The numbering of the various depicted stages refers to the left half unit of fig. 1. The stages shown in cornered brackets are transition states. R is an aspartate residue known to be the phosphoryl acceptor in the active centre (Degani and Boyer, 1973). The apical or equatorial position of and the charge distribution between the ligands of phosphorus is decided by the preference rules for apical-pseudorotation-apical displacement reactions which describe the pathway for the process with the lowest activation energy requirements. The stabilization of the pentacovalent species arises from interactions of the ligands with binding groups of the active centre in positions consistent with the geometry of pentacovalent phosphorus thus accomplishing pseudocyclization (not shown). The proton transfers (cf. stages 2 → 3 and 4" → 1) are facilitated by the much higher acidity of equatorial compared to apical hydroxy groups.

amount of free energy conserved in its conformation may amount up to -25 kcal/mole (cf. Lumry, 1971, 1974). In (Na,K)-ATPase transport function, the chemical free energy of ATP is thus gradually released and transformed into "mechanical" free energy of conformational rearrangements in the enzyme usable for doing osmotic work.

The mechanical free energy of the enzyme is essentially exploited for two simultaneous quasi-mechanical energy-requiring processes in enzyme transport function (cf. stages 2 → 3 and 4 → 1 in fig. 1 and table 2), i.e. first the translocation of sodium binding sites of the enzyme from communicating with the intracellular to communicating with the extracellular fluid as well as the translocation of the potassium ion binding sites from communicating with the extracellular to communicating with the intracellular fluid and second the change of the geometries of the cation binding sites from high sodium ion to high potassium ion affinity in the one half unit (membrane outside) and from high potassium ion to high sodium ion affinity in the other half unit (membrane inside). Of course, both mechanical processes are fundamental for (Na,K)-ATPase transport function.

On principle, the outlined reaction mechanism of ATP hydrolysis is fully reversible. However, ATP synthesis by (Na,K)-ATPase occurs only in the intact cell. Osmotic free energies able to drive the pump to run backward become transformed into mechanical free energies of (Na,K)-ATPase (3 → 1) and 1 → 4 in fig. 1) which are used for the pseudorotation of the phosphorus ligands required for the phosphorylation of both the enzyme by orthophosphate (1 → 4" → 3 in fig. 3) and of ADP by the phosphoenzyme (3 → 1).

COMMENTS

As will be summarized in the following, the proposed chemical reaction mechanism is in line with or accounts for the known submolecular processes of (Na,K)-ATPase reaction.

1. In the disruption of the terminal phosphoryl group of ATP (stages 1' → 3 in fig. 3), the more peripheral bond is broken (cf. Benkovic and Schray, 1973).
2. The source of the bridge oxygen in the enzyme phosphorylated by ATP is the phosphoryl accepting carboxyl oxygen of active centre (Dahms et al., 1973) which displaces a phosphoryl oxygen (2 → 3).
3. In the phosphoenzyme intermediate (stage 4), water attacks the phosphoryl phosphorus and not the acyl carbon (Dahms et al., 1973).
4. The isolated enzyme catalyzes neither an ATP ⇌ orthophosphate exchange nor an ATP ⇌ water exchange (Dahms and Boyer, 1973) which would require a complete reversal of the whole reaction sequence. The apparent irreversibility is accounted for by the high free-energy demands of both the pseudorotation of the phosphorus ligands (2 → 1") and the ATP release.(1' → 1).
5. During ATP cleavage, an incorporation of oxygen from water into orthophosphate occurs which is about 50% in excess that expected for ATP hydrolysis. Thus, the phosphoenzyme intermediate appears to be formed simultaneously from ATP as well as from orthophosphate under conditions optimal for ATP hydrolysis (Dahms and Boyer, 1973) (cf. the

reversibility of the second part of the reaction cycle).

6. The phosphorylation of the enzyme by orthophosphate in the presence of magnesium ions $(1 \rightarrow 4' \rightarrow 4 \rightarrow 3)$ occurs easily (cf. Dahms et al., 1973) since the involved permutation of phosphorus ligands $(4'' \rightarrow 4')$ demands less free energy than the other one $(2 \rightarrow 1'')$. Nevertheless, this phosphorylation process, too, induces a conformational change of the enzyme which is traceable by ouabain binding (Schönfeld et al., 1972).

7. The source of the bridge oxygen in the enzyme phosphorylated by orthophosphate is again the carboxylate oxygen (Dahms et al., 1973).

8. In the presence of magnesium ions and potassium ions, the enzyme catalyzes a rapid orthophosphate $\rightleftharpoons$ water exchange (Dahms and Boyer, 1973). The reversible elimination of a water molecule from incorporated phosphate occurs at the level of the phosphoenzyme intermediate and involves the displacement of a phosphoryl oxygen by a water oxygen (Dahms et al., 1973) (cf. stage $4 \rightleftharpoons 3$).

9. The orthophosphate $\rightleftharpoons$ water exchange as described before is not blocked by oligomycin (Dahms and Boyer, 1973). Thus, the inhibition of the enzymic ATP hydrolysis by oligomycin must occur at an earlier enzyme stage presumably by combination of oligomycin with the intermediate delineated in stage 2.

10. ADP is not required for the orthophosphate $\rightleftharpoons$ water exchange. The lack of ADP requirement and lack of ATP $\rightleftharpoons$ orthophosphate and ATP $\rightleftharpoons$ water exchanges make it extremely unlikely that the orthophosphate $\rightleftharpoons$ water exchange as catalyzed by (Na,K)-ATPase involves total reversal ATP hydrolysis (Dahms and Boyer, 1973).

SUMMARY AND CONCLUSIONS

1. A thermodynamical treatment of the (Na,K)-ATPase transport and ATP synthesis system confirms the challenge of a flip-flop type mechanism.

2. The free-energy changes of the single events in both half units of (Na,K)-ATPase approximately compensate each other at the four major steps forming together one complete reaction and transport cycle.

3. The resulting flatness of the net free-energy profile along the reaction coordinate appears to account for both the easy reversibility and the high thermodynamical efficiency of the (Na,K)-pump system.

4. Based on the fundamentals of phosphorus chemistry and physical chemistry of protein behaviour, a hypothetical mechanism for the interconversion of skalar chemical free energy of ATP into vectorial osmotic free energy is described.

5. The proposed chemical reaction mechanism is in line with or accounts for the known submolecular processes of (Na,K)-ATPase reactions.
6. (Na,K)-ATPase appears to be a prototype enzyme for the interconversion of osmotic and chemical free energies and was found to be suitable for the design of a functional model of mitochondrial ATPase (Repke and Schön, 1974a; Repke et al., 1974a) which seems to work both as ATP synthetase and transport ATPase, too.
7. The thermodynamical implications of the flip-flop mechanism outlined in this paper for (Na,K)-ATPase, may apply for other flip-flop enzymes, too.

ACKNOWLEDGEMENTS

The authors wish to thank Professor R. Lumry for encouragement and Dr. J. G. Reich for fruitful discussions.

REFERENCES

Alving, R. E. and Laki, K. (1972). J. theor. Biol. 34, 199.

Benkovic, S. J. and Schray, K. J. (1973). In: The enzymes, 3rd edition, vol. 8, ed. P. Boyer (Academic Press, Inc., New York and London) p. 201.

Caldwell, P. C. (1969). In: Current Topics in Bioenergetics, vol. 3, ed. D. R. Sanadi (Academic Press, Inc., New York and London) p. 251.

Caplan, S. R. (1971). In: Current Topics in Bioenergetics, vol. 4, ed. D. R. Sanadi (Academic Press, Inc., New York and London) p. 1.

Dahms, A. St. and Boyer, P. D. (1973). J. biol. Chem. 248, 3155.

Dahms, A. St., Kanazawa, T. and Boyer, P. D. (1973). J. biol. Chem. 248, 6592.

Degani, Ch. and Boyer, P. D. (1973). J. biol. Chem. 248, 8222.

Dittrich, F., Schön, R. and Repke, K. R. H. (1974). Acta biol. med. germ. 33, K17.

Garay, R. P. and Garrahan, P. J. (1973). J. Physiol. (Lond.) 231, 297.

Garrahan, P. J. and Glynn, I. M. (1967). J. Physiol. (Lond.) 192, 175.

Gillespie, P., Hoffmann, P., Klusacek, H., Marquarding, D., Pfohl, S., Ramirez, F., Tsolis, E. A. and Ugi, I. (1971). Angew. Chemie 83, 691.

Henke, W., Schön, R. and Repke, K. R. H. (1974). In preparation.

Lazdunski, M. (1972). In: Current Topics in Cellular Regulation, vol. 6, eds. B. L. Horecker and E. R. Stadtman (Academic Press, Inc., New York and London), p. 267.

Lumry, R. (1971). In: Electron and Coupled Energy Transfer in Biological Systems, vol. 1, part A, eds. T. E. King and M. Klingenberg (Marcel Dekker, Inc., New York), p. 1.

Lumry, R. (1974). Ann N. Y. Acad. Sci., in press.

Lumry, R. and Biltonen, R. (1969). In: Structure and Stability of Biological Macromolecules, vol. 2, eds. S. N. Timasheff and G. D. Fasman (Marcel Dekker, Inc., New York), p. 65.

Marquarding, D., Ramirez, F., Ugi, I. and Gillespie, P. (1973). Angew. Chemie 85, 99.

Post, R. L., Kume, S. and Rogers, F. N. (1973). In: Mechanisms in Bioenergetics, eds. G. F. Azzone, L. Ernster, S. Papa, E. Quagliariello and N. Siliprandi (Academic Press, Inc., New York and London), p. 203.

Repke, K. R. H. and Schön, R. (1973). Acta biol. med. germ. 31, K19.

Repke, K. R. H. and Schön, R. (1974a). Acta biol. med. germ. 33, K27.

Repke, K. R. H. and Schön, R. (1974b) Biochim. biophys. Acta Reviews on Biomembranes, in preparation.

Repke, K. R. H., Dittrich, F. and Schön, R. (1974a). Acta biol. med. germ. 33, K39.

Repke, K. R. H., Schön, R., Henke, W., Schönfeld, W., Streckenbach, B. and Dittrich, F. (1974b). Ann. N. Y. Acad. Sci., in press.

Rosing, J. and Slater, E. C. (1972). Biochim. biophys. Acta 267, 275.

Schön, R., Dittrich, F. and Repke, K. R. H. (1974). Acta biol. med. germ. 33, K9.

Schönfeld, W., Schön, R., Menke, K.-H. and Repke, K. R. H. (1972). Acta biol. med. germ. 28, 935.

Skou, J. C. (1972). Bioenergetics 4, 203.

Whittam, R. and Wheeler, K. P. (1970). Ann. Rev. Physiol. 32, 21.

Mitochondrial Bioenergetics

MITOCHONDRIAL BIOENERGETICS: FACTS AND IDEAS*

LARS ERNSTER

Department of Biochemistry
Arrhenius Laboratory
University of Stockholm
Stockholm, Sweden

The term bioenergetics was coined by Albert Szent-Györgyi (1957) and refers to that part of biology - primarily biochemistry and biophysics - which deals with the handling of energy in living matter. Life processes, just as all events that involve work, require energy, and it is quite natural that such activities as muscle contraction, nerve conduction, transport, growth, reproduction, as well as the synthesis of all those substances that are necessary for carrying out and regulating these activities, could not take place without an adequate supply of energy.

It is now well established that the cell is the smallest biological entity that is capable of handling energy. Common to all living cells is the ability, by means of suitable catalysts, to derive energy from their environment, to convert it into a biologically useful form, and to utilize it for driving various energy-requiring processes.

Although there is a great variety of cellular processes requiring energy, most of them utilize one single compound as the energy source (Fig. 1).

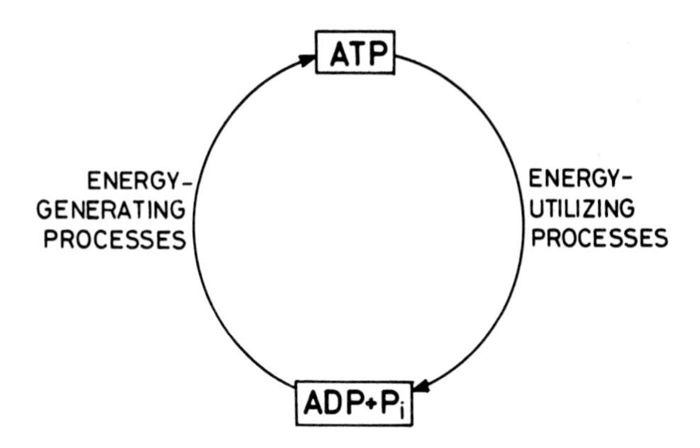

Fig. 1. The ATP Cycle

---

*Plenary Lecture, 9th FEBS Meeting, Budapest, August 25, 1974.

This universal "energy currency" is adenosine triphosphate (ATP). Relatively large amounts of energy - about 8000 calories per mole - can be stored in each of the two pyrophosphate bonds of ATP and liberated when these bonds are split. Consequently, there are many enzymes that catalyze the splitting of ATP and the utilization of the energy so liberated for various specific energy-requiring reactions; all of these reactions lead, directly or indirectly, to the formation of adenosine diphosphate (ADP) and inorganic phosphate (P_i).

In contrast, energy-generating reactions are relatively few, and they all involve an oxidoreduction as the underlying energy-yielding reaction. These reactions, commonly called oxidative or electron transport-linked phosphorylations, thus play a fundamental role in the energy supply of living cells. The highly simplified scheme in Fig. 2, describing what may be called the "Redox Cycle of the Biosphere" may serve as a general background to biological energy-conserving processes.

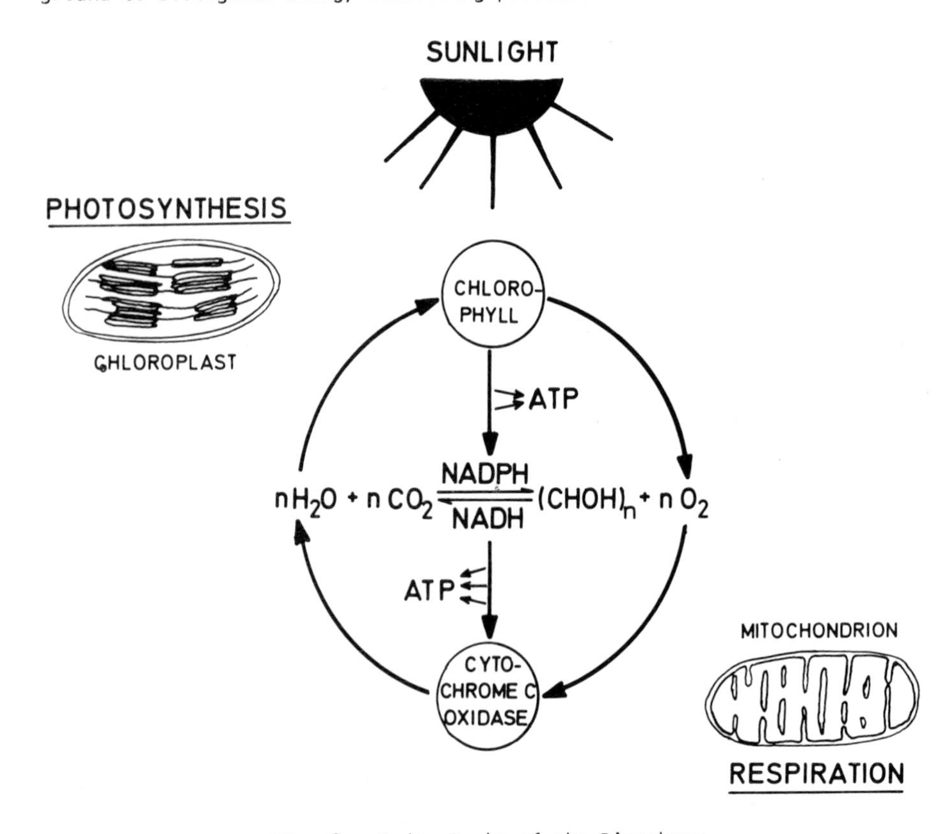

Fig. 2. Redox Cycle of the Biosphere

Sunlight is the primary source of energy for all life on Earth. Green plants can capture quanta of light-energy, photons, by means of the pigment, chlorophyll, and use this energy for splitting water into oxidizing and reducing equivalents. The oxidizing equivalents appear in the form of molecular oxygen, and this is how the oxidizing atmosphere of our Planet is thought to have evolved. The reducing equivalents are transferred through a series of redox catalysts to the nicotinamide nucleotide, $NADP^+$, to form NADPH. In the course of this electron transfer, energy is conserved and utilized for the condensation of ADP and P_i to ATP. The latter serves, in turn, as a source of energy for the manifold energy-requiring reactions of the plant cell. In the first place, through the joint action of NADPH and ATP, carbon dioxide is fixed and converted into carbohydrate and other organic material.

The whole process of photosynthesis, outlined in Fig. 2, takes place in the green-plant cell in a special organelle, the chloroplast. A similar process, although not leading to oxygen evolution, also occurs in photosynthetic bacteria and blue-green algae. However, these organisms lack chloroplasts and the photosynthetic machinery is localized in special vesicles derived from the cell membrane. It should be added that photosynthesis by plants and microorganisms accounts for the yearly production of some 10 billion tons of organic material, and that an industrial simulation of the process has been considered as a realistic possibility for facing the increasing energy crisis of mankind.

Let us now turn to the lower part of the cycle in Fig. 2. Cells of both plants and animals are capable of utilizing molecular oxygen, through a hemoprotein called cytochrome c oxidase, as an oxidant for various organic substances. The latter, consisting of carbohydrates, fats or proteins, can serve as nutrients of these cells, being split and oxidized by way of a series of enzymes to yield carbon dioxide and reducing equivalents in the form of the reduced nicotinamide nucleotide, NADH. Once formed, NADH is reoxidized to NAD^+ by way of a sequence of redox catalysts - with cytochrome c oxidase as the terminal catalyst - whereby molecular oxygen is reduced to water. This sequence of catalysts is commonly called the respiratory chain. In the course of this oxidation process, energy is conserved and utilized for the condensation of ADP and P_i to ATP, which, in turn, can serve as a source of energy for the manifold energy-requiring reactions of the plant and animal cells.

The entire process of respiration and respiration-linked phosphorylation as well as the majority of primary oxidations leading to the reduction of NAD^+ take place in the plant and animal cells in a special organelle called the mitochondrion. Respiration and respiration-linked phosphorylation also occur in aerobic bacteria, which lack mitochondria, and where the catalysts responsible for these processes are associated with the cell membrane. In spite of this difference, however, there is a striking similarity in composition and reaction mechanism between the mitochondrial and bacterial respiratory and phosphorylating systems. The same holds for the light-induced electron-transport and phosphorylating systems of plants and photosynthetic microorganisms. Moreover, there are many features in common between the light-induced and the respiratory electron-transport and ATP-synthesizing systems, indicating that they have arisen at a very early stage during evolution.

While mitochondrial respiration and phosphorylation probably represent the latest addition on the evolutionary map of cellular energy-transducing systems, they have the longest history as far as man's interest for bioenergetics is concerned. Rather than giving a historical survey, however, the purpose of the rest of this lecture is to summarize some facts and ideas relating to our present conception of the mitochondrial energy-transducing system.

* * *

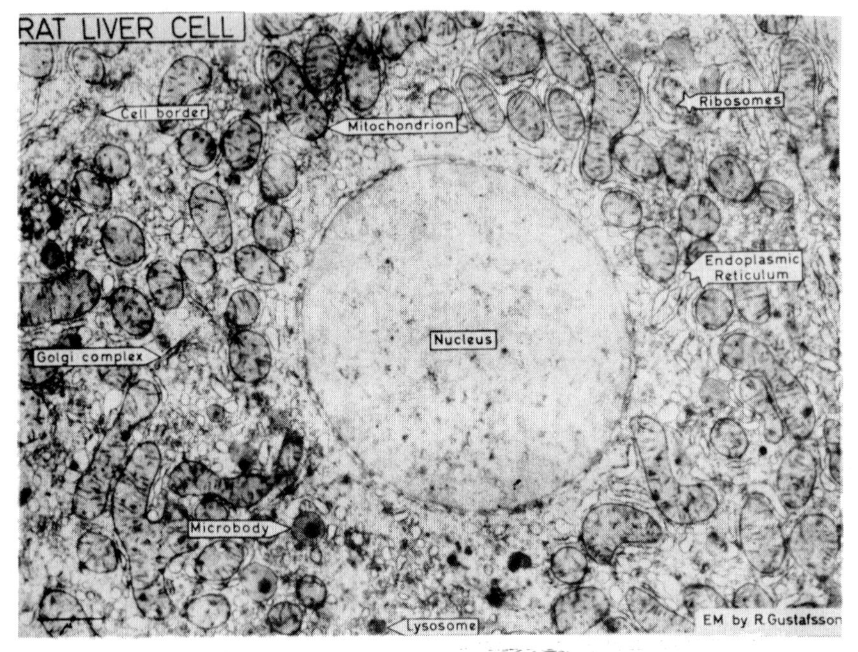

Fig. 3. Electron Micrograph of Rat Liver Cell

Before entering a discussion of various biochemical aspects of the mitochondrial energy-transducing system, it may be appropriate to show a picture of mitochondria as they occur in their natural environment, the intact cell. Fig. 3 is an electron micrograph of a liver cell. It shows numerous mitochondria, 1-2 μ in size, the next largest organelle of the animal cell after the nucleus. There are about 1,000 mitochondria in a liver cell, distributed at random over the cytoplasm.

Fig. 4 is a schematic picture of a mitochondrion, taken from a Swedish schoolbook. It shows the characteristic double-membrane structure of mitochondria, a smooth outer membrane, and a folded inner membrane giving rise to the so-called cristae (Palade, 1956).

A further detail of the inner membrane is seen in Fig. 5, which shows on the left-hand side a negatively stained specimen of a piece of isolated inner membrane. On its surface we can see a row of mushroom-like projections (Fernández-Morán, 1962). The outer membrane, on the right, contains no similar structures (Parsons et al., 1967).

Fig. 6 is a schematic picture of different parts of the mitochondrion: the outer membrane, the intermembrane space, the inner membrane (with the cristae), and the matrix. The projections are located on the inner surface of the inner membrane (Stoeckenius, 1963; Fernández-Morán et al., 1964).

The mitochondrial matrix is the site of the Krebs cycle, of fatty acid oxidation, and of the degradation of amino acids, i.e., of those processes from which the respiratory chain receives its reducing equivalents (Ernster & Kuylenstierna, 1969, 1970; Smoly et al., 1970). The respiratory chain itself, as well as its phosphorylating system, are associated with the inner membrane, with the terminal enzyme of ATP synthesis being located in the inner-membrane projections (Racker et al., 1965).

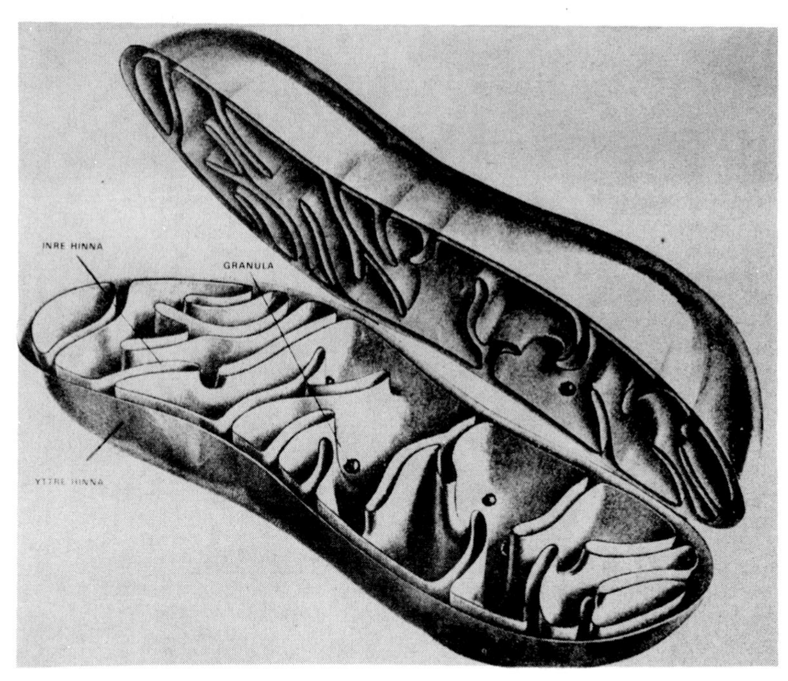

Fig. 4. Schematic Picture of Mitochondrion

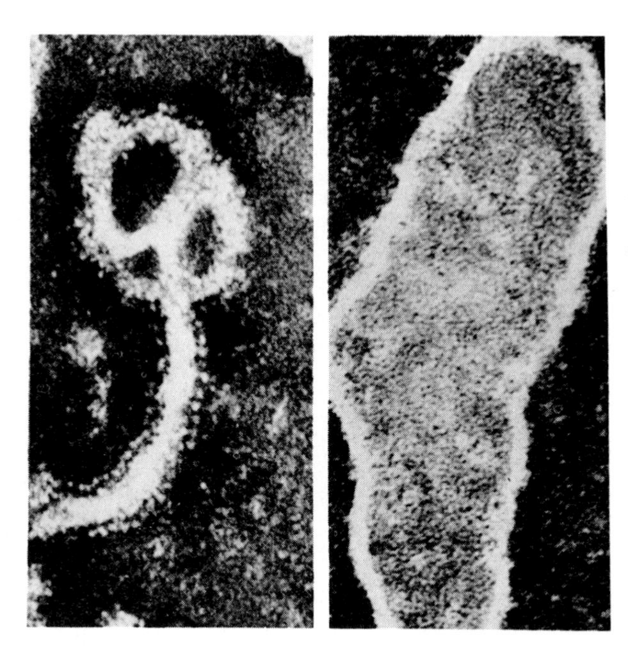

Fig. 5. Electron Micrographs of Negatively Stained Specimens of Isolated Rat-Liver Mitochondrial Membranes
Left: inner membrane
Right: outer membrane
(From Sottocasa et al., 1967)

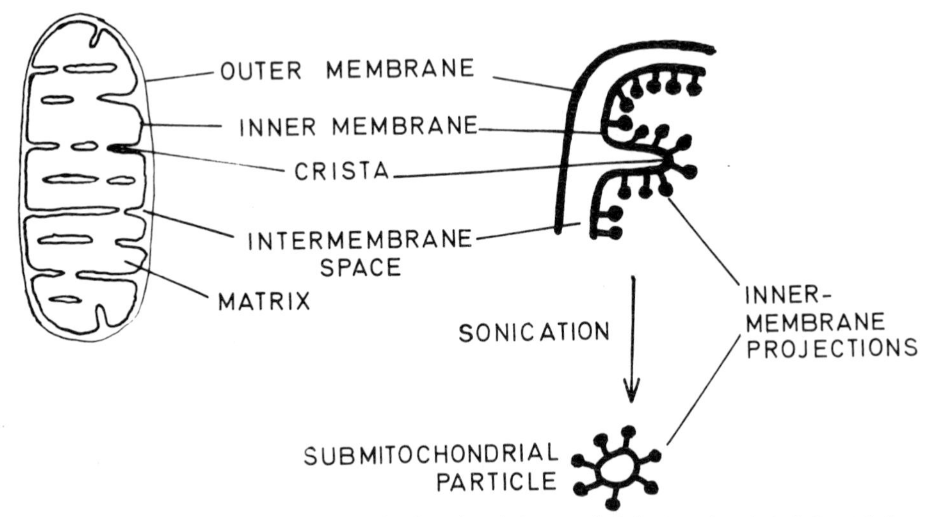

Fig. 6. Schematic Picture of Mitochondrion and Submitochondrial Particle

Fragmentation of mitochondria by sonication results in the formation of vesicles, the outer surface of which corresponds to the inner surface of the inner membrane (Ernster, 1965; Löw, 1966; Lee & Ernster, 1966; Mitchell, 1966a,b). These vesicles, often referred to as submitochondrial particles, contain the respiratory and phosphorylating systems in a concentrated and easily accessible form, and have therefore been of great significance for the study of these systems (Racker, 1970).

Fig. 7 is a schematic picture of the respiratory chain. The respiratory chain may traditionally be defined as that part of the mitochondrial electron-transport system which is firmly associated with the Keilin and Hartree heart-muscle preparation - now known to consist of submitochondrial particles - and which catalyzes the aerobic oxidation of NADH and succinate (Keilin & Hartree, 1947; Slater, 1958). It consists of two flavoproteins, a number of iron-sulfur centers, ubiquinone (or coenzyme Q), and cytochromes *b*, c_1, *c*, *a*, and a_3 (Klingenberg, 1968). Of all protein components of the respiratory chain, so far only cytochrome *c* has been accessible to detailed investigations with respect to its chemical structure; in fact, animal cytochrome *c* today is one of the best-known proteins (Margoliash, 1972).

Work in Green's laboratory has led to the isolation from submitochondrial particles of four particulate lipoprotein complexes: NADH-Q reductase (Complex I), succinate-Q reductase (Complex II), QH_2-cytochrome *c* reductase (Complex III), and cytochrome *c* oxidase (Complex IV), which, when combined in the presence of ubiquinone and cytochrome *c*, gave rise to a reconstituted NADH and succinate oxidase system with properties similar to those of the parent respiratory chain (Green, 1966; Hatefi, 1966; Hatefi *et al.*, 1974). Among these properties is the sensitivity to specific inhibitors, such as rotenone in the case of Complex I, antimycin in the case of Complex III, and cyanide in the case of Complex IV.

As indicated in Fig. 7, each complex contains multiple prosthetic groups and a number of polypeptide subunits. There are indications that some of these components may function not only as catalysts but also as regulators of electron transport through different branches of the chain. For example, in studies with ubiquinone-depleted submitochondrial particles,

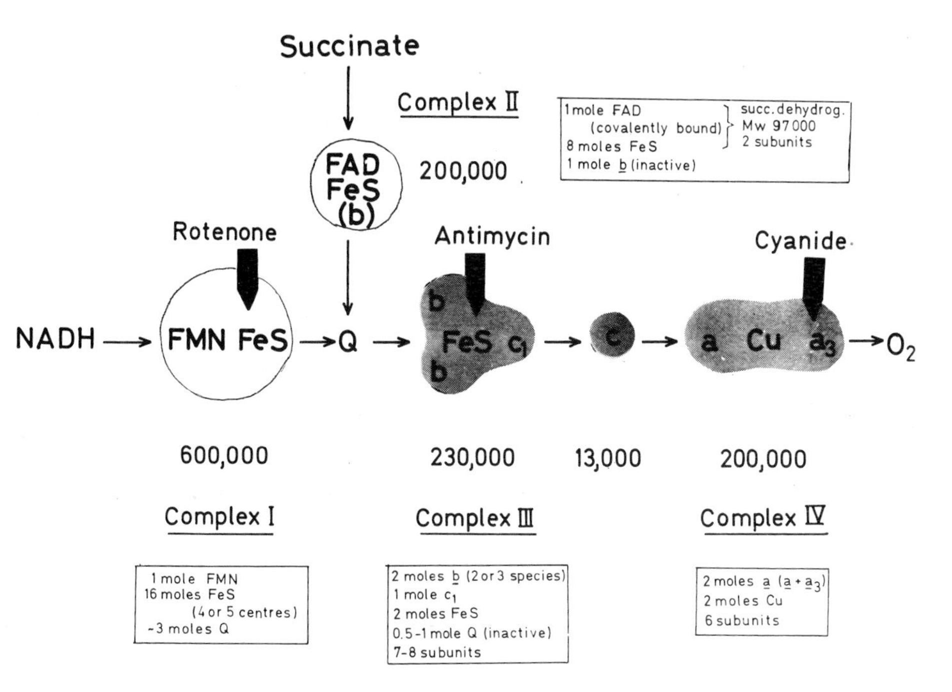

Fig. 7. The Respiratory Chain. Data regarding composition of Complexes I-IV compiled from Capaldi (1974), Hare & Crane (1974), Hatefi *et al.* (1974), Gellerfors & Nelson (1974), Gupta & Rieske (1973), Keirns *et al.* (1971), Yu *et al.* (1974).

we have obtained evidence that ubiquinone, in addition to functioning as an electron-transfer catalyst, alters certain kinetic parameters of both Complexes II and III, probably by interacting with their iron-sulfur centers (Ernster *et al.* 1969; Rossi *et al.*, 1970; Nelson *et al.*, 1971, 1972; Norling *et al.*, 1972, 1974, Głazek *et al.*, 1974).

Fig. 8 is a survey of various hydrogen-transfer catalysts that are located in the mitochondrial inner membrane and matrix and that furnish the reducing equivalents of the respiratory chain. They consist of flavin- and nicotinamide nucleotide-linked dehydrogenases and nicotinamide nucleotide transhydrogenase. Most flavin-linked dehydrogenases, including those involved in the oxidation of the CoA esters of fatty acids, of sarcosine, dimethylglycine, choline, α-glycerol phosphate, as well as a flavoprotein present in yeast and plant cells and catalyzing the oxidation of extramitochondrial NADH ($NADH_e$), probably link to the respiratory chain at the level of ubiquinone (Frisell *et al.*, 1966; Kröger & Klingenberg, 1970). An important exception is the flavoprotein lipoamide dehydrogenase (Massey, 1958) – originally Straub's diaphorase (Straub, 1939) – which serves as a catalytic link between α-ketoacid dehydrogenases and NAD^+.

It is now generally accepted that the respiratory chain contains 3 coupling-sites, at which energy made available during the transfer of a pair of electrons is conserved and can be utilized for the synthesis of ATP from ADP and P_i (Fig. 9). The three sites are located on the path of electrons from NADH to ubiquinone, from ubiquinone to cytochrome c and from

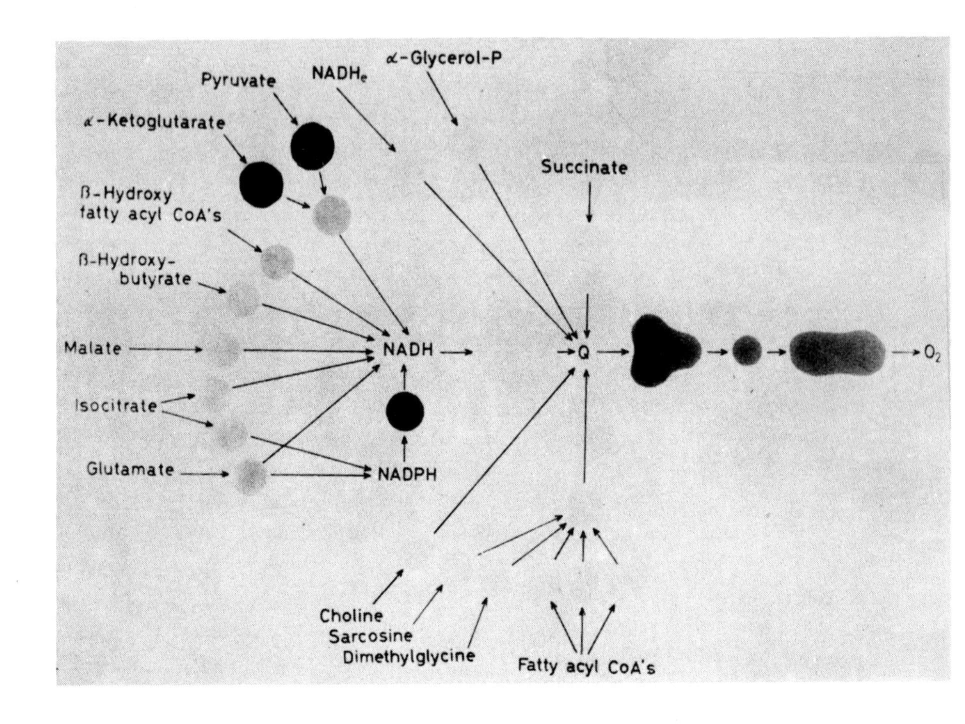

Fig. 8. Mitochondrial Dehydrogenases and Transhydrogenase

reduced cytochrome c to O_2, respectively, i.e., in Complexes I, III and IV. The evidence for this location of the energy-coupling sites rests on measurements of the phosphorylating efficiency (P/O ratio) with various substrates and artificial electron donors and acceptors (Lehninger, 1955), on studies of the steady-state kinetics of respiratory-chain catalysts in the presence and absence of ADP (State 3 and State 4) (Chance & Williams, 1956), and, more recently, on direct reconstitution experiments involving the individual complexes together with the ATP-synthesizing system (Racker, 1974).

It is also generally accepted that the 3 coupling sites utilize a common ATP-synthesizing system, and that the terminal catalyst in this system consists of the cold-labile ATPase called coupling factor F_1, isolated in the early 1960's in Racker's laboratory (Pullman et al., 1960; Racker, 1970). Evidence that F_1 indeed constitutes the enzyme responsible for the synthesis of ATP in respiratory chain-linked phosphorylation - and in light-induced phosphorylation as well - comes from reconstitution experiments (Kagawa, 1972; Racker, 1974) from studies involving the use of F_1--specific antibodies (Fessenden & Racker, 1966), and from investigations of E. coli mutants lacking F_1 (Cox et al., 1971, 1973; Gutnick et al., 1972; Nieuwenhuis et al., 1973).

Chemically, the ATP-synthesizing system consists of 3 major parts and it has been suggested that these correspond morphologically to the "base piece", the "stalk" and the "head" of the mitochondrial inner-membrane projections (Fig. 10) (Senior, 1973). The "base piece" is a hydrophobic protein (HP) complex consisting of probably four types of subunits and containing the sites of action of the phosphorylation inhibitors oligomycin (Lardy et al., 1958) and dicyclohexylcarbodiimide (DCCD) (Beechey et al.,

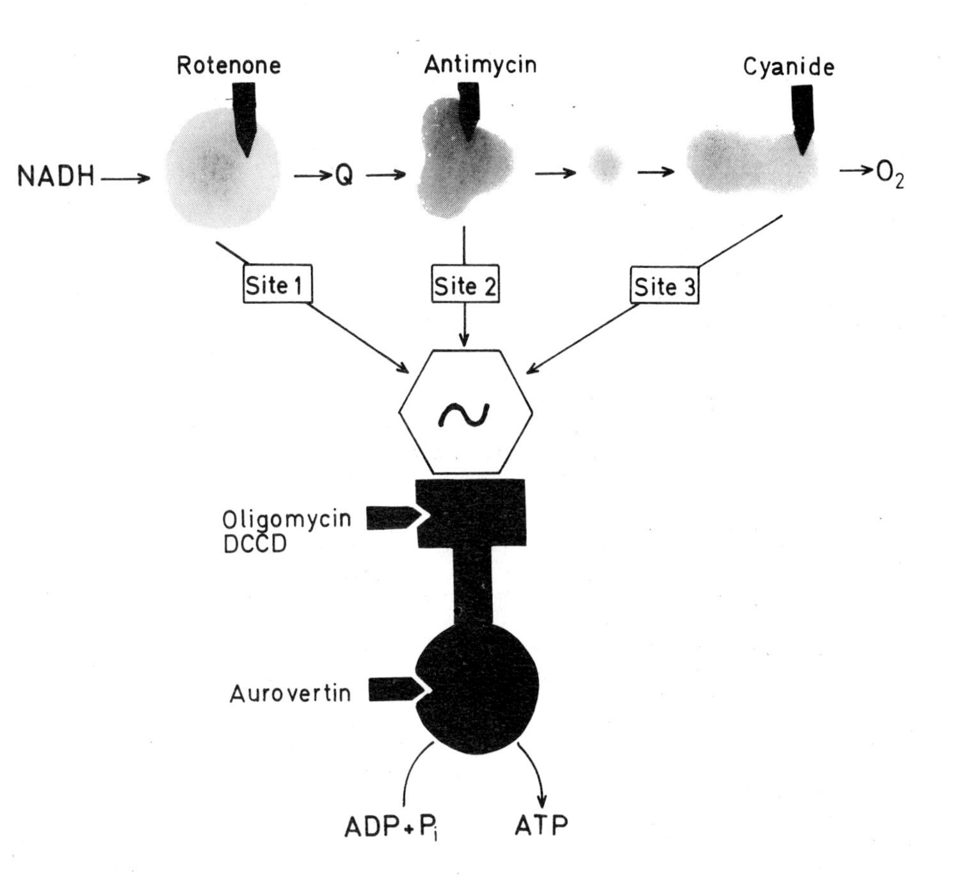

Fig. 9. Energy-Coupling Sites of the Respiratory Chain

1967). The "stalk" consists of a water-soluble protein usually referred to as the oligomycin-sensitivity conferring protein (OSCP) (MacLennan & Tzagoloff, 1968). It is required for the attachment of F_1 to the hydrophobic protein complex, thus rendering its ATPase activity sensitive to oligomycin and DCCD.

F_1 itself, the "head", is a large protein complex, consisting of five types of firmly associated subunits (Senior, 1973). Each of the two largest subunits has been concluded to occur in a number of 3 per molecule of F_1, but recent determination of the number of sulfhydryl groups indicates even numbers of subunits (Senior, 1974). The two largest subunits seem to be engaged in the catalytic activity of F_1, whereas the third subunit is required for its ability to bind to the membrane (through OSCP) and to function in oxidative phosphorylation (Nelson et al., 1973). The function of the two smallest subunits is not known. F_1 contains firmly bound adenine -nucleotides, the number and composition of which seem to vary with the source of the enzyme, and the role of which may be both catalytic, serving as primary phosphate acceptors, and regulatory (Senior, 1973).

The antibiotic aurovertin is a specific inhibitor of F_1, with the valuable feature of forming a tightly-bound, highly fluorescent complex with

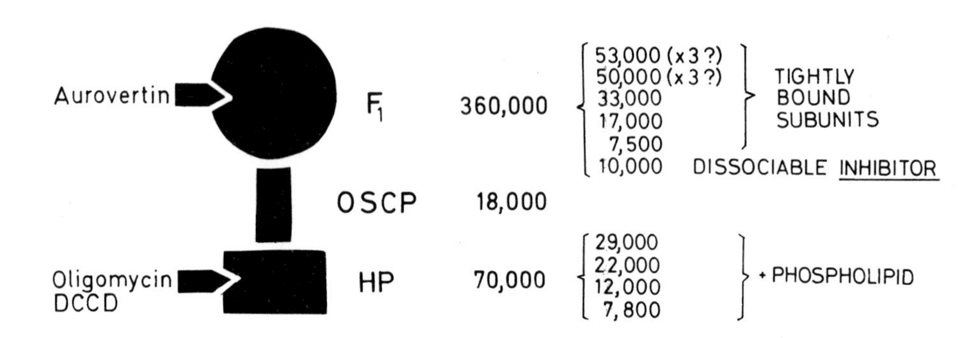

Fig. 10. The Mitochondrial ATPase System
OSCP = oligomycin-sensitivity conferring protein
HP = hydrophobic protein
(Data from Senior, 1973)

the enzyme (Lardy, 1961; Lardy & Lin, 1969; Chang & Penefsky, 1973; Bertina et al., 1973). Interestingly, aurovertin and oligomycin inhibit oxidative phosphorylation in an additive manner (Lardy et al., 1964; Lee & Ernster, 1968b), suggesting a high degree of cooperativity between various parts of the ATP-synthesizing system.

A sixth component of F_1, described by Pullman and Monroy (1963), consists of a dissociable protein of a molecular weight of about 10,000, with the interesting property of inhibiting the ATPase activity of F_1 without inhibiting oxidative phosphorylation. In 1970 it was shown in our laboratory (Asami et al., 1970) that the protein inhibits ATP-dependent energy--linked reactions of submitochondrial particles, and it was suggested that it functions as a unidirectional regulator of the ATP-synthesizing system, preventing the back-flow of energy from ATP to the respiratory chain. Recent evidence reported both from our laboratory (Ernster et al., 1972, 1974b) and by Van de Stadt and associates in Amsterdam (Van de Stadt et al., 1973) indicates that F_1 occurs in two forms, one promoted by ATP and having a high affinity for the inhibitor, and another, promoted by substrate oxidation in the presence of ADP and P_i, and having a low affinity for the inhibitor. The two forms of F_1 also seem to differ with respect to their aurovertin-binding properties (Chang & Penefsky, 1973; Bertina et al., 1973).

It is now well established that energy derived from the respiratory chain can be conserved without the participation of the phosphorylating system and utilized for driving certain energy-requiring processes other than ATP synthesis (Fig. 11). Evidence suggesting this conclusion was first obtained in our laboratory following the discovery of the reversibility of oxidative phosphorylation by Chance & Hollunger (1957; for review, cf. Ernster & Lee, 1964). It was found (Ernster, 1971, 1963a,b) that reversal of electron transport through a coupling site of the respiratory chain could be driven by energy generated at another coupling site, and that the process was insensitive to the phosphorylation inhibitor oligomycin. For example, electron transport from succinate to NADH through Coupling Site 1 could be driven by energy derived from the aerobic oxidation of succinate, i.e., from Sites 2 and 3. Thus the energy transfer between these coupling sites must have taken place without the participation of the phosphorylating system.

Another reaction of this type is the energy-linked transhydrogenase

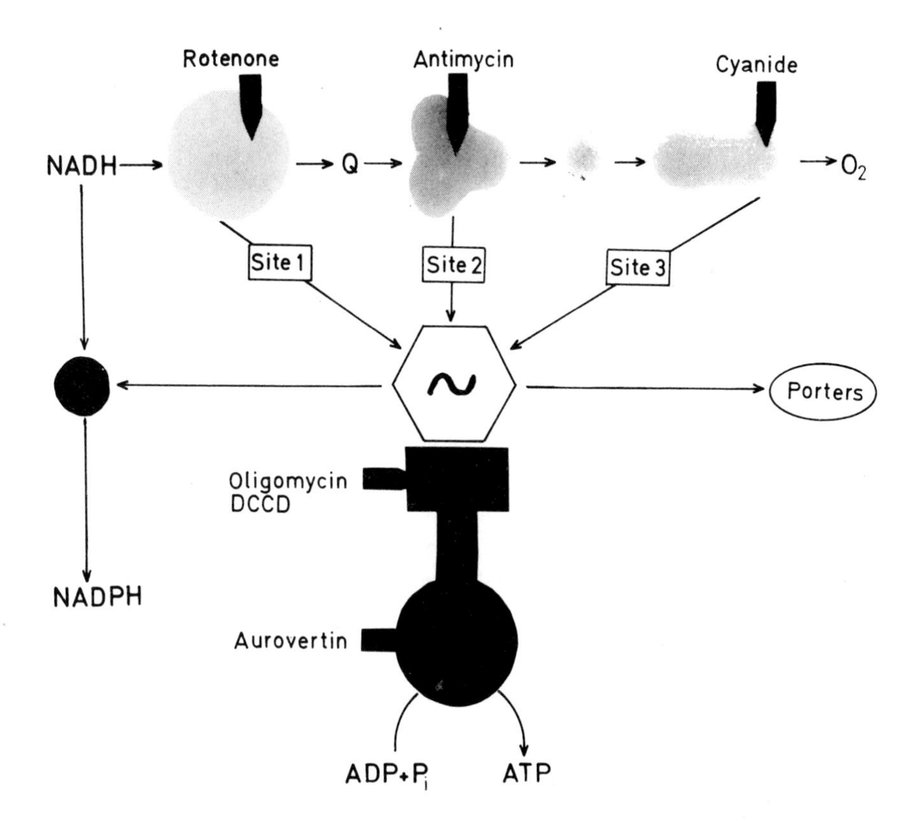

Fig. 11. Energy-Linked Transhydrogenase and Ion Transport

described by Danielson and Ernster (1963), consisting of an energy-linked shift of the equilibrium of the reaction catalyzed by the mitochondrial nicotinamide nucleotide transhydrogenase

$$NADH + NADP^+ \rightleftharpoons NAD^+ + NADPH$$

towards the formation of NAD^+ and NADPH. The reaction could be driven with either ATP or with substrate oxidation as the source of energy. In the former case the reaction was sensitive, and in the latter case insensitive, to oligomycin, indicating that it could utilize energy derived from the respiratory chain without the involvement of the phosphorylating system. Because of its very favorable equilibrium - it utilizes one high energy bond per mole of $NADP^+$ reduced - the energy-linked transhydrogenase reaction proved to be a highly sensitive assay for energy-generating systems (Lee & Ernster, 1964, 1966, 1968a). For example, we have used it to demonstrate (Lee et al., 1964) that so called nonphosphorylating submitochondrial particles may possess the capacity for respiratory chain-linked energy coupling although they have lost the ability to form or use ATP. In recent years the energy-linked transhydrogenase reaction has also been widely used as an assay for energy-coupling in bacterial mutants deficient in various parts of the energy-transducing system (Cox & Gibson, 1974).

A further type of reactions that can utilize energy derived from respiratory chain without the involvement of the phosphorylating system are those catalyzed by various ion-translocators - or porters - known to be present in the mitochondrial inner membrane (Lehninger *et al.*, 1967; Chappell, 1968; Klingenberg, 1970; Chance & Montal, 1971). The physiological function of these porters - and probably of the energy-linked transhydrogenase as well (Hoek & Ernster, 1974) - is to regulate the transfer of metabolites and reducing equivalents between mitochondria and the cytosol. A special colloquium on Ion Transport and Mitochondrial Function will be held after this Meeting under the leadership of Dr. Attila Fonyó, who has made an outstanding contribution to this field of research.

All the above reactions - as chemical reactions in general - are in principle reversible, and in fact, there is now experimental evidence for the reversibility of most reactions in Fig. 11 (*cf.* Slater, 1971). Based on measurements of the "phosphorylation potential" maintained by respiring mitochondria (Cockrell *et al.*, 1966; Slater, 1969a,b, 1971; Slater *et al.*, 1973), it has been concluded that the thermodynamic efficiency of the respiratory chain-linked phosphorylating system approaches 100 % (Slater, 1969a,b, 1971; Slater *et al.*, 1973), but the validity of this conclusion has been questioned (Klingenberg *et al.*, 1969; Ernster & Nordenbrand, 1974).

How is energy conserved in the respiratory chain and utilized for ATP synthesis and other energy-linked reactions? This has been the central problem in mitochondrial bioenergetics all since the occurrence of phosphorylations in the respiratory chain was first recognized in the 1930's (Engelhardt, 1930; Kalckar, 1937, 1939; Belitser & Tsibakova, 1939).

According to a mechanism proposed by Lipmann (1946), these phosphorylations would follow a course similar to that occurring in glycolysis, involving the formation of a phosphorylated high-energy intermediate:

$$AH_2 + B + P_i \rightleftharpoons A{\sim}P + BH_2$$
$$A{\sim}P + ADP \rightleftharpoons A + ATP$$
$$\overline{AH_2 + B + P_i + ADP \rightleftharpoons A + BH_2 + ATP}$$

(A and B denoting redox carriers). Shortly after the demonstration of succinyl-CoA as an intermediate in the phosphorylation linked to the oxidation of α-ketoglutarate (Sanadi & Littlefield, 1952), Slater (1953) proposed a mechanism involving a nonphosphorylated high-energy intermediate as the primary product of energy conservation:

$$AH_2 + B + C \rightleftharpoons A{\sim}C + BH_2$$
$$A{\sim}C + P_i + ADP \rightleftharpoons A + C + ATP$$
$$\overline{AH_2 + B + P_i + ADP \rightleftharpoons A + BH_2 + ATP}$$

Similar mechanisms were proposed by others (Lindberg & Ernster, 1954; Lehninger, 1955; Chance & Williams, 1956).

During the following years a great deal of efforts was spent in many laboratories on demonstrating and chemically defining the high-energy intermediates postulated by the above hypothesis. In fact, there is hardly any component of the respiratory chain that has not been considered on some occasion as a possible constituent of such a high-energy intermediate. All efforts, however, have been unsuccessful.

In an attempt to overcome this difficulty, Boyer (1965) proposed that the energy-yielding oxidoreductive steps in the respiratory chain might not lead to the formation of high-energy intermediates in the classical sense, i.e. compounds consisting of an electron carrier bound to a ligand by a covalent high-energy bond, but might rather give rise to conformational changes of the electron carriers involved (Fig. 12). These conformational

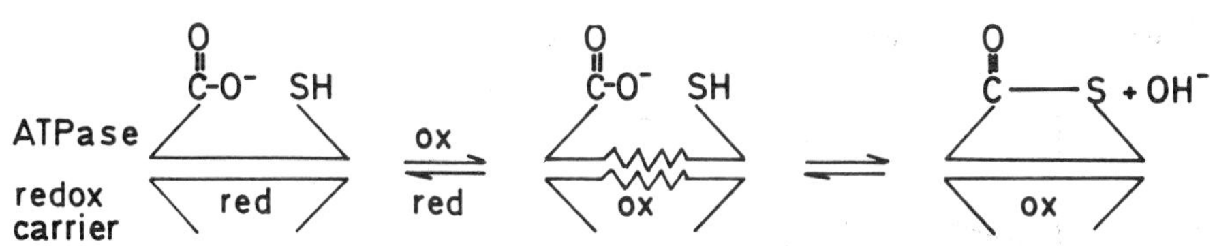

Fig. 12. Conformational Coupling Mechanism of Oxidative Phosphorylation (From Boyer, 1965)

changes of the carriers would enable them to induce the formation of a covalent high-energy bond in an adjacent ATPase molecule, with the result that the latter now would be able to synthesize ATP from ADP and P_i. An interesting recent development of this hypothesis is a proposal made by Boyer (1974) at the International Congress in Stockholm last year, according to which the terminal phase of ATP synthesis also takes place without the involvement of high-energy bonds, and the energy from the respiratory chain is needed primarily for bringing the ATPase into a conformational state so as to release ATP (Fig. 13, left). A similar mechanism has independently been proposed by Slater *et al.* (1974) (Fig. 13, right).

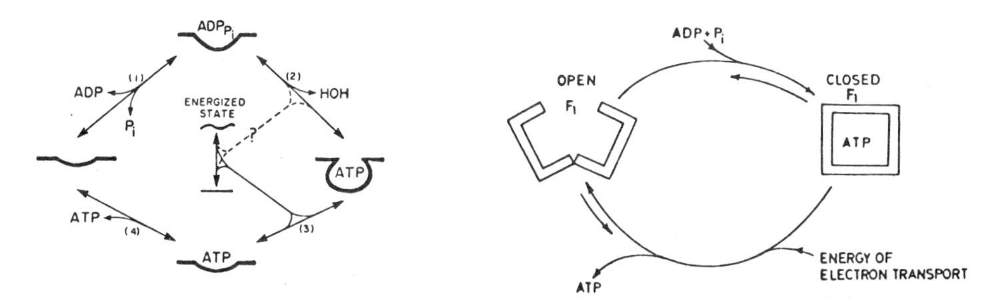

Fig. 13. Proposed Mechanisms of Oxidative Phosphorylation Involving Energy Input for ATP Release
Left: Boyer, 1974
Right: Slater *et al.*, 1974

Meanwhile, Mitchell (1961) put forward his chemiosmotic hypothesis, which he subsequently developed in great detail to represent a general mechanism for energy coupling in respiratory and photosynthetic phosphorylation (Mitchell, 1966a,b, 1968). Some essential features of the mechanism are schematically outlined in Fig. 14. It is assumed that energy-conservation in the oxidoreduction chain proceeds by way of a proton gradient across the coupling membrane that is brought about through the action of alternating hydrogen and electron carriers, which form loops across the membrane in such a way that hydrogen ions are taken up on one side of the membrane and given off on the other (Fig. 14, left). It is further assumed that the same membrane contains a proton-translocating reversible ATPase, with the same orientation as the respiratory chain, which can utilize the proton gradient generated by the latter for the synthesis of ATP from ADP and P_i (Fig. 14, right). It was postulated that the respiratory chain would consist of three loops, each translocating two protons across the membrane per two electrons transferred through the chain, and that the translocation of two protons back across the membrane would result in the

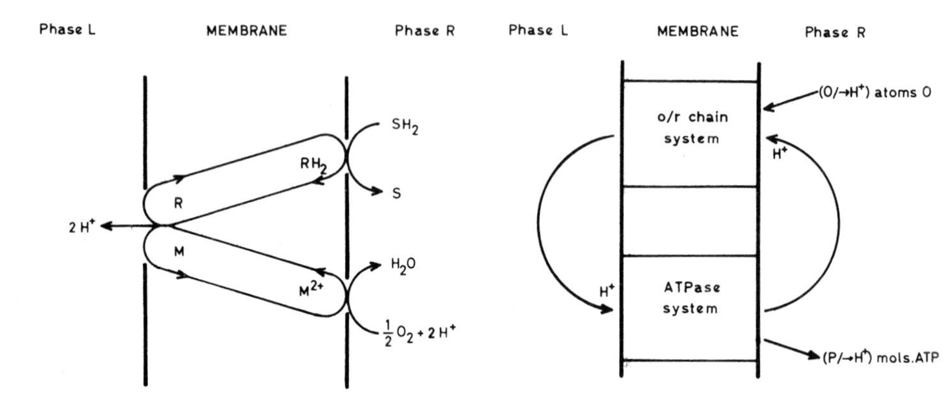

Fig. 14. Chemiosmotic Mechanism of Energy Coupling
Left: Proton-translocating oxido-reduction loop composed of a hydrogen carrier (R/RH_2) and an electron carrier (M/M^{2+})
Right: Chemiosmotic coupling between oxido-reduction (o/r) chain and ATPase system. The circulation of one proton is caused by the utilization of a certain number ($O/\rightarrow H^+$) of oxygen atoms, and causes the synthesis of a certain number ($P/\rightarrow H^+$) of ATP molecules. The P/O quotient is the product ($P/\rightarrow H^+$) x ($\rightarrow H^+/O$). (From Mitchell, 1966b)

synthesis of one molecule ATP from ADP and P_i (Fig. 15). A fourth loop was postulated for the nicotinamide nucleotide transhydrogenase. Furthermore, it was predicted that the membrane would have a low permeability not only to protons but to ions in general and would contain a substrate-specific exchange-diffusion system so that the electron transport and the ATPase system could be coupled through the sum of the electrical and the osmotic pressure differences. This sum, called the "protonmotive force", thus consists of two components, a pH difference and a membrane potential.

The chemiosmotic mechanism differs from the chemical mechanisms outlined above - both the one involving high-energy intermediates and the conformational mechanism - in two important respects (Fig. 16): (1) it involves a proton gradient rather than energized forms of redox carriers as the primary form of energy conservation; and (2) it involves an indirect interaction between the electron-transport and the ATPase systems - through the proton gradient - rather than a direct one as postulated by the chemical mechanisms. A third type of mechanism was proposed the same year by Williams (1961; cf. also Williams, 1969, 1974), which contains elements of both the chemical and chemiosmotic mechanisms (Fig. 16, right). It is similar to Mitchell's mechanism in that it postulates a proton current generated by the respiratory chain during electron transport to be the primary driving force for ATP synthesis. However, instead of establishing a gradient across the membrane, the proton current is assumed to give rise to a proton activity in the lipid phase within the membrane, localized so that it can bring about a dehydration around the active center of the ATPase, thereby promoting the synthesis of ATP. This mechanism differs from the

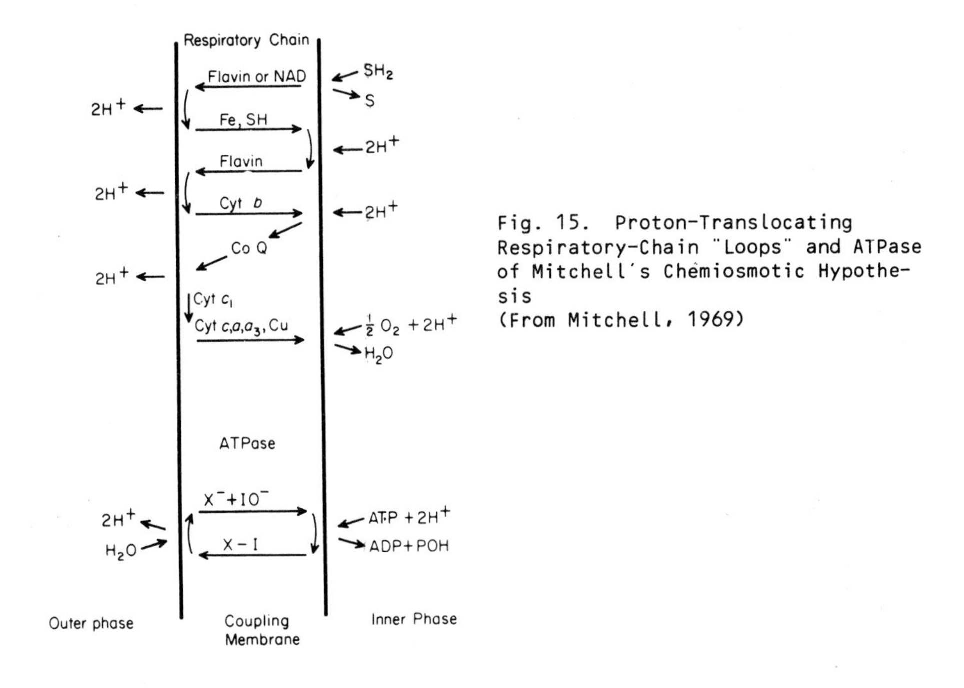

Fig. 15. Proton-Translocating Respiratory-Chain "Loops" and ATPase of Mitchell's Chemiosmotic Hypothesis
(From Mitchell, 1969)

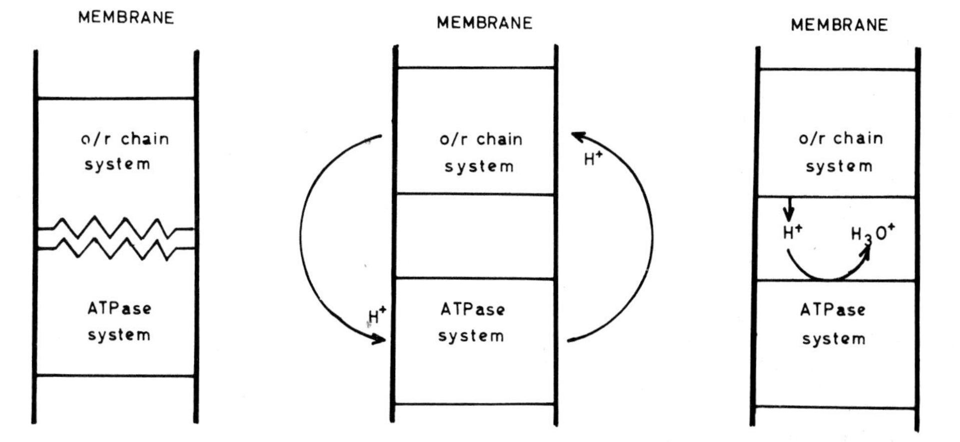

Fig. 16. Interaction of Electron-Transport (o/r chain) and ATPase Systems
Left: direct (chemical, conformational)
Middle: via proton flux across the membrane (chemiosmotic)
Right: via proton flux within the membrane (Williams, 1961)

chemical ones in that it does not require a direct interaction between components of the electron transport and the ATPase systems, but is similar to them in requiring no closed membrane and no vectorial location of the enzymes involved.

Many of Mitchell's predictions have been verified over the years (Mitchell, 1967, 1969, 1972a,b; Greville, 1969; Chance & Montal, 1971; Slater, 1971, 1974; Skulachev, 1971, 1972, 1974a,b; Kagawa, 1972; Racker, 1974), and there is no doubt that the chemiosmotic hypothesis has meant a break-through in the field of cellular bioenergetics, not least by suggesting new experiments and stimulating a great deal of discussion. There is ample evidence that the mitochondrial inner membrane - and energy-transducing membranes in general - are poorly permeable to protons and other ions, and that an intact membrane structure is a requisite for energy-transduction. The membrane contains a number of substrate-specific exchange-diffusion systems. There is also evidence that both the electron-transport system and the ATPase are located in the membrane in a vectorial fashion and are capable of establishing a proton gradient across the membrane. Agents that abolish the electrochemical gradient uncouple electron transport from phosphorylation. Also conversely, proton and other ion gradients have been shown to drive ATP synthesis and energy-linked electron-transport reactions (including the energy-linked transhydrogenase) in various preparations of coupling membranes. Significantly, reconstituted vesicles containing phospholipids and either isolated electron-transport complexes or the isolated ATPase complex (including the hydrophobic proteins) have been shown to be able to give rise to a proton gradient and/or a membrane potential, and, when combined in the same vesicle, to catalyze oxidative phosphorylation and its reversal. And finally, Racker and Stoeckenius (1974) have recently shown that bacterial rhodopsin, a light-induced generator of proton gradient from *Halobacterium halobium*, is capable of giving rise to ATP synthesis when incorporated into phospholipid vesicles together with mitochondrial ATPase.

Whereas the above results strongly support the chemiosmotic hypothesis, other observations make it difficult to accept it in its present form. I shall in the following briefly illustrate these difficulties with the energy-linked transhydrogenase reaction as an example - a reaction that has been studied extensively in our laboratory over the past years - and subsequently consider them in a more general context.

The transhydrogenase reaction represents an unusually clear-cut case in mitochondrial electron transport in the sense that both of its substrates, NAD(H) and NADP(H), are chemically well-defined, water-soluble compounds, and they are readily dissociated from the enzyme. The following lines of experimental evidence indicate that the energy-linked transhydrogenase reaction probably does not proceed by a chemiosmotic coupling mechanism:

1. NAD(H) and NADP(H) react with the enzyme on the same side of the mitochondrial inner membrane (the inside in intact mitochondria and the outside in submitochondrial particles) and there is no evidence for the involvement of transport of the nicotinamide nucleotides within or across the membrane (Lee & Ernster, 1966; Juntti *et al.*, 1969; Ernster & Kuylenstierna, 1970).

2. Experiments with tritiated substrates reveal that the reaction proceeds by a direct transfer of hydrogen between NAD(H) and NADP(H), without an exchange with the protons of water (Fig. 17) (Lee *et al.*, 1965).

3. Kinetic studies of the transhydrogenase have indicated that the enzyme occurs in two states, an active state, whose formation is promoted by NAD^+ and NADPH, and an inactive state, whose formation is promoted by NADH and $NADP^+$ (Fig. 18) (Rydström *et al.*, 1970). It was further found that in

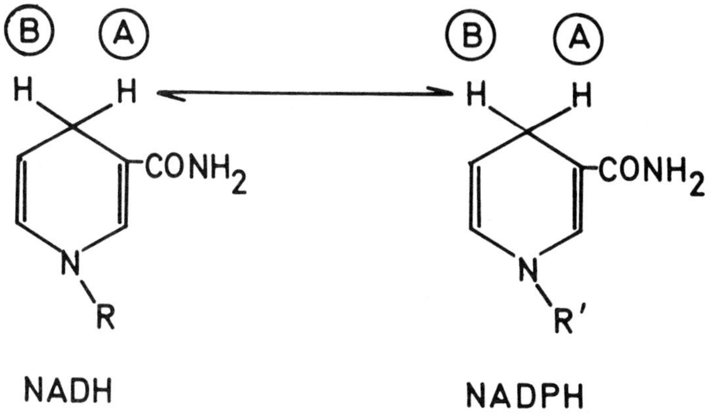

Fig. 17. Stereospecificity of Mitochondrial Nicotinamide Nucleotide Transhydrogenase
(From Lee *et al.*, 1965)

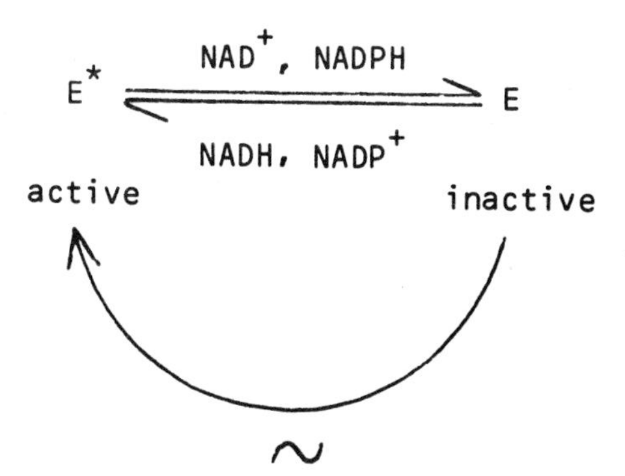

Fig. 18. Factors Influencing the Transition of Mitochondrial Nicotinamide Nucleotide Transhydrogenase between Active (E^*) and Inactive (E) Forms
∿ refers to energy derived from respiration or ATP hydrolysis
(From Rydström *et al.*, 1970)

the presence of an energy supply - from either the respiratory chain or ATP - there occurred a conversion of the enzyme from the inactive to the active state, resulting in both an increase in the reaction velocity and a shift of the equilibrium towards the formation of NAD^+ and NADPH. A detailed comparison of the steady-state kinetics of the reactions in the absence and presence of an energy supply revealed that both reactions proceeded according to a ternary-complex, Theorell-Chance mechanism, with NAD^+ and NADH as the first substrate bound to the enzyme, and that the effect of energization consisted primarily of drastic changes in the dissociation constants of the enzyme-NAD^+ and enzyme-NADH complexes, favoring the release of NAD^+ and the binding of NADH (Fig. 19) (Teixeira da Cruz *et al.*, 1971; Rydström *et al.*, 1971; Rydström, 1972a,b). Since energization is unlikely to alter the structure of NAD^+ or NADH, this change in dissociation constants must

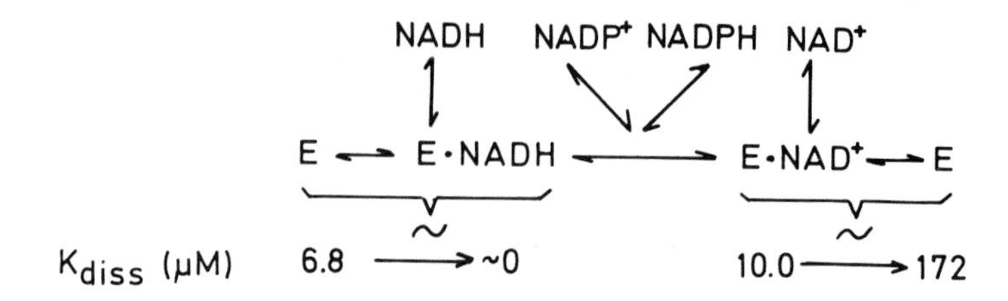

Fig. 19. Reaction Mechanism of Mitochondrial Nicotinamide Nucleotide Transhydrogenase (E) and Changes of the Dissociation Constants (K_{diss}) of the E·NADH and E·NAD$^+$ Complexes upon Energization (∿) of Beef-Heart Submitochondrial Particles
(From Teixeira da Cruz et al., 1971; Rydström et al., 1971:, Rydström, 1972b)

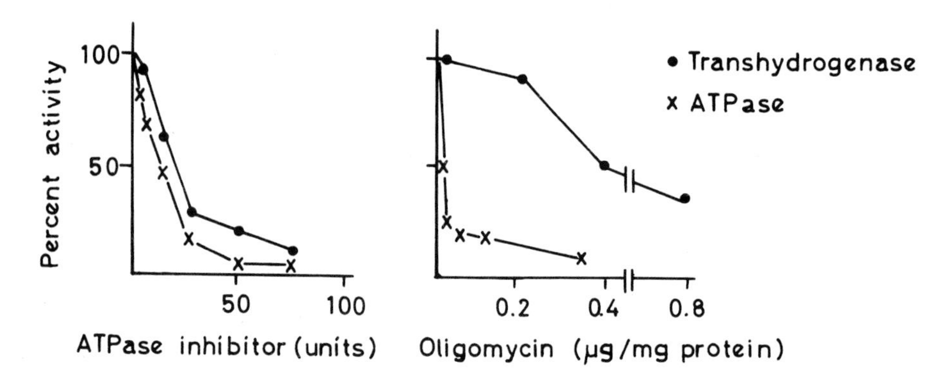

Fig. 20. Effects of ATPase Inhibitor and Oligomycin on the ATP-Driven Transhydrogenase and ATPase Activities of Submitochondrial Particles
Submitochondrial (Mg^{++}-ATP) particles from beef heart, prepared as described by Lee and Ernster (1967) were pretreated according to Van de Stadt et al. (1973) to remove endogenous ATPase inhibitor. ATPase inhibitor was purified from beef heart mitochondria according to Horstman and Racker (1970). ATPase and ATP-driven transhydrogenase activities were assayed as described by Asami et al. (1970)
(From Ernster et al., 1974b)

reflect an energy-linked change in the enzyme molecule itself, i.e., the formation of an "energized" form of the enzyme.

4. Indications of transmembrane proton movements in connection with the transhydrogenase reaction have been observed with submitochondrial particles (Grinius et al., 1970) and interpreted as being due to changes in the state of protonation of NAD(H) or NADP(H) as these traverse the membrane in the course of the reaction (Skulachev, 1970; Mitchell, 1972a). However, as pointed out above, there is no evidence for the penetration of nicotinamide nucleotides into the mitochondrial inner membrane, and it is uncertain whether these protons originate directly from components of the transhydrogenase reaction or indirectly from some proton pump(s) present in the membrane.

It has also been shown that alterations in proton concentration mimic the effect of energy on the kinetics of the transhydrogenase reaction, and it has been suggested that the energy-linked conversion of the inactive into the active form of the transhydrogenase may involve a protonation of a specific group (or groups) of the enzyme (Rydström, 1972a, 1974).

5. Titration of the ATP-driven transhydrogenase reaction of submitochondrial particles with the ATPase inhibitor protein of Pullman and Monroy (1963) resulted in an inhibition of the reaction parallel to that of the ATPase (Fig. 21, left), under conditions when the capacity of the ATPase to provide energy for the transhydrogenase was not rate-limiting as revealed by oligomycin titration (Fig. 21, right) (Ernster et al., 1972, 1974b). These results strongly suggest an assembly-like arrangement between transhydrogenase and ATPase within the membrane, with a direct, molecular interaction between the two enzymes. A possible connection between transhydrogenase and the 29,000 subunit of the hydrophobic protein component of the ATP-synthetizing system has recently been suggested by Rydström et al. (1974) on the basis of certain similarities found between the two proteins after solubilization with lysolecithin.

To summarize the above points, it is evident that the transhydrogenase (1) does not form a loop across the membrane; (2) does not give rise to a transmembrane proton gradient originating from the separation of reducing equivalents into protons and electrons; (3) may occur in an "energized" and a "de-energized" conformational state; (4) may or may not act as a proton translocator across the membrane; and (5) probably interacts with the ATPase system in a direct, molecular fashion.

Do similar considerations apply to the rest of the energy-linked electron-transport system?

Fig. 21 summarizes in a schematic form our present knowledge of the transversal topology of the electron-transport and ATPase systems in the mitochondrial inner membrane. It is based on information accumulated over the past years in several laboratories using both mitochondria and submitochondrial particles as test objects and assessing the orientation of the individual catalysts on the basis of their accessibility to substrates, inhibitors, artificial electron donors and acceptors, antibodies, proteolytic enzymes, etc. (Lee, 1970; Racker, 1970; Kröger & Klingenberg, 1970; Packer, 1973; Skulachev, 1974a). Particularly useful information in the present context has been obtained with the nonpenetrant electron acceptor, ferricyanide, which, as first shown by Copenhaver and Lardy (1952) does not interact with the respiratory chain of intact mitochondria on the substrate side of the antimycin block (cf. also Lee et al., 1967).

From the various lines of information it is now evident that the pathway of electron transfer from NADH or succinate to oxygen (indicated by the solid black line in Fig. 21) describes one "loop" over the membrane, starting on the inside through the NADH and succinate dehydrogenases, reaching the outside at the level of cytochrome c - the point where ferricyanide interacts with the respiratory chain of intact mitochondria - and turning back to the inside through cytochrome c oxidase. Those components of the chain which are known as hydrogen carriers, namely, the flavins of NADH and succinate dehydrogenases, and ubiquinone, do not interact with ferricyanide in the intact mitochondria, and thus cannot serve as electron transport-linked proton translocators. Thus, there is only one point of the chain, at the level of cytochrome c, where electrons - and thereby also any protons originating from a hydrogen carrier - can reach the outer surface of the membrane. However, the electron donor to cytochrome c, the cytochrome b-c_1 complex, is not known to contain any hydrogen-carrying prosthetic group and, thus, its postulated function as a proton-translocating oxido-

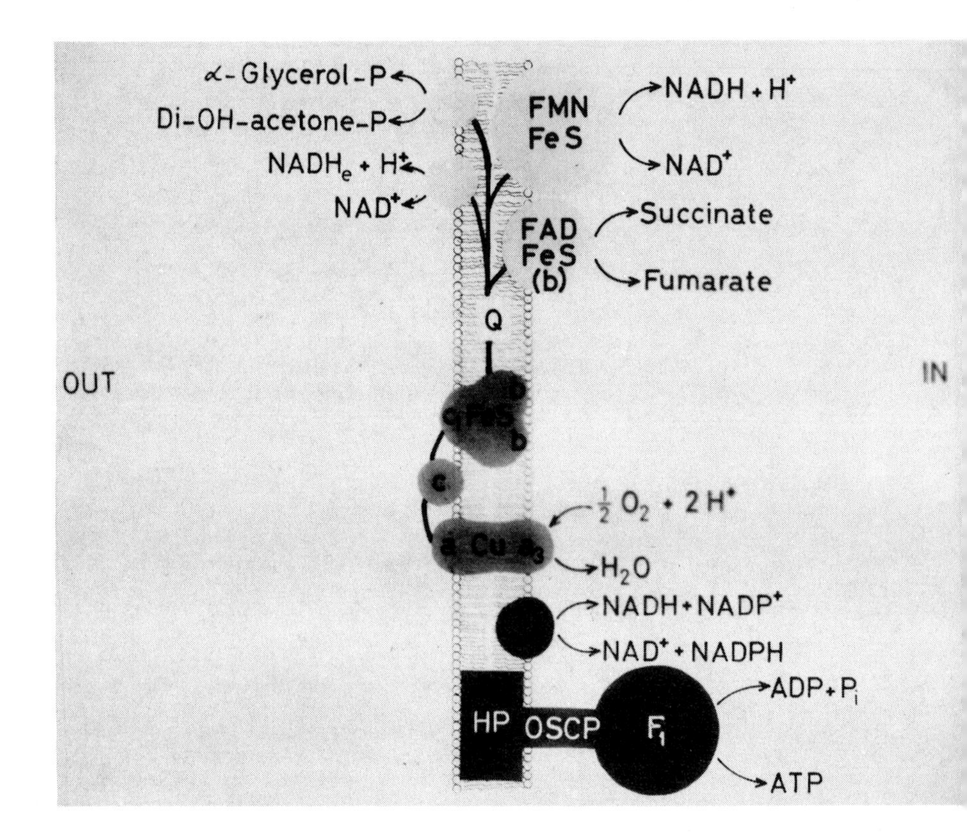

Fig. 21. Transversal Topology of Mitochondrial Inner-Membrane Catalysts. For explanation, see text.

reduction catalyst remains unproven. The same holds for cytochrome c oxidase. In other words, there seems to be no evidence for a transmembrane hydrogen carrier - i.e., a hydrogen-carrying arm of an oxidoreduction loop (cf. Fig. 14) - anywhere along the respiratory chain.

Ferricyanide does readily interact with the flavoproteins involved in the oxidation of α-glycerophosphate and of extramitochondrial NADH (the latter enzyme being present in yeast and plant mitochondria) which are located on the outer surface of the membrane (cf. Kröger & Klingenberg, 1970). In general, it appears that the transversal topology of the electron-transport system in the mitochondrial inner membrane is adapted primarily to the enzyme topology of its environment: substrates originating from the mitochondrial matrix are oxidized by enzymes located at the inner surface, and those originating from the extramitochondrial space are oxidized by enzymes located at the outer surface. This impression is further strengthened by recent data in the literature indicating a role of cytochrome c as the electron acceptor for the cytochrome b_5-like hemoprotein associated with the mitochondrial outer membrane (Sottocasa et al., 1967) as well as for the sulfite oxidase present in the intermembrane space (Ito, 1971; Wattiaux--De Coninck & Wattiaux, 1971; Cohen et al., 1972).

The lack of evidence for the occurrence of proton-translocating electron

-transfer loops corresponding to the three energy-coupling sites of the respiratory chain raises the question as to the origin and the mechanism of formation of the proton gradient observed during coupled electron transport. There is evidence for the occurrence in the membrane of both electrogenic (Papa et al., 1974) and electroneutral (Azzone et al., 1974) proton pumps, but the precise function of these in relation to the primary event of energy conservation is not yet understood. Chance and associates (Chance et al., 1970; Chance & Montal, 1971) have stressed the view that the transmembrane proton gradient may be a secondary consequence of a change in quaternary structure, resulting in altered pK_a values, of electron-transferring proteins involved in energy-conservation - a "membrane Bohr effect" - and at the International Congress in Stockholm last year Slater (1974) presented a hypothetic reaction sequence according to which such a mechanism could account for the phosphorylation occurring at Coupling Site 3 of the respiratory chain by a direct interaction between an "energized" form of cytochrome c oxidase and the ATPase. There is now evidence from several laboratories for ATP-induced spectral shifts of cytochrome c oxidase (Wikström & Saris, 1970; Nicholls et al., 1973; Wilson & Brocklehurst, 1973; Wikström, 1974), but the exact relationship of these shifts to the mechanism of energy conservation remains to be established. Concerning the ATPase, I have already referred to evidence from the laboratories of both Boyer (1974) and Slater (Slater et al., 1974) suggesting that energy is required primarily for the release - rather than the formation - of ATP. It is not clear how a proton-translocating ATPase, of the types proposed by Mitchell (1961, 1966a,b, 1973, 1974), can account for such a mechanism.

A last piece of information about mitochondria that may be relevant in the present context relates to the stoichiometry and lateral topology of their inner-membrane catalysts.

It has recently been concluded in Slater's laboratory (Bertina et al., 1973) that both rat-liver and beef-heart mitochondria contain equimolar amounts of F_1, oligomycin-binding protein and Complex III, as based on closely similar aurovertin, oligomycin and antimycin titers found with both types of mitochondria. These data, together with earlier estimates of the cytochrome contents (cf. Klingenberg, 1968), indicate that the mitochondrial inner membrane contains the ATPase complex, Complex III, cytochrome c, and Complex IV in the approximate molar proportions 1:1:2:2. Fig. 22 is a fantasy view through the plane of the membrane, showing the lateral occupancy of the membrane by the components of the respiratory chain and the ATPase system. The cytochromes and the ATPase are visualized as assemblies - an appealing possibility though yet unproven. Ubiquinone is present in the membrane at a concentration 5-10 times higher than the cytochrome chains, whereas Complexes I and II are present in concentrations 5-10 times lower. The flavoproteins probably interact with the cytochromes by way of a mobile pool of ubiquinone as proposed by Kröger and Klingenberg (1970). The respiratory chain and the ATPase system are estimated to constitute altogether about 30 to 40 % of the inner-membrane protein, the remainder - the unmarked spots in the Figure - being made up by yet unidentified protein, probably various translocators, transhydrogenase, and possibly "structural" proteins. It may be noted that the mitochondrial inner membrane - like other energy-transducing membranes - is particularly rich in protein (Branton & Deemer, 1972), a feature also revealed by the elegant freeze-fracture electron micrographs produced in recent years by Packer and his colleagues (cf. Packer, 1973). It has been calculated that less than one-half of the membrane can consist of a phospholipid bilayer. Whether, and to what extent, the protein components possess a lateral mobility, as envisaged by the fluid mosaic membrane model of Singer and Nicholson (1972),

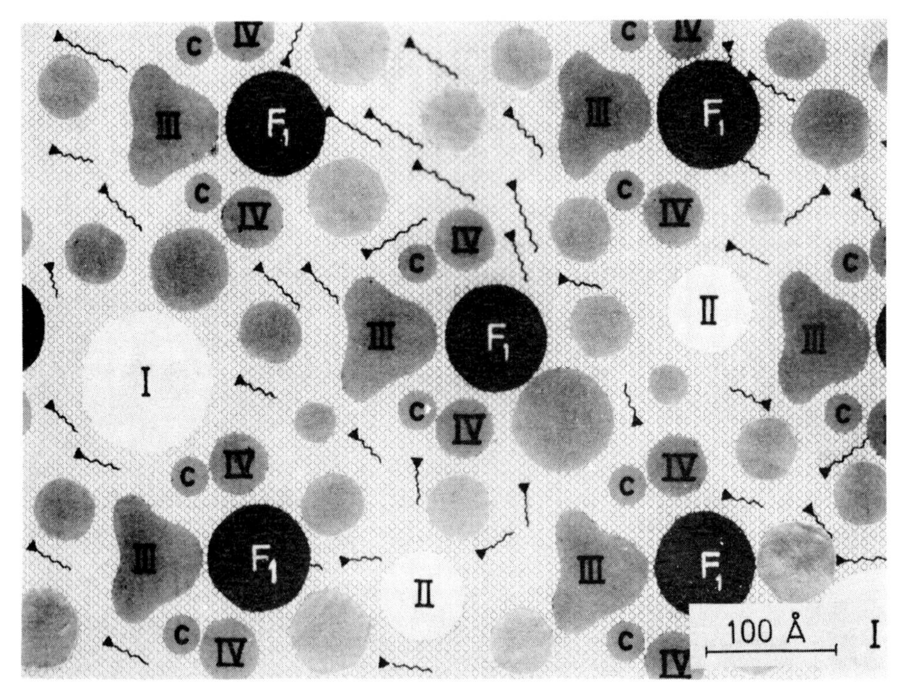

Fig. 22. Stoichiometry and Lateral Topology of Respiratory-Chain Catalysts and ATPase in the Mitochondrial Inner Membrane
o = polar heads of phospholipids
⇝ = ubiquinone
Unmarked spots are unidentified proteins
For further explanation, see text.

is not yet known.

There are now several indications, based on the use of various probes and inhibitors, of an extensive interaction among different components of the energy-transducing system, not only within the individual enzyme complexes but also between them (Lee *et al.*, 1969; Azzi *et al.*, 1969, 1974; Baum *et al.*, 1971; Ernster *et al.*, 1971; Slater, 1974). It has been shown, for example, that substrate oxidation via the respiratory chain can alter the binding properties of F_1 for aurovertin (Bertina *et al.*, 1973; Chang & Penefsky, 1973) and for the ATPase inhibitor (Ernster *et al.*, 1972; Van de Stadt *et al.*, 1973), and, conversely, that ATP can alter the antimycin--binding kinetics of Complex III (Berden & Slater, 1972). These effects, however, can be explained equally well by an indirect interaction, through a change in proton activity and/or membrane potential, as in terms of an interaction at the molecular level. Obviously, here much remains to be done, not least perhaps with reconstituted systems. A step in this direction will be presented at this Meeting by my colleague Kerstin Nordenbrand (Nordenbrand *et al.*, 1974; *cf.* also Ernster *et al.*, 1974a).

In conclusion, it appears that an urgent problem for the chemiosmotic hypothesis at the present time would be to revise its definition of proton--translocating redox mechanisms so as to take into account existing knowledge of the composition, reaction sequence and topology of the electron-

-transfer chain. Indeed, the chemiosmotic hypothesis in its present form seems to be in no better position in specifying its proton carriers than is the chemical hypothesis in specifying its high-energy intermediates. Also the role of protons in the ATPase reaction would seem to require re-evaluation to account for recent evidence indicating energy-dependence of ATP release. It is increasingly evident that conformational changes of proteins are instrumental in bringing about the observed proton movements in energy-transducing membranes. Whether their role is merely catalytic, as maintained by the chemiosmotic hypothesis, or whether they represent the primary events in both electron transport- and ATP-linked energy transduction, as envisaged by the proponents of the conformational hypothesis, remains to be decided. And finally, a question of great urgency concerns the occurrence of direct interaction between components of the electron-transport and ATPase system. Here, the burden of proof rests entirely on the side of the chemical-conformational hypotheses, although it must be remembered that the demonstration of an indirect interaction through a proton gradient, as found for example in reconstituted systems, is in itself no proof against the existence of a direct interaction as well. Clearly, the demonstration that pentose phosphate can function as an intermediate between glucose and lactate, could not be taken as proof against the existence of glycolysis.

* * *

Having come to the end of my presentation I hope I have convinced you that mitochondrial bioenergetics is a field rich in facts and certainly not poor in ideas. It is also a field with great traditions - and a traditionally controversial one. It is now about half-a-century ago that Warburg and Wieland had their famous controversy about the nature of cell respiration. The question was as to whether respiration consisted of a transport of oxygen or a transport of hydrogen between Warburg's _Atmungsferment_ and Wieland's dehydrogenases. The problem was solved by Keilin who established the sequence of cytochromes connecting _Atmungsferment_ with the dehydrogenases - the respiratory chain - and showed that both Warburg and Wieland were right; they were looking at the same problem but from two different sides.

But we can go even further back in time. About a century ago, Pasteur and Liebig had their great dispute about the chemical nature of life processes. At that time, the problem also concerned an aspect of cellular energy metabolism, although not as sophisticated as respiration; it concerned the fermentation of sugar to alcohol in yeast. The question was whether isolated catalysts - ferments - could carry out fermentation in a solution, in the absence of cell structures - a view stressed by Liebig - or whether there was something in the yeast cell beyond the ferments - a "_force vitale_" - without which no metabolism could take place - a standpoint maintained by Pasteur. And that time, too, the problem was solved by a third person, Buchner, who succeeded in 1897 to demonstrate fermentation in a press-juice from yeast. The era of vitalism seemed to be over.

Yet, 15 years later, on the 6th of December 1912, a lecture was given at the University of Heidelberg by a young _Privatdozent_, Otto Warburg, with the title "Über die Wirkung der Struktur auf chemische Vorgänge in Zellen". In this lecture (Warburg, 1913) he presented evidence that the respiratory enzymes, in contrast to those involved in fermentation, do require for their function something beyond the catalysts themselves: they are tightly associated with, and enhanced by, a structural component of the cell. Warburg concluded his lecture by referring to the just-terminated controversy between chemists and biologists by saying: "Ich glaube, Ihnen heute gezeigt zu haben, dass hier ein Gegensatz nicht existiert, dass beide, die Fermentchemiker und die Biologen, recht haben. Die Beschleunigung der energieliefernden Reaktionen in Zellen ist eine Fermentwirkung und eine Strukturwirkung, nicht so, dass erstens die Fermente und zweitens die Struktur beschleunigen, sondern die Struktur beschleunigt die Fermentwirkung." Warburg was of course unaware that the structure to which he was referring was the mitochondrial inner membrane.

Today, the discussion concerns the role of the membrane structure in energy transduction. We no longer question that a membrane structure is a requisite for energy conservation in both respiration and photosynthesis. We also all agree that the membrane serves not only as a permeability barrier, separating spaces and components from each-other, but also as the very site of the catalysts involved in energy transduction, directing and promoting - beschleunigend - their actions and interactions across and within the membrane. Whatever divergence of views that may exist, will, I believe, in the last analysis turn out to have resulted from looking at the same problem from two different angles - some among us may be focusing their interest primarily on events going on within, others on events going on across the membrane. Basically, I think, we will be able to conclude once again "dass hier ein Gegensatz nicht existiert, dass beide recht haben". There exists no conflict here: both are right.

* * *

In closing, I wish to thank the following colleagues - visitors and staff members - who have over the years made a most valuable and stimulating contribution in our laboratory to the work quoted in this presentation: Kouichi Asami, Yoram Avi-Dor, Giovanni Felice Azzone, Obi Chude, Julien Coleman, Lennart Danielson, Luciano Frigeri, Pär Gellerfors, Elzbieta Glazek, Henry Hoberman, Jan Hoek, Torill Hundal, Kerstin Juntti, Bo Kuylenstierna, Chuan-pu Lee, In-Young Lee, Buck Nelson, Kerstin Nordenbrand, Birgitta Norling, Kristina Ohlson, Aleksandr Panov, Giuseppe Paradies, Barbro Persson, Elisabetta Rossi, Jan Rydström, Nicole Simard-Duquesne, Jennie Smoly, Gian Luigi Sottocasa, Patrik Swanljung, António Teixeira da Cruz, Ulla-Britta Torndal, Eugene Weinbach. Research grants from the Swedish Cancer Society and the Swedish Medical and Natural-Science Research Councils are gratefully acknowledged.

This lecture was outlined in the summer of 1974 during the tenure of a Fogarty International Scholarship-in-Residence at the National Institutes of Health, Bethesda, Maryland, U.S.A. The final manuscript, including the figures, was prepared in collaboration with Kerstin Nordenbrand, whom I wish to thank for wholehearted interest and assistance. I also wish to acknowledge the excellent help of Maud Horovitz with the typescript and of Torill Hundal with the lantern slides.

* * *

REFERENCES

Asami, K., Juntti, K., and Ernster, L. (1970). Biochim. Biophys. Acta 205, 307.

Azzi, A., Bragadin, M., Layton, D., Graziotti, P., and Luciani, S. (1974). BBA Library 13, 303.

Azzi, A., Chance, B., Radda, G.K., and Lee, C.P. (1969). Proc. Natl. Acad. Sci. U.S. 62, 612.

Azzone, G.F., Massari, S., Colonna, R., Dell'Antone, P., Frigeri, L., and Beltrame, M. (1974). BBA Library 13, 405.

Baum, H., Hall, G.S., Nelder, J., and Beechey, R.B. (1971). In: "Energy Transduction in Respiration and Photosynthesis", E. Quagliariello, S. Papa and C.S. Rossi, eds., Adriatica Editrice, Bari, p. 747.
Beechey, R.B., Roberton, A.M., Holloway, C.T., and Knight, I.G. (1967). Biochemistry 12, 3867.
Belitser, V.A., and Tsibakova, E.T. (1939). Biokhimiya 4, 516.
Berden, J.A., and Slater, E.C. (1972). Biochim. Biophys. Acta 256, 199.
Bertina, R.M., Schrier, P.I., and Slater, E.C. (1973). Biochim. Biophys. Acta 305, 503.
Boyer, P.D. (1965). In: "Oxidases and Related Redox Systems", T.E. King, H.S. Mason and M. Morrison, eds., Wiley, New York, p. 994.
Boyer, P.D. (1974). BBA Library 13, 289.
Branton, D., and Deemer, D.W. (1972). "Membrane Structure", Springer, Vienna.
Capaldi, R.A. (1974). Arch. Biochem. Biophys. 163, 99.
Chance, B., and Hollunger, G. (1957). Fed. Proc. 16, 163.
Chance, B., and Montal, M. (1971). In: "Current Topics in Membranes and Transport", F. Bronner and A. Kleinzeller, eds., Academic Press, New York, Vol. 2, p. 99.
Chance, B., Radda, G.K., and Lee, C.P. (1970). In: "Electron Transport and Energy Conservation", J.M. Tager, S. Papa, E. Quagliariello and E.C. Slater, eds., Adriatica Editrice, Bari, p. 551.
Chance, B., and Williams, G.R. (1956). Adv. Enzymol. 17, 65.
Chang, T.-M., and Penefsky, H.S. (1973). J. Biol. Chem. 248, 2746.
Chappell, J.B. (1968). Brit. Med. Bull. 24, 150.
Cockrell, R.S., Harris, E.J., and Pressman, B.C. (1966). Biochemistry 5, 2326.
Cohen, H.J., Betcher-Lange, S., Kessler, D.L., and Rajagopalan, K.V. (1972). J. Biol. Chem. 247, 7759.
Copenhaver, J.H., and Lardy, H.A. (1952). J. Biol. Chem. 195, 225.
Cox, G.B., and Gibson, F. (1974). Biochim. Biophys. Acta 346, 1.
Cox, G.B., Gibson, F., McCann, L.M., Butlin, J.D. and Crane, F.L. (1973). Biochem. J. 132, 689.
Cox, G.B., Newton, N.A., Butlin, J.D., and Gibson, F. (1971). Biochem. J. 125, 489.
Danielson, L., and Ernster, L. (1963). Biochem. Z. 338, 188.
Engelhardt, W.A. (1930). Biochem. Z. 227, 16.
Ernster, L. (1961). In: "Biological Structure and Function", T.W. Goodwin and O. Lindberg, eds., Academic Press, London, Vol. 2, p. 139.
Ernster, L. (1963a). In: "Symposium on Intracellular Respiration: Phosphorylating and Non-Phosphorylating Oxidation Reactions. Proc. 5th Intern. Congr. Biochem., Moscow, 1961", E.C. Slater, ed., Pergamon, Oxford, Vol. 5, p. 115.
Ernster, L. (1963b). In: "Funktionelle und Morphologische Organisation der Zelle", P. Karlson, ed., Springer, Berlin, p. 98.
Ernster, L. (1965). Fed. Proc. 24, 1222.
Ernster, L., Juntti, K., and Asami, K. (1972). J. Bioenergetics 4, 149.
Ernster, L., Juntti, K., Asami, K., and Coleman, J. (1974b). Verbal communications, ASBC Symp. on Energy Coupling Mechanisms, Minneapolis.
Ernster, L., and Kuylenstierna, B. (1969). In: "Mitochondria - Structure and Function", L. Ernster and Z. Drahota, eds., Academic Press, London, p. 5.
Ernster, L., and Kuylenstierna, B. (1970). In: "Membranes of Mitochondria and Chloroplasts", E. Racker, ed., Van Nostrand Reinhold Co., New York, p. 172.
Ernster, L., and Lee, C.P. (1964). Annu. Rev. Biochem. 33, 729.

Ernster, L., Lee, I.-Y., Norling, B., and Persson, B. (1969). Eur. J. Biochem. 9, 299.
Ernster, L., and Nordenbrand, K. (1974). BBA Library 13, 283.
Ernster, L., Nordenbrand, K., Chude, O., and Juntti, K. (1974a). In: "Membrane Proteins in Transport and Phosphorylation", G.F. Azzone, M.E. Klingenberg, E. Quagliariello, and N. Siliprandi, eds., North-Holland, Amsterdam, p. 29.
Ernster, L., Nordenbrand, K., Lee, C.P., Avi-Dor, Y., and Hundal, T. (1971). In: "Energy Transduction in Respiration and Photosynthesis", E. Quagliariello, S. Papa and C.S. Rossi, eds., Adriatica Editrice, Bari, p. 57.
Fernández-Morán, H. (1962). Circulation 26, 1039.
Fernández-Morán, H., Oda, T., Blair, P.V., and Green, D.E. (1964). J. Cell Biol. 22, 63.
Fessenden, J.M., and Racker, E. (1966). J. Biol. Chem. 241, 2483.
Frisell, W.R., Cronin, J.R., and Mackenzie, C.G. (1966). BBA Library 8, 367.
Gellerfors, P., and Nelson, B.D. (1974). Submitted for publication.
Glazek, E., Norling, B., Nelson, B.D., and Ernster, L. (1974). FEBS Lett. 46, 123.
Green, D.E. (1966). Comprehensive Biochem. 14, 309.
Greville, G.D. (1969). In: "Current Topics in Bioenergetics", D.R. Sanadi, ed., Academic Press, New York, Vol. 3, p. 1.
Grinius, L., Jasaitis, L., Kadzianskas, Yu.P., Liberman, E.A., Skulachev, V.P., Topali, V.P., Tsofina, L.M., and Vladimirova, M.A. (1970). Biochim. Biophys. Acta 216, 1.
Gupta, U.D., and Rieske, J.S. (1973). Biochem. Biophys. Res. Commun. 54, 1247.
Gutnick, D.L., Kanner, B.I., and Postma, P.W. (1972). Biochim. Biophys. Acta 283, 217.
Hatefi, Y. (1966). Comprehensive Biochem. 14, 199.
Hatefi, Y., Hanstein, W.G., Davis, K.A., and You, K.S. (1974). Ann. N.Y. Acad. Sci. 277, 504.
Hare, J.F., and Crane, F.L. (1974). Subcell. Biochem. 3, 1.
Hoek, J.B., and Ernster, L. (1974). In: "Alcohol and Aldehyde Metabolizing Systems", R.G. Thurman, T. Yonetani, J.R. Williamson and B. Chance, eds., Academic Press, New York, p. 351.
Horstman, L.L., and Racker, E. (1970). J. Biol. Chem. 245, 1336.
Ito, A. (1971). J. Biochem. 70, 1061.
Juntti, K., Torndal, U.-B., and Ernster, L. (1969). In: "Electron Transport and Energy Conservation", J.M. Tager, S. Papa, E. Quagliariello and E.C. Slater, eds., Adriatica Editrice, Bari, p. 257.
Kagawa, Y. (1972). Biochim. Biophys. Acta 265, 297.
Kalckar, H. (1937). Enzymologia 2, 47.
Kalckar, H. (1939). Biochem. J. 33, 631.
Keilin, D., and Hartree, E.F. (1947). Biochem. J. 41, 500.
Keirns, J.J., Yang, C.S., and Gilmour, M.U. (1971). Biochem. Biophys. Res. Commun. 45, 835.
Klingenberg, M. (1968). In: "Biological Oxidations", T.P. Singer, ed., Interscience, New York, p. 3.
Klingenberg, M. (1970). FEBS Lett., 6, 145.
Klingenberg, M., Heldt, H.W., and Pfaff, E. (1969). In: "The Energy Level and Metabolic Control in Mitochondria", S. Papa, J.M. Tager, E. Quagliariello and E.C. Slater, eds., Adriatica Editrice, Bari, p. 237.
Kröger, A., and Klingenberg, M. (1970). Vitamins and Hormones 28, 533.
Lardy, H.A. (1961). In: "Biological Structure and Function", T.W. Goodwin and O. Lindberg, eds., Academic Press, London, Vol. 2, p. 265.
Lardy, H.A., Connelly, J.L., and Johnson, D., 1964, Biochemistry 3, 1961.

Lardy, H.A., Johnson, D., and McMurray, W.C. (1958). Arch. Biochem. Biophys. 78, 587.
Lardy, H.A., and Lin, C.H.C. (1969). In: "Inhibitors - Tools in Cell Research", T. Bücher and H. Sies, eds., Springer, Berlin, 255.
Lee, C.P. (1970). In: "Electron Transport and Energy Conservation", J.M. Tager, S. Papa, E. Quagliariello, and E.C. Slater, eds., Adriatica Editrice, Bari, p. 291.
Lee, C.P., Azzone, G.F., and Ernster, L. (1964). Nature 201, 152.
Lee, C.P., and Ernster, L. (1964). Biochim. Biophys. Acta 81, 187.
Lee, C.P., and Ernster, L. (1966). BBA Library 7, 218.
Lee, C.P., and Ernster, L. (1967). Meth. Enzymol. 10, 543.
Lee, C.P., and Ernster, L. (1968a). Eur. J. Biochem. 3, 385.
Lee, C.P., and Ernster, L. (1968b). Eur. J. Biochem. 3, 391.
Lee, C.P., Ernster, L., and Chance, B. (1969). Eur. J. Biochem. 8, 153.
Lee, C.P., Simard-Duquesne, N., Ernster, L., and Hoberman, H.D. (1965). Biochim. Biophys. Acta 105, 397.
Lee, C.P., Sottocasa, G.L., and Ernster, L. (1967). Meth. Enzymol. 10, 33.
Lehninger, A.L. (1955). Harvey Lectures 49, 174.
Lehninger, A.L., Carafoli, E., and Rossi, C.S. (1967). Adv. Enzymol. 29, 259.
Lindberg, O., and Ernster, L. (1954). "Chemistry and Physiology of Mitochondria and Microsomes", Springer, Vienna.
Lipmann, F. (1946). In: "Currents in Biochemical Research", D.E. Green, ed., Interscience, New York, p. 137.
Löw, H. (1966). BBA Library 7, 25.
MacLennan, D.H., and Tzagoloff, A. (1968). Biochemistry 7, 1603.
Margoliash, E. (1972). Harvey Lectures 66, 177.
Massey, V. (1958). Biochim. Biophys. Acta 30, 205.
Mitchell, P. (1961). Nature 191, 144.
Mitchell, P. (1966a). Biol. Rev. 41, 445.
Mitchell, P. (1966b). "Chemiosmotic Coupling in Oxidative and Photosynthetic Phosphorylation", Glynn Research, Bodmin, Cornwall, England.
Mitchell, P. (1967). Fed. Proc. 26, 1370.
Mitchell, P. (1968). "Chemiosmotic Coupling and Energy Transduction", Glynn Research, Bodmin, Cornwall, England.
Mitchell, P. (1969). In: "Mitochondria - Structure and Function", L. Ernster and Z. Drahota, eds., Academic Press, London, p. 219.
Mitchell, P. (1972a). J. Bioenergetics 3, 5.
Mitchell, P. (1972b). Fed. Eur. Biochem. Soc. Symp. 28, 353.
Mitchell, P. (1973). FEBS Lett. 33, 267.
Mitchell, P. (1974). FEBS Lett. 43, 189.
Nelson, B.D., Norling, B., Persson, B., and Ernster, L. (1971). Biochem. Biophys. Res. Commun. 44, 1312, 1321.
Nelson, B.D., Norling, B., Persson, B., and Ernster, L. (1972). Biochim. Biophys. Acta 267, 205.
Nelson, N., Deters, D.W., Nelson, H., and Racker, E. (1973). J. Biol. Chem. 248, 2049.
Nicholls, P., Erecinska, M., and Wilson, D.F.(1973).In:"Mechanisms in Bioenergetics", G.F. Azzone, L. Ernster, S. Papa, E. Quagliariello, and N. Siliprandi, eds., Academic Press, New York, p. 561.
Nieuwenhuis, F.J.R.M., Kanner, B.I., Gutnick, D.L., Postma, P.W., and van Dam, K. (1973). Biochim. Biophys. Acta 325, 62.
Nordenbrand, K., Chude, O., and Ernster, L. (1974). Abstr., 9th FEBS Meeting, Budapest, p. 272.
Norling, B., Glazek, E., Nelson, B.D., and Ernster, L. (1974). Eur. J. Biochem. 47, 475.

Norling, B., Nelson, B.D., Nordenbrand, K., and Ernster, L. (1972). Biochim. Biophys. Acta 275, 18.
Packer, L. (1973). In: "Mechanisms in Bioenergetics", G.F. Azzone, L. Ernster, S. Papa, E. Quagliariello, and N. Siliprandi, eds., Academic Press, New York, p. 33.
Palade, G.E. (1956). In: "Enzymes: Units of Biological Structure and Function", O.H. Gaebler, ed., Academic Press, New York, p. 185.
Papa, S., Guerrieri, F., and Lorusso, M. (1974). BBA Library 13, 417.
Parsons, D.F., Williams, G.R., Thompson, W., Wilson, D.F., and Chance, B. (1967). In: "Round Table Discussion on Mitochondrial Structure and Compartmentation", E. Quagliariello, S. Papa, E.C. Slater and J.M. Tager, eds., Adriatica Editrice, Bari, p. 29.
Pullman, M.E., and Monroy, G.C. (1963). J. Biol. Chem. 238, 3762.
Pullman, M.E., Penefsky, H.S., Datta, A., and Racker, E. (1960). J. Biol. Chem. 235, 3322.
Racker, E. (1970). In: "Membranes of Mitochondria and Chloroplasts", E. Racker, ed., Van Nostrand Reinhold, New York, p. 127.
Racker, E. (1974). BBA Library 13, 269.
Racker, E., and Stoeckenius, W. (1974). J. Biol. Chem. 249, 662.
Racker, E., Tyler, D.D., Estabrook, R.W., Conover, T.E., Parsons, D.F., and Chance, B. (1965). In: "Oxidases and Related Redox Systems", T.E. King, H.S. Mason and M. Morrison, eds., Wiley, New York, p. 1077.
Rossi, E., Norling, B., Persson, B., and Ernster, L. (1970). Eur. J. Biochem. 16, 508.
Rydström, J. (1972a). Ph.D. Thesis, University of Stockholm, Chem. Commun., No. VII.
Rydström, J. (1972b). Eur. J. Biochem. 31, 496.
Rydström, J. (1974). Eur. J. Biochem. 45, 67.
Rydström, J., Hoek, J.B., and Hundal, T. (1974). Biochem. Biophys. Res. Commun. 60, 448.
Rydström, J., Teixeira da Cruz, A., and Ernster, L. (1970). Eur. J. Biochem. 17, 56.
Rydström, J., Teixeira da Cruz, A., and Ernster, L. (1971). Eur. J. Biochem. 23, 212.
Sanadi, D.R., and Littlefield, J.W. (1952). Science 116, 327.
Senior, A.E. (1973). Biochim. Biophys. Acta 301, 249.
Senior, A.E. (1974). Verbal communication, ASBC Meeting, Minneapolis.
Singer, S.J., and Nicholson, L. (1972). Science 175, 720.
Skulachev, V.P. (1970). FEBS Lett. 11, 301.
Skulachev, V.P. (1971). In: "Current Topics in Bioenergetics", D.R. Sanadi, ed., Academic Press, New York, Vol. 4, p. 127.
Skulachev, V.P. (1972). J. Bioenergetics 3, 25.
Skulachev, V.P. (1974a). Ann. N.Y. Acad. Sci. 227, 188.
Skulachev, V.P. (1974b). BBA Library 13, 243.
Slater, E.C. (1953). Nature 172, 975.
Slater, E.C. (1958). Adv. Enzymol. 20, 147.
Slater, E.C. (1969a). In: "The Energy Level and Metabolic Control in Mitochondria", S. Papa, J.M. Tager, E. Quagliariello and E.C. Slater, eds., Adriatica Editrice, Bari, p. 255.
Slater, E.C. (1969b). In: "Mitochondria - Structure and Function", L. Ernster and Z. Drahota, eds., Academic Press, London, p. 205.
Slater, E.C. (1971). Q. Rev. Biophys. 4, 35.
Slater, E.C. (1974). BBA Library 13, 1.
Slater, E.C., Rosing, J., Harris, D.A., Van de Stadt, R.J. and Kemp, A., Jr. (1974). In: "Membrane Proteins in Transport and Phosphorylation", G.F. Azzone, M.E. Klingenberg, E. Quagliariello and N. Siliprandi, eds.,

North-Holland, Amsterdam, p. 137.
Slater, E.C., Rosing, J., and Mol, A. (1973). Biochim. Biophys. Acta 292, 534.
Smoly, J.M., Kuylenstierna, B., and Ernster, L. (1970). Proc. Natl. Acad. Sci. U.S. 66, 125.
Sottocasa, G.L., Kuylenstierna, B., Ernster, L., and Bergstrand, A. (1967). J. Cell Biol. 32, 415.
Stoeckenius, W. (1963). J. Cell Biol. 17, 443.
Straub, F.B. (1939). Biochem. J. 33, 787.
Szent-Györgyi, A. (1957). "Bioenergetics", Academic Press, New York.
Teixeira da Cruz, A., Rydström, J., and Ernster, L. (1971). Eur. J. Biochem. 23, 203.
Van de Stadt, R.J., De Boer, B.L. and van Dam, K. (1973). Biochim. Biophys. Acta 292, 338.
Warburg, O. (1913). "Über die Wirkung der Struktur auf chemische Vorgänge in Zellen", Gustav Fischer, Jena.
Wattiaux-De Coninck, S., and Wattiaux, R. (1971). Eur. J. Biochem. 19, 552.
Wikström, M.K.F. (1974). Ann. N.Y. Acad. Sci. 227, 146.
Wikström, M.K.F., and Saris, N.-E.L. (1970). In: "Electron Transport and Energy Conservation", J.M. Tager, S. Papa, E. Quagliariello, and E.C. Slater, eds., Adriatica Editrice, Bari, p. 77.
Williams, R.J.P. (1961). J. Theoret. Biol. 1, 1.
Williams, R.J.P. (1969). In: "Current Topics in Bioenergetics", D.R. Sanadi, ed., Academic Press, New York, Vol. 3, p. 79.
Williams, R.J.P. (1974). Ann. N.Y. Acad. Sci. 227, 98.
Wilson, D.F., and Brocklehurst, E.S. (1973). Arch. Biochem. Biophys. 158, 200.
Yu, C.A., Yu, L. and King, T.E. (1974). J. Biol. Chem. 249, 4905.

CARRIER MEDIATED TRANSPORT OF PHOSPHATE IN MITOCHONDRIA

A. Fonyó, E. Ligeti, F. Palmieri and E. Quagliariello

Experimental Research Department, Semmelweis University Medical School 1082 Budapest, Hungary and C.N.R. Unit for the Study of Mitochondria and Bioenergetics, 70126 Bari, Italy

INTRODUCTION

The mitochondrial inner membrane contains several carriers for the transport of metabolites through the otherwise impermeable membrane. Some of these carriers require for their function free SH-groups. It was demonstrated first in case of the phosphate carrier that low levels of organic mercurials, NEM and DTNB inhibit its function (Fonyó, 1968; Fonyó and Bessman, 1968; Tyler, 1968, 1969; Haugaard et al., 1969; Guerin et al., 1970). The phosphate transport was so sensitive to organic mercurials that these compounds were considered to be specific inhibitors of it. It was shown later that also the dicarboxylate, the tricarboxylate and the oxoglutarate carriers are inhibited by organic mercurials, but not by NEM (Robinson and Oei, 1970; Meijer et al., 1970; Quagliariello and Palmieri, 1972; Johnson and Chappell, 1973; Palmieri et al., 1974). The glutamate carrier is not sensitive to organic mercurials but it is readily inhibited by lipid

Abbreviations: ASPM: N- N-acetyl-4-sulfamoyl-phenyl maleimide; CMS: p-chloro-mercuriphenyl-sulfonic acid; DTNB: 5,5'-dithio-bis 2-nitrobenzoic acid ; EGTA: ethyleneglycol-bis beta-aminoethyl ether -N,N'-tetraacetic acid; NEM: N-ethylmaleimide

soluble SH-group reagents, NEM, DTNB and fuscin (Meijer et al., 1972; Vignais and Vignais, 1973; Meyer and Vignais, 1973). The adenine nucleotide carrier is not sensitive to organic mercurials; a sensitivity for NEM and for fuscin develops in it however if mitochondria are pretreated with low concentrations of ADP (Leblanc and Clauser, 1972; Vignais and Vignais, 1972, 1973; Vignais et al., 1973, 1974). It follows from these that the reactivity of the SH-groups shows wide variation from one carrier to the other. In each carrier the reactivity of its thiol groups is determined by their location within the protein molecule, by their environment.

The SH-groups of the P_i carrier are probably the most exposed ones to the action of various SH-group reagents. It was anticipated that some information on the location of its SH-groups and on the structure of the carrier can be gained by following changes in thiol reactivity under various conditions. In the first part of the present paper the methods are summarized by which P_i carrier activity can be assayed quantitatively. In the second part results of the titration of the carrier activity with different inhibitors are presented together with some suggestions on the possible structure of the P_i carrier.

ASSAY OF THE PHOSPHATE CARRIER ACTIVITY

For titration of the P_i carrier with inhibitors it is an absolute requirement that the P_i transport measured should depend solely on the activity of the carrier and not on electron transport or energy conservation. In mitochondria a number of P_i requiring processes can be monitored the rate of which parallels the rate of P_i transport: State 3 respiration and the coupled synthesis of ATP, the swelling during valinomycin-induced ion uptake. In these processes the rate of P_i transport is limited either by the rate of electron transport or by the rate of energy conservation (Figs. 1 and 2); in ATP synthesis even adenine nucleotide transport may be rate limiting. If the maximum capacity of the P_i carrier is in excess as compared to any of the capacities of the previously mentioned processes then a significant part of the carrier can be titrated by the inhibitor without any apparent inhibition of the actual rate of P_i transport.

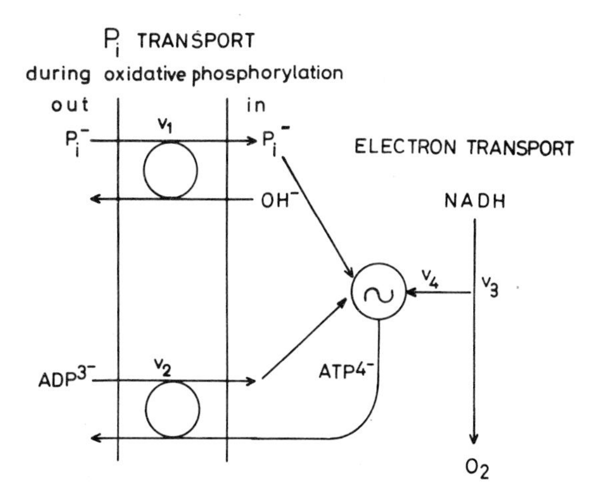

Fig. 1. Scheme of the rate limiting steps of oxidative phosphorylation. The symbols above the arrows indicate the velocities.

The rate of P_i transport is dependent only on P_i carrier activity in the following processes: 1/, the exchange between $^{32}P_i$ and $^{31}P_i$ in mitochondria loaded previously with phosphate; 2/, the exchange between P_i and acetate in mitochondria which had accumulated acetate salts; 3/, the swelling of mitochondria in iso-osmotic ammonium phosphate, and in the presence of carboxylic ionophores, iso-osmotic K-phosphate solution and 4/, the efflux of phosphate in the presence of carboxylic ionophores from mitochondria which had accumulated K-phosphate.

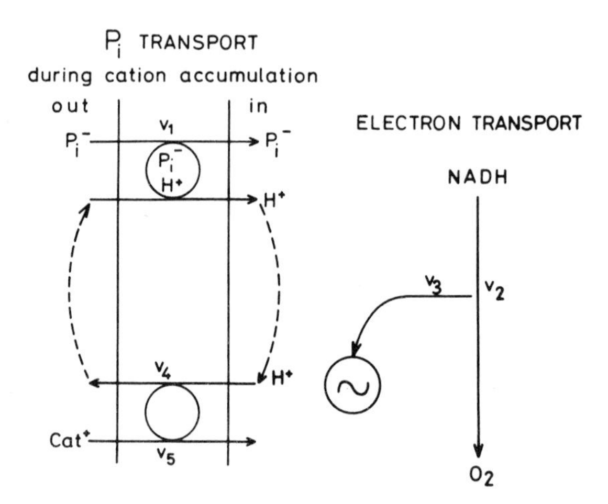

Fig. 2. Scheme of the rate limiting steps in induced ion uptake. The symbols above the arrows indicate the velocities.

The $^{32}P_i$ - $^{31}P_i$ exchange

Phosphate is transported through the inner mitochondrial membrane via 2 carrier systems. The phosphate carrier exchanges P_i either for hydroxyl or for P_i ions (exchange against hydroxyl ions cannot be distinguished from co-transport with protons, see Mitchell and Moyle, 1969). The dicarboxylate carrier catalyzes exchange either between phosphate and dicarboxylate or between phosphate and phosphate ions (Fig. 3.). The phosphate carrier is inhibited by organic mercurials and by NEM while the dicarboxylate carrier is inhibited by organic mercurials and n-butylmalonate but not by NEM. In order to measure the rate of $^{32}P_i$ - $^{31}P_i$ exchange exclusively through the phosphate carrier, the dicarboxylate carrier has to be inhibited by n-butylmalonate. Coty and Pedersen (1974) preloaded mitochondria with inorganic phosphate and measured the rate of exchange of the subsequently added $^{32}P_i$ in the presence of n-butylmalonate. This direct measurement of the P_i carrier ac-

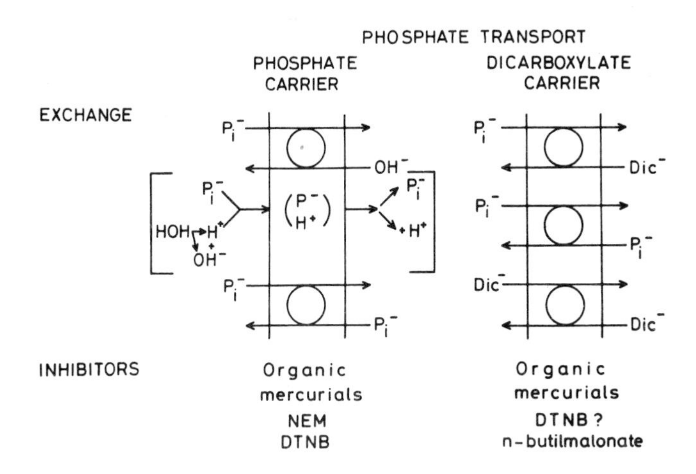

Fig. 3. Scheme of phosphate transport via the phosphate and the dicarboxylate carrier.

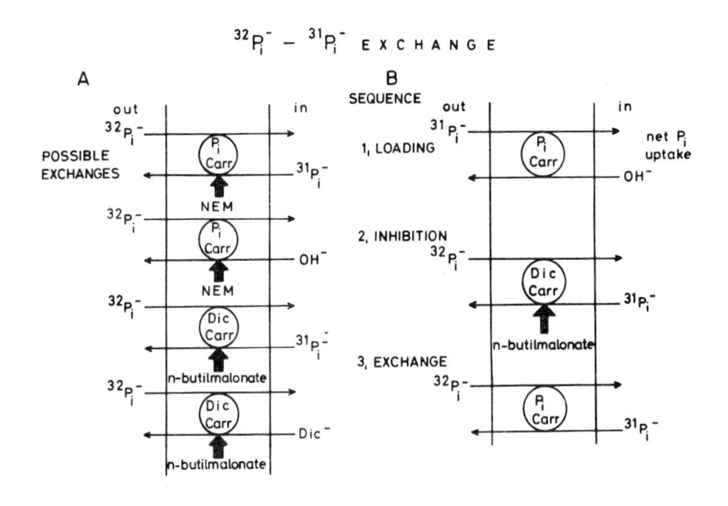

Fig. 4. Scheme of measuring $^{32}P_i$ exchange through the phosphate carrier.

tivity (Fig. 4.) is clearly independent on energy supply or cation transport.

The indirect assay systems of P_i carrier activity are based on monitoring the volume changes that occur during P_i transport and which can be followed optically.

The exchange between P_i and acetate after calcium-acetate uptake

Respiring mitochondria accumulate calcium together with acetate and swell proportionately to the uptake in a medium which contains acetate but not phosphate; the swollen state is maintained as long as the energy supply of the mitochondria is intact (Chappell and Crofts, 1966) . Addition of P_i to the swollen mitochondria causes the exchange of P_i with the intramitochondrial acetate: this gives an intramitochondrial precipitation of calcium phosphate and rapid shrinkage of the mitochondria (Fig. 5.; Rasmussen et al., 1965). The P_i - acetate exchange is sensitive to all known inhibitors of the P_i carrier including NEM which does not affect the dicarboxylate carrier(Fonyó, 1969, 1974) . The rate of shrinkage of the mitochondria is an indicator of the rate of P_i transport. Kinetic parameters of the exchange were determined under various conditions (Fonyó et al., 1974). The exchange requires that the mitochondria should be in the energized state but the rate of exchange is not dependent on the rate of energy conservation.

The exchange is composed of two processes: the carrier mediated influx of phosphate and the efflux of acetate or acetic acid by free diffusion. The rate of the overall exchange process may be either limited by the maximal rate of the carrier mediated phosphate influx or by the maximal rate of diffusion of acetate through the membrane. The movement of acetate consists of three consecutive steps: the protona-

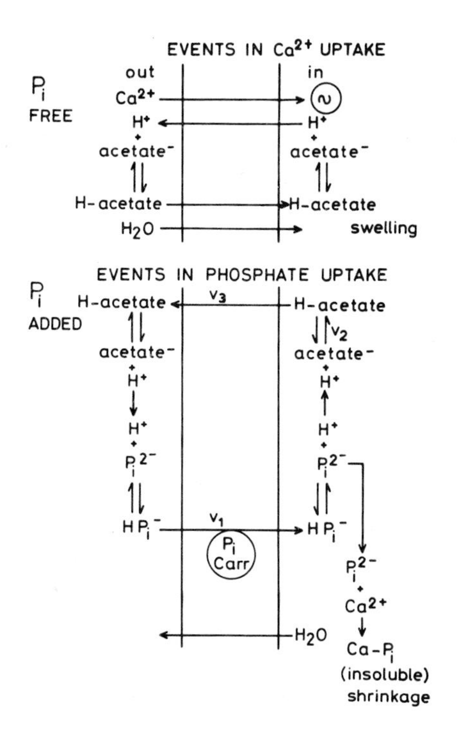

Fig. 5. Scheme of the volume changes during calcium acetate uptake and consecutive phosphate-acetate exchange in mitochondria

tion of acetate to acetic acid at the membrane-matrix interface, the diffusion of acetic acid through the membrane and the deprotonation of acetic acid to acetate at the other interface. The problem of applicability of the phosphate-acetate exchange for assaying the P_i carrier activity is whether all the three steps in the movement of acetate are faster than the movement of phosphate or not.

The evidence that acetate outflow is not rate limiting is only indirect. Were the maximal carrier capacity in excess over the maximal rate of acetate movement then titration of carrier activity with any inhibitor would at first not change the rate of P_i transport. Inhibition would appear only when carrier activity balanced acetate efflux. It was found however that titration of the exchange process with some (but not all) inhibitors yielded hyperbolic inhibition curves. It is assumed therefore that the rate of the phosphate-acetate exchange is limited by the capacity of the P_i carrier.

Most results described in the present paper were obtained by recording the rate of shrinkage of mitochondria optically.

Swelling of mitochondria in ammonium phosphate solution

Mitochondria swell rapidly in iso-osmotic ammonium phosphate solution in the presence of respiratory inhibitors and EDTA or EGTA (Chappell and Crofts, 1966). The swelling depends on the activity of the phosphate carrier (Fonyó and Bessman, 1968; Tyler, 1968, 1969; Williams and Orr, 1974). This simple method has been used for studies of the inhibition characteristics of the P_i carrier (Guerin et al., 1970; Fonyó et al., 1973; Klingenberg et al., 1974). The swelling indicates the penetration of ammonium phosphate into the mitochondria and the osmotic equilibration of water. Ammonium ions cannot penetrate the membrane in their charged form, but they are in protonation-deprotonation equilibrium with ammonia. This latter, being lipid soluble, penetrates the membrane easily by diffusion. Following the ammonia, phosphate is transported through the membrane by the P_i carrier either in protonated form (Mitchell and Moyle, 1969) or exchanged against hydroxyl ions (Fig. 6.). Similar osmotic swelling takes place in mitochondria suspended in iso-osmotic K-phosphate solution if K-ion permeability together with proton permeability is induced by carboxylic ionophores as nigericin (Mitchell and Moyle, 1969; Johnson and Chappell, 1973). In this osmotic swelling it is the activity of the P_i carrier that is rate

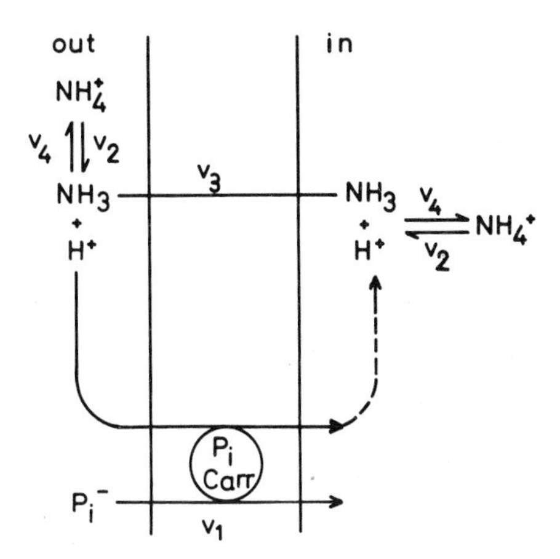

Fig. 6. Scheme of ammonium phosphate penetration into mitochondria. The symbols above the arrows indicate the velocities.

limiting. This assumption is based on two facts: the inhibition curve of some but not all SH-group reagents is of the hyperbolic type, furthermore the swelling of mitochondria in iso-osmotic ammonium acetate solution is much faster than in ammonium phosphate solution (Johnson and Chappell, 1973) .

The rate of swelling in both ammonium- and K-phosphate solution depends strongly on the pH: at acid pH the swelling is slow and with increasing pH it becomes much faster (Mitchell and Moyle, 1969; Fonyó, 1972) . The reason for this is poorly understood. The original suggestion of Mitchell and Moyle was that the carrier had an alkaline pH optimum. It was found however later that the apparent V_{max} of the carrier as measured by the phosphate-acetate exchange was unchanged between the pH values 6.5 and 7.4 (Fonyó et al., 1974). The P_i carrier is subject to inactivation in strong salt solutions (Williams and Orr, 1974); a reversible inactivation may occur in the 0.1 M phosphate solutions used to measure the swelling.

Efflux of phosphate from phosphate-loaded mitochondria

In the processes listed above phosphate influx into mitochondria was measured. The opposite movement of phosphate, its outflow is also suitable under controlled conditions to measure the activity of the P_i carrier. First K-phosphate uptake has to be induced in respiring mitochondria by addition of valinomycin: this results in swelling. The addition of a carboxylic ionophore, e.g. nigericin results in the efflux of K-ions and phosphate and mitochondrial shrinkage. If saturating levels of nigericin are employed and thus the membrane became completely permeable for K-ions and protons, the rate of efflux and thus the rate of shrinkage depends on the maximal capacity of the phosphate carrier.

TITRATION OF THE PHOSPHATE CARRIER WITH SH-GROUP REAGENTS

SH-group reagents react with a number of mitochondrial thiol groups of which only a small fraction belongs to the P_i carrier (Guerin and Guerin, 1972; Fonyó et al., 1973; Klingenberg et al., 1974). The effect of a given amount of reagent added depends on the amount of mitochondrial SH-groups which reacted under the conditions employed and on the relative affinities of the carrier SH-groups as compared to those of other proteins.

Titration of phosphate carrier activity with NEM.

NEM is a penetrant SH-group-reagent (Vignais et al., 1974): its action is rather slow and the inhibition by NEM depends on both the concentration of the reagent and the time available for reaction. NEM does not dissociate in the pH range between 6 and 8: the effect of changing the pH on the inhibition caused by NEM is the result of change of ionization of the SH-groups of the P_i carrier.

Ionization of the SH-groups of the carrier favoured their reaction with NEM (Fig. 7.). Inhibition occured at lower NEM concentration at pH 7.4 than at 6.5. In these experiments NEM was allowed to react for 60 seconds with the mitochondria after which period an excess of 2- mercaptoethanol was added to remove the unreacted portion of NEM. Similar results were published by Klingenberg et al. (1974) who used a different assay technique for the P_i carrier. Ionization of the thiol groups of the P_i carrier increased the rate of their reaction with NEM (Fig. 8.): that amount of NEM which required about 240 seconds to inhibit the carrier at pH 6.5 inhibited carrier activity at pH 7.4 in less than 10 seconds. The reaction rate of simple, low molecular weight thiol compounds with NEM increases also on ionization (Klingenberg et al., 1974).

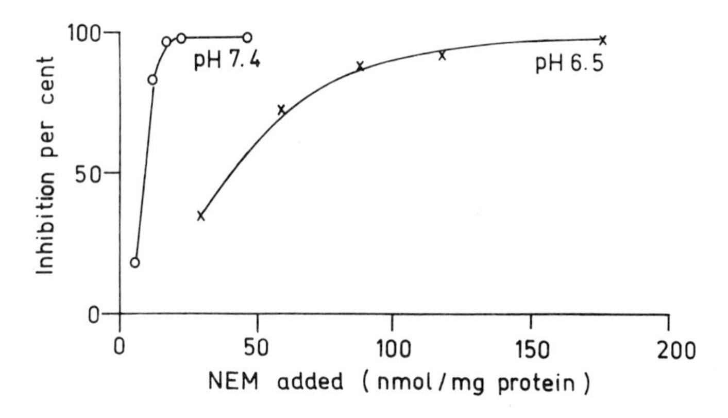

Fig. 7. Inhibition of phosphate carrier activity by NEM. Carrier activity was assayed by following absorbance changes on addition of phosphate to mitochondria that were swollen because of uptake of calcium acetate. Temperature: 11° C. NEM was added when the swelling was complete: after a further 60 seconds excess 2-mercaptoethanol was added, followed by 2 mM tris-phosphate.

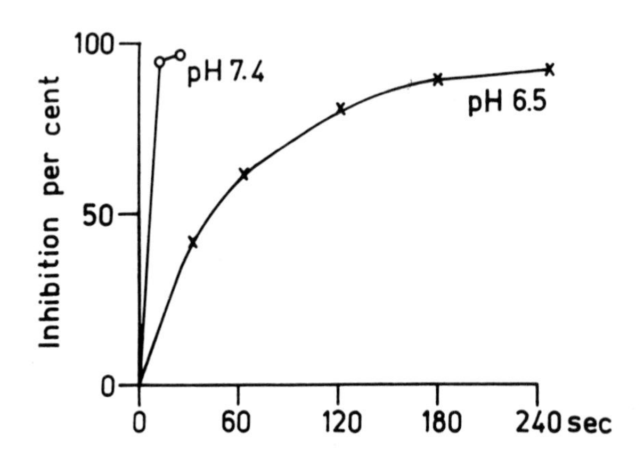

Fig. 8. The rate of inactivation of the phosphate carrier by NEM. Experimental details were as in Fig. 7. with the difference that the time of addition of 2-mercaptoethanol was varied.

Titration of phosphate carrier activity with DTNB

Similarly to NEM, DTNB is also a penetrant SH-group reagent (Vignais et al., 1974)but it differs from NEM that it is capable of ionization. The P_i carrier is inhibited by lower amount of DTNB at pH 6.5 than at pH 7.4 (Fig. 9, see also Klingenberg et al, 1974). Simple thiols react faster with DTNB at higher than at lower pH. Two opposing factors determine the rate of reaction of the SH-groups of the P_i carrier with DTNB: the ionization of the SH-groups of the carrier accelerates the reaction while the ionization of the DTNB molecule itself has a strong rate decreasing effect.

Titration of the phosphate carrier with organic mercurials

The reaction of organic mercurial compounds, CMS and mersalyl is much faster with thiols than the reaction with either NEM or DTNB.

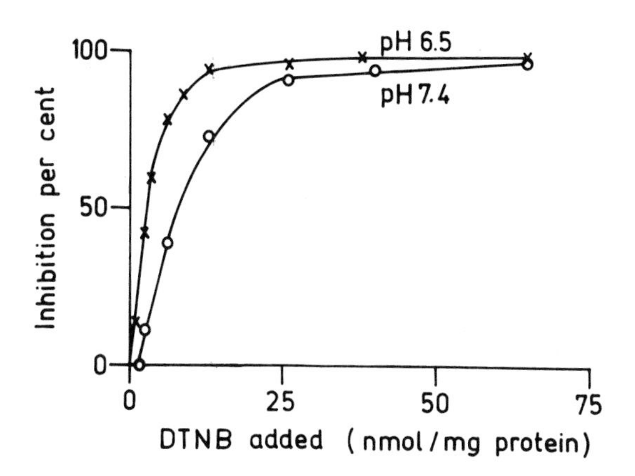

Fig. 9. Inhibition of phosphate carrier activity by DTNB. Experimental details were as in Fig. 7.

In contrast to the experiments reported in Figs 7-9, time was not a limiting factor in the experiments with the mercurials.

With both CMS and mersalyl less added reagent was required for inhibition of the P_i carrier at pH 6.5 than at pH 7.4 (Figs 10. and 11). This means that ionization of the mercurials, similarly to the ionization of DTNB and of ASPM (Klingenberg et al., 1974) decreases their reactivity with the SH-groups of the P_i carrier: the ionized form of the reagent has less accessibility to the SH-groups of the carrier than the undissociated form. Either are anionic groups of the carrier close to the SH-groups and act electrostatically, or the SH-groups are in a hydrophobic environment which the ionized form of the SH-group reagent can reach only with difficulties.

It is apparent in Figs. 10 and 11 that addition of a few nmoles of CMS or mersalyl to mitochondria does not inhibit P_i transport at all. Similar results were also presented by Papa et al. (1973) on the basis

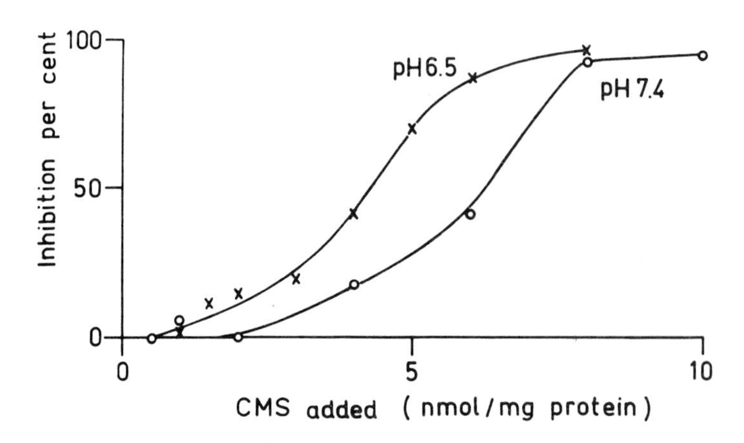

Fig. 10. Inhibition of phosphate carrier activity by CMS. Experimental details were as in Fig. 7. but 2-mercaptoethanol was not added. The addition of phosphate followed 60 seconds the addition of CMS.

of $^{32}P_i$ exchange experiments. The reason of this non-inhibition of transport may be either that the mercurials if added at low levels did not react with the carrier, or that they reacted with the carrier but carrier activity was not rate limiting under the experimental conditions or, that although the mercurials did react with the carrier the reaction did not result in inhibition.

The binding of mersalyl to the SH-groups of the P_i carrier was tested by making use of the reversibility of the mersalyl and the irreversibility of the NEM binding and inhibition (Fig. 12.). On adding a low amount of mersalyl to mitochondria one part of the SH-groups will be bound by it and the rest left free. Further addition of an excess of NEM will bind all those groups which remained free but not those which were already bound by the mercurial. Addition of a thiol compound as 2-mercaptoethanol in excess releases the mercurial-bound SH-groups but

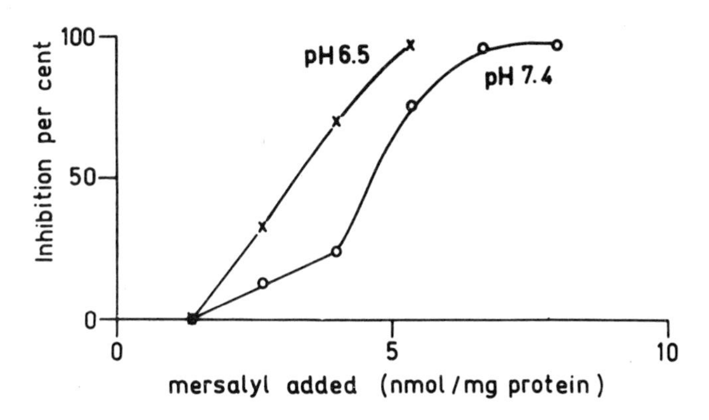

Fig. 11. Inhibition of phosphate carrier activity by mersalyl. Experimental details were as in Fig. 10.

not those bound by NEM. Thus the mercurial reagent "protected" the SH-groups from being irreversibly bound by NEM. It follows that 2-mercapto-ethanol will reactivate transport only if the SH-groups bound by the mercurial were involved in function of the carrier.

In the phosphate - acetate exchange system mersalyl, at a level which did not inhibit P_i transport in itself, "protected" the carrier from the irreversible inhibition of NEM. This means that mersalyl added at a non-inhibitory level reacted indeed with the carrier.

The P_i carrier activity was also titrated with mersalyl using the ammonium phosphate swelling system (Fig. 13.). In this Figure both the "direct" titration of carrier activity and the "protective" effect of the same amount of mersalyl were plotted. It is seen that those amounts of mersalyl which caused little if any inhibition of swelling had a progressive "protecting" effect if added before NEM. In the experiment shown in Fig. 13. addition of 2 nmoles mersalyl per mg protein had hardly any inhibitory action on swelling i.e. on P_i carrier activity, but if the same amount of mersalyl was added before NEM then 2-mercaptoethanol reactivated more than half of the original carrier activity.

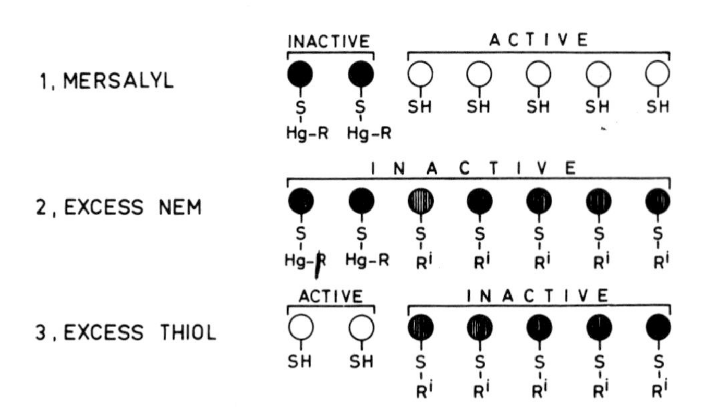

Fig. 12. Scheme of sequential reversible and irreversible partial block of SH-groups and their release by thiol compounds.

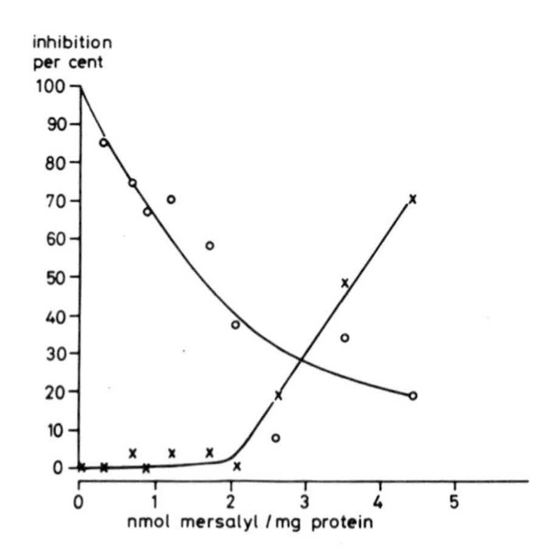

Fig. 13. Titration of swelling of mitochondria in ammonium phosphate with mersalyl and the "protective" effect of mersalyl against NEM inhibition. The crosses indicate the addition of mersalyl alone; the circles the addition of mersalyl before NEM, followed by 2-mercaptoethanol.

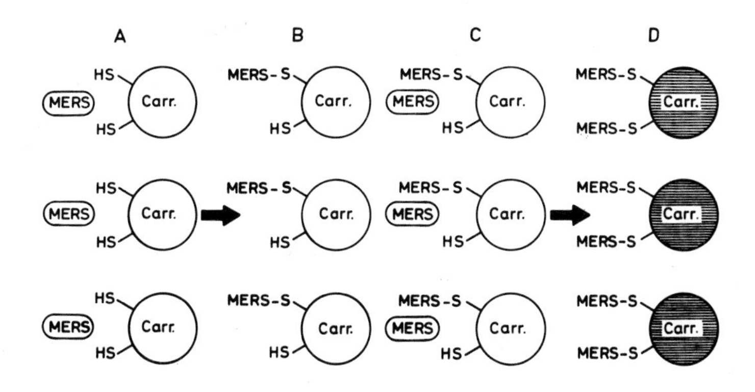

Fig. 14. Scheme of the proposed structure of the phosphate carrier and the sequential reaction of mersalyl with its two SH-groups. The open circles indicate functionally active, the hatched inactive carriers.

It is theoretically possible that the low amounts of mersalyl did not inhibit P_i transport because the maximal capacity of the carrier was in excess to the actual rate of transport. This explanation of results, although it cannot be ruled out at present completely, seems unlikely. The movement of ammonium acetate into mitochondria is much faster than that of ammonium phosphate. This fact indicates that in ammonium phosphate penetration the carrier mediated transport of P_i is rate limiting (Johnson and Chappell, 1973) .

A further possibility is that the mersalyl applied at low level reacted with the SH-groups of the carrier without inhibiting carrier function. In this case the "direct" and the "indirect" titration of P_i transport in Fig. 13. would mean that the same SH-groups of the carrier which could be blocked without loss of carrier activity could be operative and were able to support carrier function if released and all the other SH-groups irreversibly blocked. This explanation would imply that the P_i carrier has two equivalent SH-groups from which only one is required for activity. We presume that if the carrier is titrated with mersalyl those SH-groups will react faster which did not yet bind a molecule of the mercurial. With limiting amounts of mersalyl each carrier molecule will bind a single molecule of reagent, as binding of the first molecule decreases affinity for the second one. If more mersalyl is added the second molecule will be bound too (Fig. 14.) . The carrier is active with a single SH-group bound and is inactivated on binding the second SH-group. The equivalence of the SH-groups suggests that the carrier might have a symmetrical structure.

ACKNOWLEDGEMENTS

The expert and devoted technical assistance of Mrs. Julianna Szilassy and Mr. Luigi Anzivino is gratefully acknowledged.

SUMMARY

The activity of the phosphate carrier of liver mitochondria was titrated with SH-group binding inhibitors: NEM, DTNB, p-chloro-mercuriphenyl sulfonic acid and mersalyl. Ionization of the SH-groups of the carrier increased the rate of reaction with NEM. Ionization of the dissociable reagents, DTNB, CMS and mersalyl decreased their reactivity with the SH-groups of the carrier. After binding of a part of the SH-groups of the carrier by organic mercurials phosphate transport remains intact. The carrier has probably two, functionally equivalent SH-groups per unit: only one of them is required for transport function.

REFERENCES

Chappell, J.B. and Crofts, A.R. (1966) . In: Regulation of Metabolic Processes in Mitochondria. J.M. Tager, S. Papa, E. Quagliariello and E.C. Slater Eds., Elsevier Publ. Co., Amsterdam, London and New York p. 293.

Coty, W.A. and Pedersen, P.L. (1974) . J. Biol. Chem. 249, 2593.

Fonyó, A. (1968). Biochem. Biophys. Res. Comm. 32, 624.

Fonyó, A. (1969). In: Biochemistry of Intracellular Structures. L. Wojtczak, W. Drabikowski and H. Strzelecka-Golaszewska Eds., PWN Warszawa, p. 21.

Fonyó, A. (1972). In: The Biochemistry and Biophysics of Mitochondrial Membranes. G.F. Azzone, E. Carafoli, A.L. Lehninger, E. Quagliariello and N. Siliprandi Eds., Academic Press, New York and London, p. 701.

Fonyó, A. (1974). Biochem. Biophys. Res. Commun. 57, 1069.

Fonyó, A. and Bessman, S.P. (1968). Biochem. Med. 2, 143.

Fonyó, A., Csillag, A., Ligeti, E. and Ritvay, J. (1973). In: Mechanisms in Bioenergetics. G.F. Azzone, L. Ernster, S. Papa, E. Quagliariello and N. Siliprandi Eds., Academic Press, New York and London, p. 347.

Fonyó, A., Palmieri F., Ritvay, J. and Quagliariello, E. (1974). In: Membrane Proteins in Transport and Phosphorylation. G.F. Azzone, M. Klingenberg, E. Quagliariello and N. Siliprandi Eds., North Holland, Amsterdam and London, p. 283.

Guerin, B., Guerin M. and Klingenberg, M. (1970). FEBS Letters 10, 265.

Guerin, M. and Guerin, B. (1972). Abstr. Commun. Meet. Fed. Eur. Biochem. Soc. 8. p. 173.

Haugaard, N., Lee, N.H., Kostrzewa, R., Horn, R.S. and Haugaard, E.S. (1969). Biochim. Biophys. Acta 172, 198.

Johnson, R.N. and Chappell, J.B. (1973). Biochem. J. 134, 769.

Klingenberg, M., Durand, R. and Guerin, B. (1974). Eur. J. Biochem. 42, 135.

Leblanc, P. and Clauser, H. (1972). FEBS Letters 23, 107.

Meijer, A.J., Groot, G.S.P. and Tager, J.M. (1970). FEBS Letters 8, 41.

Meijer, A.J., Brouwer, A., Reingoud, D.J. and Tager, J.M. (1972). Biochim. Biophys. Acta 283, 421.

Meyer, J. and Vignais, P.M. (1973). Biochim. Biophys. Acta 325, 375.

Mitchell, P. and Moyle, J. (1969). Eur. J. Biochem. 9, 149.

Palmieri, F., Passarella, S., Stipani, I. and Quagliariello, E. (1974). Biochim. Biophys. Acta 333, 195.

Papa, S., Kanduc, D. and Lofrumento, N.E. (1973). FEBS Letters 36, 9.

Quagliariello, E. and Palmieri, F. (1972). In: Biochemistry and Biophysics of Mitochondrial Membranes. G.F. Azzone, E. Carafoli, A.L. Lehninger, E. Quagliariello and N. Siliprandi Eds., Academic Press, New York and London, p. 659.

Rasmussen, H., Chance, B. and Ogata, E. (1965). Proc. Natl. Acad. Sci. U.S. 53, 1069.

Robinson, B.H. and Oei, J. (1970). FEBS Letters 11, 200.

Tyler, D.D. (1968). Biochem. J. 107, 121.

Tyler, D.D. (1969). Biochem. J. 111, 665.

Vignais, P.V. and Vignais, P.M. (1972). FEBS Letters 26, 27.

Vignais, P.M. and Vignais, P.V. (1973). Biochim. Biophys. Acta 325, 357.

Vignais, P.V., Vignais, P.M., Lauquin, G. and Morel, F. (1973). Biochimie 55, 763.

Vignais, P.M., Chabert, J. and Vignais, P.V. (1974). This volume.

Williams, G.R. and Orr, J.L. (1974). In: Dynamics of Energy Transducing Membranes, BBA Library 13. L. Ernster, R.W. Estabrook and E.C. Slater Eds., Elsevier Publ. Co. Amsterdam, London and New York. p. 497.

THE USE OF SULPHYDRYL REAGENTS TO IDENTIFY PROTEINS UNDERGOING ADP-DEPENDENT CONFORMATIONAL CHANGES IN MITOCHONDRIAL MEMBRANE

P. M. VIGNAIS, J. CHABERT and P. V. VIGNAIS

DRF/Biochimie, CEN-G, B.P. 85, 38041 - Grenoble-Cédex (France)

-SH groups play a role in many mitochondrial functions. They are involved in electron transfer of the respiratory chain, in some anion and cation transports and have been assumed to contribute to the formation of "high energy" intermediates in energy transducing reactions (cf Gautheron, 1973 for review). It is therefore a difficult enterprise to attempt to locate and characterize the specific -SH groups directly involved in the different functions of the inner mitochondrial membrane.

In an effort of simplification and clarification, we have first looked for a test which could give a rough indication as to the localization of the -SH groups with regard to the sidedness of the mitochondrial membrane. Mitochondria are closed organelles containing in the matrix space large amounts of highly reactive free glutathione which can conveniently be titrated. The penetrability of the most currently used -SH reagents listed on Figure 1 was assessed by the determination of free internal glutathione.

<u>Penetrability of -SH reagents in the mitochondrial membrane</u> -

Rat liver mitochondria were preincubated with -SH reagents in isotonic conditions, centrifuged, and their content in reduced and oxidized glutathione analyzed (Table 1). The decrease in the amount of internal free glutathione was taken as criterion of penetrability of the inner mitochondrial membrane by the reagent. The test is easily interpretable when the reagent gives irreversible addition derivatives. It may not be so clear cut with mercurials which give ionic reversible derivatives. Furthermore mercurials have a tendency to induce a swelling of mitochondria which may

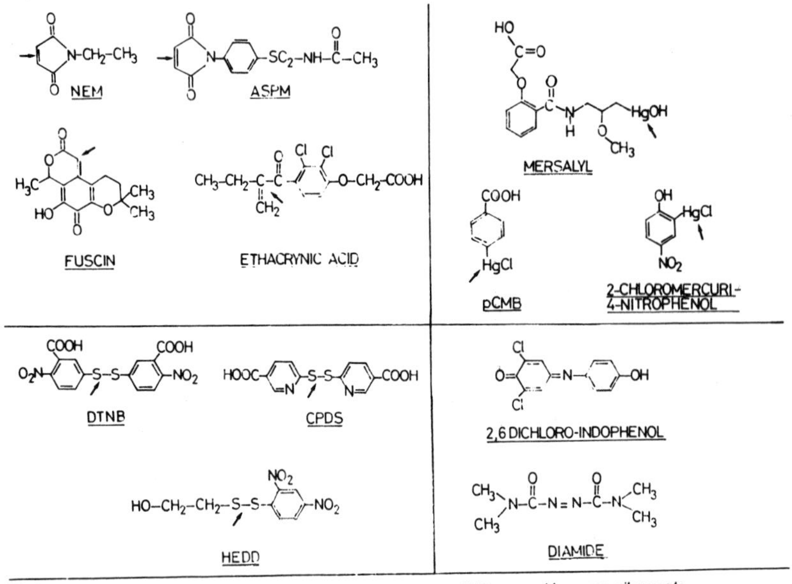

NEM : N-ethyl maleimide
ASPM : N-(N-acetyl-4-sulfamoyl phenyl) maleimide
DTNB : 5,5'-dithio-bis (2-nitrobenzoic acid) (Ellman reagent)
CPDS : 6,6'-dithionicotinic acid
pCMB : p.chloromercuribenzoate
HEDD : β-hydroxyethyl-2,4-dinitrophenyl disulphide
DIAMIDE : N,N,N',N'-tetramethylazoformamide
DCI : 2,6-dichloro-indophenol

Fig. 1. Structure of commonly used -SH reagents

result either in a leakage of internal glutathione or an increased permeability of the membrane toward the reagent.

As shown in Table 1 when NEM, fuscin, avenaciolide and dichloroindophenol are added to mitochondria the content of internal GSH is markedly decreased. The formation of addition derivatives of GSH with the above-mentioned reagents has been checked by paper chromatography (States and Segal, 1969 ; Vignais and Vignais, 1973). It may be then concluded that NEM, fuscin, avenaciolide and dichloroindophenol are penetrant reagents. Other reagents such as the mercurials mersalyl and pCMB (used at low concentration to avoid swelling of mitochondria), DTNB, CPDS, diamide and ferricyanide are not penetrant.

It must be emphasized that the distinction between "penetrant" and "non-penetrant" is merely valid in the conditions of the test (pH, concentration, temperature, duration of incubation) and may not apply to all types of biological membranes. For example the red cell membrane is more permeable

Table 1

Free glutathione content of rat liver mitochondria treated with SH reagents

Expt. Nb	-SH reagent (μM)		GSH (nmoles/mg protein)	GSSG	Comments[4]
1) 10 min incubation 20°		None	5.1	0.2	
	20	Mersalyl	5.0	0.2	NP
	20	DCI	ND[1]	2.2	P
	20	NEM	0.3	0.2	P
	20	Fuscin	1.3	0.3	P
	20	Avenaciolide	1.3	0.1	P
	40	CMNP[2]	3.3	0.4	S
	50	pCMB	3.9	0.3	S
	100	DTNB	5.1	0.3	NP
	4000	Ferricyanide	4.9	0.7	NP
2) 5 min incubation 20°		None	6.0	0.2	
	100	Diamide	5.4	0.3	NP
	100	CPDS	5.5	0.4	NP
	100	HEDD	4.2	0.8	S
	100	HEDD (pH 6.8)	0.1	1.3[3]	
3) 2 min incubation 20°		None	5.3	0.2	
	100	ASPM	5.6	0.4	NP
	150	ASPM	3.8	0.2	
	300	ASPM	2.1	0.6	
	100	ASPM (pH 6.8)	2.3	0.7	P

[1]ND = not detected ; [2]CMNP = 2, chloromercuri-4, nitrophenol

[3]Adduct compound, $GSSCH_2 CH_2OH$, may be titrated as GSSG

[4]NP = non penetrant, P = penetrant, S = swelling inducer.

Glutathione was determined by the method of Tietze (1969). (For details : cf Vignais and Vignais, 1973).

than the inner mitochondrial membrane (for review see Chappell and Haarhoff, 1967). This may explain why diamide is able to produce oxidized glutathione in the red cell (Kosower et al., 1969) while it does not in mitochondria (Table 1). The non-penetrant character of diamide may be related to the reversibility of the inhibitory effect of diamide on some mitochondrial functions (Siliprandi et al., 1974). As far as ASPM is concerned, it is non-penetrant at relatively low concentration (100μM) and at pH 7.4, but penetrant at pH 6.8. When the concentration is raised it becomes penetrant even at pH 7.4 (Table 1). This may be linked to the weak acid property of ASPM (Merz et al., 1965). The inhibition of phosphate transport by ASPM at pH lower than 7.4 (Klingenberg et al., 1974) may be related to the penetration of ASPM.

As mentioned above, some of the reagents (in particular HEDD, pCMB, and 2-chloromercuri-4-nitrophenol) present the disadvantage of inducing a rapid swelling of mitochondria. As a consequence they cannot be used safely to study the localization of mitochondrial -SH groups.

Other factors such as kinetic factors guide the choice for a reagent. A fast reacting compound may be used for short time periods either to detect more reactive -SH groups or to avoid interaction with chemical groups other than -SHs. Table 2 shows that diamide and ASPM react nearly as fast while for ferricyanide longer periods of incubation (20 min) would be necessary to titrate 5 to 6 nmoles of external -SH / mg protein.

Table 2

Amount of membrane -SH groups altered by non-penetrant -SH reagents (Diamide, ASPM or Ferricyanide)

Incubation medium[1]	Preincubation with[2]	Bound (^{14}C)NEM[3] (nmoles /mg protein)	Δ
ETK	Nil	23.4	
	Diamide	16.1	7.3
	ASPM	17.0	6.4
	Ferricyanide	22.4	1.0
ETK	Nil	19.4	
+ Phosphate	Diamide	13.2	6.2
+ Succinate	ASPM	14.4	5.0
	Ferricyanide	18.0	1.4

[1] 0.1mM EDTA, 20mM Tris,HCl, 120mM KCl (ETK) 20μM ATR and when indicated 4mM phosphate and 4mM succinate, final pH 7.4
[2] 100μM diamide or ASPM, or 2mM ferricyanide (3 min at 20°)
[3] 110μM (^{14}C)NEM incubated for 10 min at 0° with 4.3mg protein (rat liver mitochondria) in 5 ml of incubation medium.

Topology of -SH groups unmasked in mitochondria upon addition of ADP -

When mitochondria are incubated for a short period of time with ADP and a low concentration of a penetrant -SH reagent such as NEM their ability to further carry out the transport of ADP is markedly decreased (Vignais and Vignais, 1972 ; Vignais et al., 1973). There are several lines of evidence which indicate that ADP in the preincubation medium interacts with the ADP carrier : 1) the effect is ADP or ATP specific. Other nucleotides such as UDP, CDP or GDP which are not transported are not effective. 2) the concentration of ADP required in the preincubation medium to obtain 50% inhibition of the ADP transport is 1 to 2μM, that is the same value as the K_M for ADP. 3) the inhibitory effect of ADP is overcome by atractyloside which is an inhibitor of the ADP transport.

The traces of ADP not only potentiate the inhibitory effect of NEM on ADP transport but also unmask additional -SH groups. Titration with (^{14}C)-NEM shows that the addition of ADP during preincubation slightly increases the titer of -SH groups (1 to 2 moles/mole cytochrome a). The increase is more easily detected in respiring than in non-respiring mitochondria (Table 3). As illustrated in Figure 2 it is steadily maintained after a 5 min incubation with (^{14}C)NEM (Fig. 2A) and is observed over a wide range of (^{14}C)NEM concentration (Fig. 2B) and of pH (Fig. 2C).

The unmasking of -SH groups brought about by ADP has been interpreted as reflecting an ADP-induced conformational modification of the membrane structure (Vignais et al., 1973). The unmasked -SH groups are able to react with NEM or fuscin which are penetrant but not with mersalyl which is not. One may therefore speculate that these -SH groups are located either within the inner membrane or on the matrix face of the membrane.

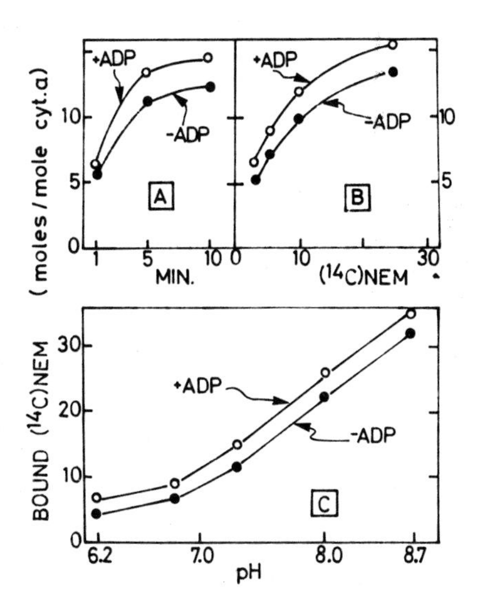

Fig. 2. Unmasking of -SH groups by ADP. The preincubation with (^{12}C)-NEM and the incubation with (^{14}C)-NEM were essentially the same as in Table 3. The medium used was made of 0.12M KCl, 0.01M Tris, HCl, 0.01M morpholinoethane sulfonic acid, 0.1mM EDTA, 4mM phosphate and 4mM succinate.

A) Incubation at pH 7.4 with 20μM (^{14}C)NEM

B) Incubation at pH 7.4 for 10 min with (^{14}C)NEM

C) Incubation for 10 min with 20μM (^{14}C)NEM.

Table 3

Binding of (^{14}C)NEM to rat liver mitochondria partially saturated with unlabelled NEM

Additions	Bound (^{14}C)NEM (moles/mole cyt.a)	
	Non-respiring mitochondria (A)	Respiring mitochondria (B)
10 μM ADP	9.1	15.2
10 μM atractyloside	8.7	12.4
	Δ: 0.4	Δ: 2.8

Mitochondria (10mg protein/ml) in 0.27M sucrose were preincubated for 15 min at 0° with 100μM unlabelled NEM. They were centrifuged and washed once before being resuspended in cold 0.27M sucrose and divided into media A and B (50ml each). Medium A : ETK medium, pH 7.4, 7.5μM rotenone, 4mM phosphate and 4mM malate. Medium B : same as A but 4mM malate was replaced by 4mM succinate. The A and B mitochondrial suspensions were kept for one min at 20° and 4 min at 0°. Then 5ml aliquots (4.4mg protein) of each suspension were added to 5ml of ETK medium containing 10μM ADP or atractyloside as indicated and 20μM (^{14}C)NEM, incubated for 5 min at 0° and centrifuged.

Table 4

Inhibition of (^{35}S)atractyloside binding to rat liver mitochondria by preincubation with ADP and permeant -SH reagents

Additions	Bound (^{35}S)ATR (mole/mole cyt.a)	
	- ADP	+ ADP[1]
None	1.06	1.03
50 μM NEM	1.07	0.36
50 μM Fuscin	1.05	0.20
50 μM Mersalyl	1.07	1. 01
100 μM ASPM	0.95	1. 02
200 μM ASPM	0.95	0.57
100 μM CPDS	1.04	1. 06
200 μM DTNB	1.02	1. 00
200 μM Diamide	1.04	0.98
2 mM Ferricyanide	1.02	1. 05

[1]10 μM ADP (Same conditions as in Vignais and Vignais, 1972).

A convenient way to relate the unmasking of -SH groups to the functioning of the ADP carrier is to study the binding of (^{35}S)atractyloside (Vignais and Vignais, 1972 ; Vignais et al., 1973). As shown in Table 4, the binding of (^{35}S)atractyloside is inhibited by preincubation of mitochondria with both permeant -SH reagents and traces of ADP (which are ineffective when added separately). Non-penetrant -SH reagents are ineffective.

In summary, typical permeant -SH reagents such as NEM or fuscin appear to inhibit irreversibly, by -SH covalent binding, the change of membrane conformation which is brought about by external ADP and which results in an unmasking of -SH groups. This structural change not only prevents the function of the ADP carrier but also the binding of (^{35}S)-atractyloside. However, our data do not allow us to decide yet whether the -SH groups which are unmasked belong to the translocator itself or to proteins in the close neighbourhood.

Acknowledgements -

We thank for their generous gifts of -SH reagents : Prof. D.H.R. Barton, London, (fuscin), Prof. E.M. Kosower, New-York, (diamide), Prof. G. Pfleiderer, Bochum, (ASPM), Dr. A. Wotjczak, Warsaw, (HEDD). This work was supported by research grants from the CNRS (ERA n° 36) and the DGRST.

References -

Chappell, J.B. and Haarhoff, K.N. (1967), in "Biochemistry of Mitochondria", eds. E.C. Slater, Z. Kaniuga, and L. Wojtczak (Acad. Press, London and New-York) pp. 75-91.

Gautheron, D.C. (1973), Biochimie, 55, 727-745.

Klingenberg, M., Durand, E. and Guérin, B. (1974), Eur. J. Biochem., 42, 135-150.

Kosower, N.S., Kosower, E.M. and Wertheim, B. (1969), Biochem. Biophys. Res. Commun., 37, 593-596.

Merz, H., Pfleiderer, G. and Wieland, Th. (1965), Biochem. Z., 342, 68-75.

Siliprandi, D., Scutari, G., Zoccarato, F. and Siliprandi, N. (1974), FEBS-Lett., 42, 197-199.

States, B. and Segal, S. (1969), Anal. Biochem., 27, 323-329.

Tietze, F. (1969), Anal. Biochem., 27, 502-522.

Vignais, P.V. and Vignais, P.M. (1972), FEBS-Lett., 26, 27-31.

Vignais, P.M. and Vignais, P.V. (1973), Biochim. Biophys. Acta, 325, 357-374.

Vignais, P.V., Vignais, P.M., Lauquin, G. and Morel, F. (1973), Biochimie, 55, 763-778.

SUBJECT INDEX

***FEBS* Proceedings**

Proceedings of the seventh meeting Varna 1971
Volume 24
BIOCHEMISTRY OF CELL DIFFERENTIATION
edited by A.Monroy and R. Tsanev
1973, x + 192 pp.

Volume 23
FUNCTIONAL UNITS IN PROTEIN BIOSYNTHESIS
edited by R. A. Cox and A. A. Hadjiolov
1972, xii + 430 pp.

Volume 22
VIRUS-CELL INTERACTIONS AND VIRAL ANTIMETABOLITES
edited by D. Shugar
May 1972, viii + 232 pp.

Proceedings of the sixth meeting Madrid 1969
Volume 21
MACROMOLECULES
Biosynthesis and Function
edited by C. Asensio, C. F. Heredia, D. Nachmansohn and S. Ochoa
1970, viii + 374 pp.

Volume 20
MEMBRANES
Structure and Function
edited by J. R. Villanueva and F. Ponz
1970, viii + 154 pp.

Volume 19
METABOLIC REGULATION AND ENZYME ACTION
edited by A. Sols and S. Grisolia
1970, x + 382 pp.

Proceedings of the fifth meeting Prague 1968
Volume 18
ENZYMES AND ISOENZYMES
Structure, Properties and Function
edited by D. Shugar
1970, x + 362 pp.

Volume 17
MITOCHONDRIA Structure and Function
edited by L. Ernster and Z. Drahota
1969, viii + 394 pp.

Volume 16
BIOCHEMICAL ASPECTS OF ANTIMETABOLITES AND OF DRUG HYDROXYLATION
edited by D. Shugar
1969, viii + 296 pp.

Volume 15
GAMMA GLOBULINS Structure and Biosynthesis
edited by F. Franěk and D. Shugar
1969, viii + 242 pp.

These volumes are available from ACADEMIC PRESS LONDON AND NEW YORK

Proceedings of the fourth meeting Oslo 1967

Volume 14
STRUCTURE AND FUNCTION THE ENDOPLASMIC RETICUL
IN ANIMAL CELLS
edited by F. C. Gran
1968, viii + 100 pp.

Volume 13
CONTROL OF GLYCOGEN METABOLISM
edited by W. J. Whelan
1968, viii + 222 pp.

Volume 12
CELLULAR COMPARTMENTALIZATION AND CONTROL OF FATTY ACID METABOLISM
edited by F. C. Gran
1968, viii + 116 pp.

Volume 11
STRUCTURE AND FUNCTION (
TRANSFER RNA AND 5S-RNA
edited by L. O. Fröholm and S. G. Laland
1968, viii + 186 pp.

Volume 10
THE BIOCHEMISTRY OF VIRUS REPLICATION
edited by S. G. Laland and L. O. Fröholm
1968, viii + 108 pp.

Volume 9
REGULATION OF ENZYME ACTIVITY AND ALLOSTERIC INTERACTIONS
edited by E. Kvamme and A. Pihl
1968, viii + 170 pp.

Proceedings of the third meeting Warsaw 1966
Volume 8
BIOCHEMISTRY OF BLOOD PLATELETS
edited by E. Kowalski and S. Niewiarowski
1967, viii + 192 pp.

Volume 7
BIOCHEMISTRY OF MITOCHONDRIA
edited by E. C. Slater, Z. Kaniuga and L. Wojtczak
1967, viii + 122 pp.

Volume 6
GENETIC ELEMENTS
Properties and Function
edited by D. Shugar
1967, x + 362 pp.

Proceedings of the first meeting London 1964
Volume 1
STRUCTURE AND ACTIVITY C
ENZYMES
edited by T. W. Goodwin, J. I. Harr
and B. S. Hartley
1964, viii + 190 pp.

ASP BIOLOGICAL AND MEDICAL PRESS B.V.

FEBS volumes published under the imprint of North-Holland Publishing Company:

Proceedings of the 8th FEBS Meeting, Amsterdam, 1972:

Vol. 25
ANALYSIS AND SIMULATION OF BIOCHEMICAL SYSTEMS
Organized by H. C. Hemker and B. Hess
1972. 460 pp.

Vol. 26
IMMUNOGLOBULINS: CELL BOUND RECEPTORS AND HUMORAL ANTIBODIES
Organized by R. E. Ballieux, M. Gruber and H. G. Seyen
1972. 104 pp.

Vol. 27
RNA VIRUSES: REPLICATION AND STRUCTURE
RIBOSOMES: STRUCTURE, FUNCTION AND BIOGENESIS
Organized by H. Bloemendal, E. M. J. Jaspars, A. van Kammen and R. J. Planta
1972. 310 pp.

Vol. 28
MITOCHONDRIA: BIOGENESIS AND BIOENERGETICS
BIOMEMBRANES: MOLECULAR ARRANGEMENTS AND TRANSPORT MECHANISMS
Organized by S. G. van den Bergh, P. Borst, L. L. M. van Deenen, J. C. Riemersma, E. C. Slater and J. M. Tager
1972. 414 pp.

Vol. 29
ENZYMES: STRUCTURE AND FUNCTION
Organized by J. Drenth, R. A. Oosterbaan and C. Veeger
1972. 241 pp.

Proceedings of the FEBS Special Meeting, Dublin, 1973:

Vol. 30
INDUSTRIAL ASPECTS OF BIOCHEMISTRY
Part 1: 1974. 547 pp.
Part 2: 1974. 424 pp.
Edited by B. Spencer